CORTICAL DEFICITS IN SCHIZOPHRENIA
From Genes to Function

CORTICAL DEFICITS IN SCHIZOPHRENIA
From Genes to Function

Edited by

Patricio O'Donnell

University of Maryland, Baltimore, MD, USA

Springer

Patricio O'Donnell
Depts. Anatomy & Neurobiology and Psychiatry
University of Maryland School of Medicine
20 Penn Street, Room S251
Baltimore, MD 21201

Library of Congress Control Number: 2007933066

ISBN-13: 978-0-387-74350-9   e-ISBN-13: 978-0-387-74351-6

Printed on acid-free paper.

9 8 7 6 5 4 3 2 1

springer.com

Contents

List of Contributors

Deborah Bauer
University of Alabama at Birmingham, Department of Psychiatry
1530 3rd Avenue South, Birmingham, AL 35294, USA

Francine Benes
McLean Hospital, Program in Structural and Molecular Neuroscience
115 Mill St., Belmont, MA 02478, USA

Katherine E. Burdick
Zucker Hillside Hospital, Division of Psychiatry Research
Glen Oaks, NY, 11004, USA

Gabriel Corfas
Harvard University, Children's Hospital, Neurobiology Program
300 Longwood Ave, Boston, MA 02115, USA

Pamela DeRosse
Zucker Hillside Hospital, Division of Psychiatry Research
Glen Oaks, NY, 11004, USA

Robert Freedman
University of Colorado, Department of Psychiatry
Campus Box C249-32, Denver, CO 80111, USA

Subroto Ghose
University of Texas Southwestern, Department of Psychiatry
5323 Harry Hines Blvd.
Dallas, TX 75390, USA

Robert W. Greene
University of Texas Southwestern, Department of Psychiatry
5323 Harry Hines Blvd.
Dallas, TX 75390, USA

Paul J. Harrison
University of Oxford, Department of Psychiatry, Warneford Hospital
Oxford OX3 7JX, United Kingdom

Takanori Hashimoto
University of Pittsburgh, Department of Psychiatry
3811 O'Hara St., Pittsburgh, PA 15213, USA

Todd E. Kippin
University of California at Santa Barbara, Dept Psychology
Santa Barbara, CA 93106, USA

David A. Lewis
University of Pittsburgh, Department of Psychiatry
3811 O'Hara St., Pittsburgh, PA 15213, USA

Benjamin Lim
McLean Hospital, Program in Structural and Molecular Neuroscience
115 Mill St., Belmont, MA 02478, USA

Jaime G. Maldonado-Aviles
University of Pittsburgh, Department of Psychiatry
3811 O'Hara St., Pittsburgh, PA 15213, USA

Anil K. Malhotra
Zucker Hillside Hospital, Division of Psychiatry Research
Glen Oaks, NY, 11004, USA

David Matzilevich
McLean Hospital, Program in Structural and Molecular Neuroscience
115 Mill St., Belmont, MA 02478, USA

Robert E. McCullumsmith
University of Alabama at Birmingham, Department of Psychiatry
1530 3rd Avenue South, Birmingham, AL 35294, USA

James H. Meador-Woodruff
University of Alabama at Birmingham, Department of Psychiatry
1530 3rd Avenue South, Birmingham, AL 35294, USA

Harvey Morris
University of Pittsburgh, Department of Psychiatry
3811 O'Hara St., Pittsburgh, PA 15213, USA

Patricio O'Donnell
University of Maryland School of Medicine
20 Penn St., Baltimore, MD 21286, USA

Christine Roy
Harvard University, Children's Hospital, Neurobiology Program
300 Longwood Ave, Boston, MA 02115, USA

Karen K. Szumlinski
University of California at Santa Barbara, Dept Psychology
Santa Barbara, CA 93106, USA

Carol A. Tamminga
University of Texas Southwestern, Department of Psychiatry
5323 Harry Hines Blvd.
Dallas, TX 75390, USA

Jason R. Tregellas
University of Colorado, Department of Psychiatry
Campus Box C249-32, Denver, CO 80111, USA

John Walsh
McLean Hospital, Program in Structural and Molecular Neuroscience
115 Mill St., Belmont, MA 02478, USA

Preface

The genetics of schizophrenia have taken the driver seat in the research related to this disorder in recent years. For a long time, it had been known that sharing genes with someone affected with schizophrenia conferred a 50% chance of acquiring the disease. Although this is fifty times the incidence in the normal population, it was evident that genes could not explain all factors involved. At best, one could think of a combination of gene traits conferring predisposition for the disease. Now, it is evident that although there are no genes that individually confer a strong liability, several candidate gene alleles do confer some risk, and perhaps a combination of genes could cause the emergence of symptoms in this multi-faceted psychiatric condition. As the research on predisposing genes soars, it is becoming clear that we need to be able to identify cellular processes the genes control and to provide pathophysiological scenarios that could link those cellular phenomena with symptoms. In this book, we have chosen to elaborate on that link.

The contributions to this book have been requested to reflect the current state of the connection between schizophrenia genetics and pathophysiology. The book opens with a chapter by Paul Harrison highlighting the several convergence points among the diverse schizophrenia-related genes. This is followed by chapters reviewing several different specific genes, including Neuregulin, DISC-1, dysbindin, and Homer, among others, and by several chapters presenting information on physiological cortical processes that could be affected by those genes. These include glutamatergic, GABAergic and cholinergic neurotransmission.

Current hot topics in schizophrenia research are also highlighted in the book. A concept that is gaining strength in the field, for example, is the role of inhibitory neurotransmission in schizophrenia pathophysiology, with converging (and convincing) data from *post-mortem* and imaging studies. The possibility of deficits in cortical inhibition (causing the emergence of "noisy" cortical activity) is discussed throughout the book in relationship with possible alterations in GABA and/or glutamate activity in several cortical areas including the prefrontal cortex and the temporal lobe. Thus, the chapters in this book summarize the current state of affairs in schizophrenia pathophysiology, with a consideration of genetic elements that may contribute to it. This timely book could prove a useful tool for those interested in a better understanding of schizophrenia. Even if more predisposing genes (not included in this book) become identified, the work laid out here will prove critical for assigning a functional (or dysfunctional) role to those genes. Although this is a fast-evolving field, the information and concepts covered in this book will likely be useful for years to come.

Patricio O'Donnell, MD, PhD

Foreword

The genetics of schizophrenia have taken the driver seat in recent years in the research related to this disorder. For a long time, it had been known that sharing genes with someone affected with schizophrenia conferred a 50% chance of acquiring the disease. Although this is 50 times the incidence in the normal population, it was evident that genes could not explain all factors involved. At best, one could think of a combination of gene traits conferring predisposition for the disease. Now, it is evident that although there are no genes conferring a strong liability independently, several candidate gene alleles do confer risk, and perhaps a combination of several genes could cause the emergence of symptoms in this multi-faceted psychiatric condition. As the research on predisposing genes soars, it is becoming clear that we need to be able to provide pathophysiological scenarios that could link the cellular phenomena the genes control with symptoms. In this book, we have chosen to elaborate on that link.

Chapter 1

Schizophrenia Genes: Searching for Common Features, Functions, and Mechanisms

Paul J. Harrison

Abstract. As chapters in this book make clear, substantial progress has been made in the past five years towards identifying susceptibility genes that underlie the 80% heritability of schizophrenia (Sullivan et al. 2003; Harrison and Weinberger 2005; Owen et al. 2005). At first sight, the various genes appear to have little in common. However, the theme of this chapter is convergence; the notion that the genes may share important features, in terms of the nature of genetic association, the mediating molecular mechanisms, and their pathogenic roles. If any of these convergences prove even partially correct, the hope is that it may facilitate the ongoing search for additional genes and, in particular, how they operate to confer schizophrenia susceptibility.

1.1 The Nature of Genetic Associations with Schizophrenia

At the outset, it is worth noting that the evidence concerning schizophrenia susceptibility genes is, and will for several reasons likely remain, both statistically and molecularly ambiguous. Three aspects of this uncertainty are briefly outlined, since they set the scene for the discussion of convergence.

1.1.1 Statistics: Necessary but not Sufficient

Meta-analyses of the genome-wide linkage scans of schizophrenia were influential in moving the field forwards from skepticism to reasonable confidence that several loci harboring schizophrenia genes had been identified (Badner and Gershon 2002; Lewis et al. 2003). Similarly, meta-analysis can help establish the robustness of genetic associations, and are now being reported for several of the genes.

However, the fact that multiple alleles within each susceptibility gene are reportedly associated with schizophrenia makes this task difficult. Should the meta-analysis be for one or more polymorphism, or for haplotypes? Should family and case control studies be combined? Such issues, along with methodological factors concerning how the meta-analysis is conducted, can lead to discrepant results. A good example is provided by the two meta-analyses of Neuregulin (NRG1) and schizophrenia, published almost simultaneously, and using virtually the same datasets. Munafo et al. (2006) found no significant association with any single allele, but a significant (p=0.02) association when the most significant haplotype p-value from each study was combined. They concluded that their analysis 'provides support for the association of NRG1 with schizophrenia, but...firmly establishing the role...requires much larger sample sizes.' Li et al. (2006), on the other hand, found association of 6 individual NRG1 SNPs with schizophrenia (0.00003<p<0.014), and the haplotype analysis was highly significant (p=8x10^{-10}). Their interpretation was correspondingly more upbeat, as reflected in the title of the paper.

The meta-analytic findings illustrate two points *vis à vis* convergence. First, that statistical genetics alone is unlikely to provide an unequivocal and unambiguous answer; it is necessary, but it is not sufficient. Thus, complementary evidence, particularly plausible molecular and biological mechanisms, will also be required, in order to provide convergent evidence to establish the candidacy of a gene. The same will apply to the search for gene-gene and gene-environment interactions. Second, the two NRG1 meta-analyses do agree on one important point: the specific variant(s) associated with schizophrenia has not been identified. What explains the consistent but unfocused association signals originating from the NRG1 locus? The same question pertains to all other putative schizophrenia genes; the issue is whether it has the same answer across genes, in which case a convergent molecular basis for association may exist. With regard to this possibility, two issues need consideration: allelic heterogeneity, and identification of the genuine risk variants.

1.1.2 Allelic heterogeneity or false positives?

The first possible explanation for the association of multiple polymorphisms within a gene with schizophrenia is allelic heterogeneity. For example, Li et al. (2006) found evidence that different SNPs and haplotypes within NRG1 are associated in Caucasian and Asian populations. Allelic heterogeneity is plausible, and arguably even probable, given findings in Mendelian disorders, of which cystic fibrosis (>1400 different mutations in the CFTR gene) is a striking example (Rowe et al. 2005). In this allelic heterogeneity scenario, two or more of the associated NRG1 polymorphisms are 'true' associated variants, in different individuals and populations. On the other hand, in complex disorders, the relationship between genetic variation and phenotype is inherently less clear, both statistically and functionally, than in Mendelian disorders (reflected in the differing terminologies for genetic variation in the two kinds of disease – 'mutation' versus 'polymorphism'). In consequence, there is a much higher prior likelihood that an individual polymorphism reported as being associated with schizophrenia is a false positive. Given this, it is a moot point whether the null hypothesis for schizophrenia genes is for or against allelic heterogeneity. (Note that this problem, and indeed the whole question of replication and proof of genetic association is not unique to schizophrenia; it is being grappled with in other complex disorders too (Botstein and Risch 2003; Bertram and Tanzi 2004; Cavalleri et al. 2005; Rosand et al. 2006).

1.1.3 Identifying the causal variant(s)

The association of multiple alleles within a gene with schizophrenia may not result from allelic heterogeneity, or false positives, but reflect the fact that the various polymorphisms are marker SNPs in linkage disequilibrium (LD) with the causal variant(s). This scenario does not lend itself to simple empirical testing. Identifying the 'true' variant may be difficult, either because (1) it is a rare mutation, or (2) because it is distant from the marker SNPs, and may itself be non-coding (see Rebbeck et al. 2004 for review).

Regarding rare but causal (coding) variants, in the case of NRG1 and dysbindin, sequencing of many dozens of chromosomes has not identified any novel coding mutations in schizophrenia, rendering this possibility unlikely (although see Walss-Bass et al. 2006).

The alternative scenario – that the reported SNP associations are in LD with another, but still non-coding, variant – is hard to quantify or disprove, in part because of the range of possibilities. For example, the true risk allele may reside in an adjacent gene, a gene on the opposite strand, or in an intronic microRNA. Apolipoprotein E provides a precedent for the range of genetic complexity of this kind that may exist, even for a gene where a single, coding, disease-associated polymorphism has been identified (Seitz et al. 2005). Bioinformatics can help estimate the prior probabilities of these molecular nuances, but ultimately there will be no substitute for empirical data – yet discovering them or ruling them out will be a considerable task. Again, NRG1 is the exemplar: it is a large gene, with over twenty exons, several large introns, six putative promoters, extensive alternative splicing, and potential microRNAs within the locus (Harrison and Law 2006).

In summary, the genetic and 'intragenic' architecture of schizophrenia remains frustratingly elusive. Despite convincing signals from several loci, the basis for, and precise source of, genetic association is unknown. Causal, coding mutations seem increasingly unlikely but cannot be dismissed. It is not clear if the currently associated polymorphisms are directly functional, or, if not, the identity, location and number of those that are. Prioritizing the options to pursue, and how far to pursue them, will become a major decision for research groups to ponder. Against this background of uncertainty and the escalating number of research questions regarding schizophrenia genes and how they operate, the potential value of some common or unifying themes, or convergences, becomes apparent.

1.2 Types of Convergence

The remainder of this chapter discusses various possible convergences concerning the mechanism of genetic association, and the nature and properties of schizophrenia genes.

Table 1.1. Examples of other disease genes reported to confer susceptibility via altered gene expression

Disease	Polymorphism	Effect	Reference
Neuropsychiatric Disorders			
Alzheimer's disease	Promoter presenilin 1 SNP	Alters abundance	Theuns et al 2003
Alzheimer's disease	Intronic ubiquitin SNP	Alters splicing	Bertram et al 2005
Progressive supranuclear palsy	Intronic tau SNP	Alters splicing	Rademakers et al 2005
Parkinson's disease	Promoter α-synuclein length polymorphism	Alters abundance	Maragonore et al 2006
Major depression	Promoter 5-HT$_{1A}$R SNP	Alters abundance	Lemonde et al 2003
Other Diseases			
Rheumatoid arthritis	Intronic SLC22A4 SNP	Alters splicing	Tokuhiro et al 2003
Sarcoidosis	Exonic non-coding BTNL-2 SNP	Alters splicing	Valentonyte et al 2005
Cystic fibrosis	Intronic CFTR SNP	Alters splicing	Rowe et al 2005

1.2.1 Genetic Associations Explained by Effects on Gene Expression?

As noted, the polymorphisms currently associated with schizophrenia in all susceptibility genes are non-coding (with the notable if controversial exception of the Val[158]Met substitution in catechol-O-methyltransferase [COMT]; see Tunbridge et al. 2006). Most polymorphisms occur in promoter regions, introns, or the 3' untranslated region; the remainder are conservative exonic substitutions. The possibility that the 'true' variants are in fact coding variants yet to be discovered was mentioned above. However, the continuing failure to identify these makes it increasingly likely that non-coding variants are actually responsible for the associations. This implies strongly that the functionality of the polymorphisms must reside in their effect on gene expression, not upon the sequence and thence the structure and function of the encoded protein. Indeed, in the case of RGS4, it was the decreased mRNA abundance observed in schizophrenia brain which led to its investigation as a candidate gene (see Talkowski et al. 2006); a similar rationale applied to Akt1 (Emamian et al. 2004). There are plenty of precedents in other diseases for risk alleles affecting gene expression (Table 1), including some of the many CTFR gene mutations (Rowe et al. 2005). See Schadt et al. (2005) for discussion of how gene expression and disease may be causally related.

Altered gene expression may manifest in a range of ways: changes in the overall amount of mRNA or protein synthesized; altered abundance or proportion of specific isoforms; differences in the timing of expression; or a change in the cell type or subcellular compartments wherein the gene products are expressed. Such differences can be caused by comparably diverse mechanisms (e.g. Duan et al. 2003; Jaenisch and Bird 2003; Wilkins and Haig 2003; Shen et al. 1999; Pastinen and Hudson 2004; Matlin et al. 2005; Kleinjau and van Heyningen 2005). It is even possible that the allelic variants affect expression of neighboring or functionally related genes rather than that of the susceptibility gene itself (Lipska et al. 2006a). Again, the problem that arises is deciding which of these multiple facets of gene expression and putative underlying mechanisms to study, and how to do so.

Although at a very early stage of this kind of research, empirical evidence is growing that the expression of schizophrenia susceptibility genes is indeed altered, in one or other of the above ways (for review, see Harrison and Weinberger 2005). For example, dysbindin mRNA (Weickert et al. 2004) and protein (Talbot et al. 2004) are both decreased in schizophrenia. For DISC1, mRNA expression is unchanged in schizophrenia (Lipska et al. 2006a), but there is a redistribution of DISC1 protein between cytoplasm and nucleus (Sawamura et al. 2005). In the case of NRG1, two kinds of altered expression have been reported. Firstly, an increase in mRNA encoding the type I NRG1 isoform in schizophrenia (Hashimoto et al. 2004; Law et al. 2006). Secondly, type IV NRG1 mRNA is increased in individuals carrying the NRG1 risk haplotype, independent of diagnosis (Law et al. 2006). These NRG1 expression findings – one related to genotype, the other to diagnosis, as well as the isoform selectivity of both changes – illustrate important points to consider when evaluating the notion that schizophrenia risk gene variants mediate their effects via altered gene expression.

1.2.2 Genotype Effects on Gene Expression

If a SNP (or one with which it is in LD) is functional because it affects gene expression, one would expect this to occur independent of diagnosis. That is, the SNP is functional *per se*; the fact that it is a risk allele for a disease is presumably because the alteration in expression that it influences is a pathophysiological risk factor. An increasing number of examples of schizophrenia risk alleles or haplotypes that alter gene expression are being reported, including the NRG1 type IV finding mentioned above. Some studies report overall mRNA abundance differences between genotype groups (e.g. Xu et al. 2005; Law et al. 2006), whereas a complementary approach is to measure differential allelic expression, in which the proportion of mRNAs encoded by each chromosome is measured in informative heterozygotes, as for dysbindin (Bray et al. 2005), CNPase (Peirce et al. 2006) and DGR2 (Shifman et al. 2006).

1.2.3 Altered Expression in Schizophrenics vs. Controls

Examples of altered expression of several susceptibility genes in schizophrenia compared to controls (independent of genotype) were noted above, and in other recent reports (e.g. Eastwood et al. 2005a; Eastwood and Harrison 2007). Interpretation of these findings is somewhat more difficult than differences related to genotype. Most known schizophrenia risk alleles are only carried by a minority of cases (and with a modest odds ratio compared to controls). For example, the NRG1 risk haplotype in the original two studies had a frequency of ~14% in cases and ~7% in controls. Thus, a group difference in expression between cases and controls cannot be driven just by a modestly greater proportion of risk allele-carrying subjects in the schizophrenia group. At the extreme, a diagnostic difference in expression may be completely separate from, and irrelevant to, the fact the gene may be a susceptibility gene; its altered expression in schizophrenia could be secondary to many other factors, including differential confounding by smoking (Mexal et al. 2005) or antipsychotic medication. For this reason, a diagnosis-related change in expression does not necessarily and of itself add to the evidence that the gene may be a susceptibility gene. At the other extreme, however, a diagnosis-related difference in expression *may* in fact be directly related to its candidacy since, as discussed below, another aspect of convergence is that the various genes converge on common biochemical pathways. As part of this convergence, there may be shared effects on expression of genes within the pathway, regardless of which risk alleles (or, for that matter, epigenetic or environmental factors) happen to be acting in each individual. A similar suggestion has been made by Mirnics et al. (2006). For instance, maybe the up-regulation of type I NRG1 in schizophrenia reflects the fact that this isoform is regulated by a shared functional effect of the risk variants of NRG1 and of its ErbB4 receptor – and other genes within this pathway that prove to be susceptibility genes (see below).

1.2.4 Altered Splicing as a Basis for Altered Expression

The NRG1 data also illustrate the important point that the altered expression may affect specific isoforms of a gene, either via effects on promoters or on splicing, depending in part on the intragenic location of the associated SNPs. Other examples are also being reported for other genes. However, whether the findings are viewed as increasing the likelihood that there *is* genetic association with schizophrenia, is debatable; it illustrates the point made earlier regarding the relevance of biological plausibility and a molecular mechanism when deciding if a particular polymorphism or gene is a susceptibility factor.

1.2.5 Convergence of the Genes on Synapses, Glutamate and NMDA Receptors?

A separate aspect of convergence relates to the putative function of the genes. For all the current susceptibility genes, except DISC1 and G72 (DAOA), there was information about their biology that preceded the link to schizophrenia, allowing an initial consideration as to the pathophysiological implications of the genetic association. Moreover, there has been a subsequent surge of experimental interest in each gene, ranging from their cell biology (e.g. Porteous et al. 2006) to their impact on brain structure and function (e.g. Callicott et al. 2005).

The functional convergence that has received the most interest is the suggestion that the various genes share an impact upon glutamatergic synaptic transmission, especially that mediated by NMDA receptors (Harrison and Owen, 2003; Moghaddam, 2003). This is an attractive notion not least because it chimes with the prominent 'pre-genetic' theories as to the role of glutamate, NMDA receptors, and synaptic dysfunction in schizophrenia (Olney and Farber 1995; Goff and Coyle 2001; Harrison and Eastwood 2001; Frankle et al. 2003). This mode of convergence is described in detail in Harrison and Weinberger (2005).

Although attractive and plausible, this proposed convergence is relatively weak: it is not a convergence upon a specific molecule, or a defined molecular pathway. Thus, comparisons with the established convergence of genetic and other factors upon β-amyloid processing in Alzheimer's disease, or upon α-synuclein and Parkinson's disease, though illustrative, are somewhat misleading. In consequence, how meaningful, or falsifiable, is the notion? Much of the genome is expressed in the brain, and many genes are involved in one way or another with synapses or glutamate. The proposal therefore has a non-trivial prior probability of being true and, as such, the convergence may be apparent rather than real. As a first step to address this issue, we performed a bioinformatic and literature search to establish the proportion of randomly selected genes that met pre-determined criteria for being called 'synaptic', 'glutamatergic' or 'NMDA receptor modulating', and then compared this frequency to that for the leading schizophrenia susceptibility genes. We found that all three categories were over-represented amongst the schizophrenia genes (Harrison and West, 2006). In this respect, therefore, the proposal has some statistical merit. It remains to be seen if this continues to be the case as new schizophrenia genes, and new functions for existing genes, are discovered, and whether the proposal can be more fully specified in molecular terms.

In summary, a functional convergence of schizophrenia susceptibility genes on synapses, glutamate and NMDA receptors seems reasonable, but the ultimate validity and utility of this convergence requires much further research. All the same caveats apply to other comparable proposals, such as that the genes converge via effects on glia, plasticity, or oxidative stress (e.g. Moises et al. 2002; Carter 2006).

1.2.6 Convergence of the Genes upon Neurodevelopment?

The idea that schizophrenia genes work because they impact on brain maturation is attractive, fitting in as it does with the prevailing neurodevelopmental view of the disorder. Moreover, it is true that NRG1, arguably the leading gene, is unequivocally involved in multiple aspects of neurodevelopment (Falls 2003; Harrison and Law 2006). There are also persuasive data that a truncated form of DISC1, as seen in the Scottish pedigree wherein DISC1 was identified, impairs cortical development (Kamiya et al. 2005). More speculatively, Schmidt-Kastner et al. (2006) note that expression of over half of the current susceptibility genes are affected by hypoxia, and argue that developmental hypoxia may be a mediating mechanism to explain their involvement in the disease.

As with the other convergences, the notion of a neurodevelopmental pathogenesis for the susceptibility genes has weaknesses (Harrison 2007). First, the ontogenic profile of expression of the gene is sometimes taken to bear upon its candidacy. However, most genes are expressed both during neurodevelopment (whenever that is said to end) and thereafter, and it is not clear whether particular temporal changes in expression level have functional implications. For example, COMT expression and activity rises steadily until early adulthood and then increases more markedly in the fourth decade (Tunbridge et al. 2007). Does this support an involvement of the gene in the pathogenic processes hypothesised to occur during development, or the factors associated with the onset of illness? Does the fact that some genes, such as PPP3CC (Eastwood et al. 2005b) and DISC1 (Lipska et al. 2006a) have peak expression much earlier in life and then decline support the pre or perinatal origins of the disorder? Second, if a 'neurodevelopmental' gene continues to be expressed at moderate or high levels in the brain throughout life (as is the case for NRG1 (Law et al. 2004) and, as far as is known, all other susceptibility genes), it is reasonable to assume that it continues to have a function. Whether this function is still the same as that during developmental is an empirical question; equally, if the gene has different functions at different ages, how could one decide which function was relevant to its involvement in schizophrenia?

In summary, at least some of the pathogenicity of susceptibility genes may well operate during, and impact on, brain maturation. However, there are no convincing data at present to suggest this will be a key convergence.

1.2.7 Convergence between Genes?

Gene-gene interactions (epistasis), both statistical and biological, are beginning to be reported. For example, statistically, interactions exist between the COMT Val[158] allele and SNPs in RGS4, GRM3, DISC1 and G72 (Nicodemus et al. 2006). Biologically, there are links between RGS4 alleles and COMT activity (Lipska et al. 2006b) and between COMT and PRODH in a mouse model (Paterlini et al. 2005).

The relative importance of epistatic versus single gene effects on schizophrenia susceptibility is not clear, and will require much more extensive datasets (and statistical approaches) to be clarified.

1.2.8 Convergence of Genes within a Biochemical Pathway?

A specific example of gene-gene interaction concerns genes that act in the same biochemical pathway. A good precedent for this is provided by APP and presenilin, the genes causing familial Alzheimer's disease, in that presenilin is part of the γ-secretase complex that cleaves APP. For schizophrenia, there are several emerging examples of this kind. First, as noted above, NRG1 signals through ErbB4 to activate its effector kinase pathways. As well as the strong evidence for NRG1 as a schizophrenia gene, there are now positive reports for ErbB4 (Silberberg et al. 2006), and for epistasis between NRG1 and ErbB4 (Norton et al. 2006), as well as a functional interaction between them in schizophrenic brain (Hahn et al. 2006). The second example is dysbindin, which participates in the lysosomal BLOC1 complex; at least two other genes in the BLOC1 complex (Muted and Palladin) may also be associated with schizophrenia (Straub and Weinberger, 2006). A final example in this category is the convergence between D-amino acid oxidase (DAO) and G72. Chumakov et al. (2002) reported association of the novel gene G72 with schizophrenia; in the same study, a biochemical interaction of G72 with DAO was identified, complemented by evidence for genetic association of DAO with schizophrenia. Both these genes also typify the convergence on NMDA receptors described above, since G72 (now renamed D-amino acid oxidase activator [DAOA]) regulates DAO, which metabolizes the endogenous NMDA receptor modulator, D-serine. Indeed, adding to the genetic focus on this pathway, SNPs in the D-serine synthetic enzyme, serine racemase, have also been reported to be associated with schizophrenia (Goltsov et al. 2006), as has variation in the serine racemase interacting protein PICK1 (Fujii et al. 2006).

In summary, there is tantalizing evidence for the association of multiple genes within a biochemical pathway with schizophrenia. However, the existence and magnitude of this convergence cannot be determined at present.

1.2.9 Convergence of Genes within Loci?

The final convergence to be discussed is that schizophrenia genes cluster at particular genomic regions, a possibility first hinted at by the fact that the effect sizes for individual genes did not seem sufficient to explain the linkage signal at that locus. This is true for dysbindin on chromosome 6p, and NRG1 on chromosome 8p, amongst others. Subsequently, evidence has begun to emerge that in fact these linkage signals may conceal two (or more) genes. Thus, dysbindin, Muted, and MRDS1 (OFCC1) are all located under the broad 6p peak (Straub and Weinberger 2006); similarly, PPP3CC (Emamian et al. 2004), PCM1 (Gurling et al. 2006) and FZD3 (Yang et al. 2003) are all close to NRG1 at chromosome 8p.

A particular and striking example of this genomic convergence is that of chromosome 22q11 (Bassett and Chow 1999), a region not only implicated by the meta-analyses, but also by virtue of the manifold increase in schizophrenia incidence seen in subjects with a hemizygous deletion of 22q11. Within 22q11, four genes have now been associated with schizophrenia – COMT, itself a candidate gene (see Tunbridge et al. 2006a), ZDHHC8 (Mukai et al. 2004), PRODH (Liu et al. 2002), and DGCR2 (Shifman et al. 2006).

It is not clear why there should be genomic clustering of otherwise unrelated schizophrenia genes. In any event, no explanation is required until the phenomenon is confirmed.

1.3 Conclusions

This chapter has reviewed a miscellany of potential common features or convergences that may characterize schizophrenia susceptibility genes and how they operate. The justification for doing so is that, should such principles exist, their identification may help guide and prioritize future approaches. Without any such convergence, the number of susceptibility genes and alleles already reported results in an almost unmanageable complexity.

On the other hand, the seductive attractiveness of the concept of convergence should not mask the vagueness and lack of positive evidence for any of the specific suggestions. One sign of this weakness is to reframe each proposition in reverse. For example, how likely is it that the genes do *not* have a role in neurodevelopment, or in synaptic function, or impact some aspect of NMDA receptor signaling? Or that there *are* in fact causal coding variants hidden in all the genes? Thus, there is a problem with both the prior probability, and the falsifiability, that constrain all attempts to make sense of, or generalize from, the current situation regarding the identity of schizophrenia genes and the associated variants within them. This is probably inevitable, given the genetic architecture (i.e. lots of small effect size genes, with few or no coding mutations) and the lack of unambiguous biological markers or known pathophysiological elements with which the genes can be integrated; rather, the genes are needed to get a fix on the biology. Advances will therefore probably remain incremental, and the process will need to be iterative between molecular genetics, statistics, and functional studies of the expression and function of the genes. No single approach on its own will reveal the full picture.

Opening this book optimistically, let us recall that in 2001 there were no widely accepted chromosomal loci for schizophrenia, and no well replicated susceptibility genes. Now, in 2006, there is at least one gene in this category (NRG1) and arguably several more. Their identification has provided a new impetus to the field and is already paying dividends in terms of unraveling the pathophysiology of schizophrenia, and with the realistic prospect of substantial further progress. We should not lose sight of this goal, however intricate the molecular neurobiology of schizophrenia genes proves to be.

Acknowledgements

Work supported by the UK Medical Research Council, the Stanley Medical Research Institute, Wellcome Trust, and NARSAD. I thank the many group members and colleagues whose data and intellectual insights have helped shape the ideas presented here, especially Phil Burnet, Sharon Eastwood, Amanda Law, Liz Tunbridge, and Danny Weinberger. Valerie West kindly helped prepare the manuscript.

References

Badner JA, Gershon ES (2002) Meta-analysis of whole-genome linkage scans of bipolar disorder and schizophrenia. Mol Psychiatry 7:405-411.

Bassett AS, Chow EWC (1999) 22q11 deletion syndrome: A genetic subtype of schizophrenia. Biol Psychiatry 46:882-891.

Bertram L, Tanzi RE (2004) Alzheimer's disease: one disorder, too many genes? Hum Mol Genet 13 (R1): R135-141.

Bertram L, Hiltunen M, Parkinson M, Ingelsson M, Lange C, et al (2005) Family-Based Association between Alzheimer's Disease and Variants in UBQLN1. N Engl J Med 352:884-894.

Botstein D, Risch N (2003) Discovering genotypes underlying human phenotypes: past successes for mendelian disease, future prospects for complex disease. Nat Genet 33 Suppl: 228-237.

Bray NJ, Preece A, Williams NM, Moskvina V, Buckland PR, et al (2005) Haplotypes at the dystrobrevin binding protein 1 (DTNBP1) gene locus mediate risk for schizophrenia through reduced DTNBP1 expression. Hum Mol Genet 14:1947-1954.

Callicott JH, Straub RE, Pezawas L, Egan MF, Mattay VS, et al (2005) Variation in DISC1 affects hippocampal structure and function and increases risk for schizophrenia. Proc Natl Acad Sci USA 102:8627-8632.

Carter CJ (2006) Schizophrenia susceptibility genes converge on interlinked pathways related to glutamatergic transmission and long-term potentiation, oxidative stress and oligodendrocyte viability. Schiz Res 86:1-14.

Cavalleri GL, Lynch JM, Depondt C, Burley M-W, Wood NW et al (2005) Failure to replicate previously reported genetic associations with sporadic temporal lobe epilepsy: where to from here? Brain 128:1832-1840.

Chumakov I, Blumenfeld M, Guerassimenko O, Cavarec L, Palicio M, et al (2002) Genetic and physiological data implicating the new human gene G72 and the gene for D-amino acid oxidase in schizophrenia. Proc Natl Acad Sci USA 99:13675-13680.

Duan J, Wainwright MS, Comeron JM, Saitou N, Sanders AR, et al (2003) Synonymous mutations in the human dopamine receptor D_2 (DRD2) affect mRNA stability and synthesis of the receptor. Hum Mol Genet 12:205-216.

Eastwood SL, Burnet PWJ, Harrison PJ (2005a) Decreased hippocampal expression of the susceptibility gene PPP3CC and other calcineurin subunits in schizophrenia. Biol Psychiatry 57:702-710.

Eastwood SL, Salih T, Harrison PJ (2005b) Differential expression of calcineurin A subunit mRNA isoforms during rat hippocampal and cerebellar development. Eur J Neurosci 22:3017-3024.

Eastwood SL, Harrison PJ (2007) Decreased mRNA expression of Netrin-G1 and Netrin-G2 in the temporal lobe in schizophrenia and bipolar disorder [abstract]. Schizophr Res (in press).

Egan MF, Straub RE, Goldberg TE, Yakub I, Callicott JH, et al (2004) Variation in GRM3 affects cognition, prefrontal glutamate, and risk for schizophrenia. Proc Natl Acad Sci USA 101:12604-12609.

Emamian ES, Hall D, Birnbaum MJ, Karayiorgou M, Gogos JA (2004) Convergent evidence for impaired AKT1-GSK3β signaling in schizophrenia. Nat Genet 36:131-137.

Falls DL (2003) Neuregulins: functions, forms, and signaling strategies. Exp Cell Res 284:14-30.

Frankle WG, Lerma J, Laruelle M (2003) The synaptic hypothesis of schizophrenia. Neuron 39:205-216.

Fujii K, Maeda K, Hikida T, Mustafa A, Balkissoon R et al (2006) Serine racemase binds to PICK1: potential relevance to schizophrenia. Mol Psychiatry 11:150-157.

Goff DC, Coyle JT (2001) The emerging role of glutamate in the pathophysiology and treatment of schizophrenia. Am J Psychiatry 158:1367-1377.

Goltsov AY, Loseva JG, Andreeva TV, Grigorenko AP, Abramova LI, et al (2006) Polymorphism in the 5'-promoter region of serine racemase gene in schizophrenia. Mol Psychiatry 11:325-326.

Gurling HMD, Critchley H, Datta SR, McQuillin A, Blaveri E, Thirumalai S et al (2006) Genetic association and brain morphology studies and the chromosome 8p22 pericentriolar material 1 (PCM1) gene in susceptibility to schizophrenia. Arch Gen Psychiatry 63:844-854.

Hahn CG, Wang HY, Cho DS, Talbot K, Gur RE, et al (2006) Altered neuregulin 1-erbB4 signaling contributes to NMDA receptor hypofunction in schizophrenia. Nat Med 12:824-828.

Harrison PJ (2007) Schizophrenia susceptibility genes and neurodevelopment. Biol Psychiatry (in press).

Harrison PJ, Eastwood SL (2001) Neuropathological studies of synaptic connectivity in the hippocampal formation in schizophrenia. Hippocampus 11:508-519.

Harrison PJ, Law AJ (2006) Neuregulin 1 and schizophrenia: genetics, gene expression, and neurobiology. Biol Psychiatry 60:132-140.

Harrison PJ, Owen MJ (2003) Genes for schizophrenia? Recent findings and their pathophysiological implications. Lancet 361:417-419.

Harrison PJ, Weinberger DR (2005) Schizophrenia genes, gene expression, and neuropathology: on the matter of their convergence. Mol Psychiatry 10:40-68.

Harrison PJ, West VA (2006) Six degrees of separation: on the prior probability that schizophrenia susceptibility genes converge on synapses, glutamate, and NMDA receptors. Mol Psychiatry 11: 981-983.

Hashimoto R, Straub RE, Weickert CS, Hyde TM, Kleinman JE, Weinberger DR (2004) Expression analysis of neuregulin-1 in the dorsolateral prefrontal cortex in schizophrenia. Mol Psychiatry 9:299-307.

Jaenisch R, Bird A (2003) Epigenetic regulation of gene expression: how the genome integrates intrinsic and environmental signals. Nat Genet 33:245-254.

Kamiya A, Kubo K, Tomoda T, Takaki M, Youn R, et al (2005) A schizophrenia-associated mutation of DISC1 perturbs cerebral cortex development. Nat Cell Biol 7:1067-1078.

Kleinjan DA, van Heyningen V (2005) Long-range control of gene expression: emerging mechanisms and disruption in disease Am J Hum Genet 76:8-32.

Law AJ, Weickert CS, Hyde T, Kleinman J, Harrison PJ (2004) Neuregulin-1 (NRG1) mRNA and protein in the adult human brain. Neuroscience 127:125-136.

Law AJ, Lipska BK, Weickert CS, Hyde TM, Straub RE, et al (2006) Neuregulin 1 transcripts are differentially expressed in schizophrenia and regulated by 5' SNPs associated with the disease. Proc Natl Acad Sci USA 103:6747-6752.

Lemonde S, Turecki G, Bakish D, Du L, Hrdina PD, et al (2003) Impaired repression at a 5-hydroxytryptamine 1A receptor gene polymorphism associated with major depression and suicide. J Neurosci 23:8788087-99.

Lewis CM, Levinson DF, Wise LH, DeLisi LE, Straub RE, et al (2003) Genome scan meta-analysis of schizophrenia and bipolar disorder, part II: Schizophrenia. Am J Hum Genet 73:34-48.

Li D, Collier DA, He L (2006) Meta-analysis shows strong positive association of the neuregulin 1 gene with schizophrenia. Hum Mol Genet 15:1995-2002.

Lipska BK, Peters T, Hyde TM, Halim N, Horowitz C, et al (2006a) Expression of DISC1 binding partners is reduced in schizophrenia and associated with DISC1 SNPs. Hum Mol Genet 15:1245-1258.

Lipska BK, Mitkus S, Caruso M, Hyde TM, Chen J, et al (2006b) RGS4 mRNA expression in postmortem human cortex is associated with COMT Val158Met genotype and COMT enzyme activity. Hum Mol Genet 15:2804-2812.

Liu H, Abecasis GR, Heath SC, Knowles A, Demars S, et al (2002) Genetic variation in the 22q11 locus and susceptibility to schizophrenia. Proc Natl Acad Sci USA 99:16859-16864.

Maraganore DM, de Andrade M, Elbaz A, Farrer MJ, Ioannidis JP, et al (2006) Collaborative Analysis of {alpha}-Synuclein Gene Promoter Variability and Parkinson Disease. JAMA 296:661-670.

Matlin AJ, Clark F, Smith CWJ (2005) Understanding alternative splicing: towards a cellular code. Nat Rev Mol Cell Biol 6:386-398.

Mexal S, Frank M, Berger R, Adams CE, Ross RG, et al (2005) Differential modulation of gene expression in the NMDA postsynaptic density of schizophrenic and control smokers. Mol Brain Res 139:317-332.

Mirnics K, Levitt P, Lewis DA (2006) Critical appraisal of DNA microarraysin psychiatric genomics. Biol Psychiatry 60: 163-176.

Moghaddam B (2003) Bringing order to the glutamate chaos in schizophrenia. Neuron 40:881-884.

Moises HW, Zoetga T, Gottesman II (2002) The glial growth factors deficiency and synaptic destabilization hypothesis of schizophrenia. BMC Psychiatry 2.

Mukai J, Liu H, Burt RA, Swor DE, Lai WS, et al (2004) Evidence that the gene encoding ZDHHC8 contributes to the risk of schizophrenia. Nat Genet 36:725-731.

Munafò MR, Thiselton DL, Clark TG, Flint J (2006) Association of the NRG1 gene and schizophrenia: a meta-analysis. Mol Psychiatry 11:539-546.

Nicodemus KK, Kolachana BS, Vakkalanka R, Straub RE, Giegling I, et al (2006) Evidence for statistical epistasis between Catechol-O-Methyltransferase (COMT) and polymorphisms in RGS4, G72 (DAOA), GRM3 and DISC1: Influence on risk of schizophrenia. Hum Genet (in press).

Norton N, Moskvina V, Morris D, Bray NJ, Zammit S, et al (2006) Evidence that interaction between neuregulin 1 and its receptor erbB4 increases susceptibility to schizophrenia. Am J Med Genet B Neuropsychiatr Genet 141:96-101.

Olney JW, Farber NB (1995) Glutamate receptor dysfunction and schizophrenia. Arch Gen Psychiatry 52:998-1007.

Owen MJ, Craddock N, O'Donovan MC (2005) Schizophrenia: genes at last? Trends Genet 21:518-525.

Pastinen T, Hudson TJ (2004) Cis-acting regulatory variation in the human genome. Science 306:647-650.

Paterlini M, Zakharenko SS, Lai WS, Qin J, Zhang H, et al (2005) Transcriptional and behavioral interaction between 22q11.2 orthologs modulates schizophrenia-related phenotypes in mice. Nat Neurosci 8:1586-1594.

Peirce TR, Bray NJ, Williams NM, Norton N, Moskvina V, et al (2006) Convergent evidence for 2',3'-cyclic nucleotide 3'-phosphodiesterase as a possible susceptibility gene for schizophrenia. Arch Gen Psychiatry 63:18-24.

Porteous DJ, Thomson P, Brandon NJ, Millar JK (2006) The genetics and biology of DISC-1-An emerging role in psychosis and cognition. Biol Psychiatry 60:123-131.

Rademakers R, Melquist S, Cruts M, Theuns J, Del-Favero J, et al (2005) High-density SNP haplotyping suggests altered regulation of tau gene expression in progressive supranuclear palsy. Hum Mol Genet 14:3281-3292.

Rebbeck TR, Spitz M, Wu X (2004) Assessing the function of genetic variants in candidate gene association studies. Nat Rev Genet 5:589-597.

Rosand J, Bayley N, Rost N, de Bakker PI (2006) Many hypotheses but no replication for the association between PDE4D and stroke. Nat Genet 38:1091-1092.

Rowe S, Miller S, Sorscher E (2005) Cystic fibrosis. N Eng J Med 352:1992-2001.

Sartorius LJ, Nagappan G, Lipska BK, Lu B, Sei Y, et al (2006) Alternative splicing of human metabotropic glutamate receptor 3. J Neurochem 96:1139-1148.

Sawamura N, Yamamoto T, Ozeki Y, Ross C, Sawa A (2005) A form of DISC1 enriched in nucleus: altered subcellular distribution in orbitofrontal cortex in psychosis and substance/alcohol abuse. Proc Natl Acad Sci USA 102:1187-1192.

Schadt EE, Lamb J, Yang X, Zuh J, Edwards S, et al (2005) An integrative genomics approach to infer causal associations between gene expression and disease. Nat Genet 37:710-717.

Schmidt Kastner R, van Os J, W M Steinbusch H, Schmitz C, Kamiya A, et al (2006) Gene regulation by hypoxia and the neurodevelopmental origin of schizophrenia. Schizophr Res 84:253-271.

Seitz A, Gourevitch D, Zhang X-M, Clark L, Chen P, et al (2005) Sense and antisense transcripts of the apolipoprotein E gene in normal and ApoE knockout mice, their expression after spinal cord injury and corresponding human transcripts. Hum Mol Genet 14:2661-2670.

Shen LX, Bassilion JP, Stanton VP (1999) Single-nucleotide polymorphisms can cause different structural folds of mRNA. Proc Natl Acad Sci USA 96:7871-7876.

Shifman S, Levit A, Chen ML, Chen CH, Bronstein M, et al (2006) A complete genetic association scan of the 22q11 deletion region and functional evidence reveal an association between DGCR2 and schizophrenia. Hum Genet 120:160-170.

Silberberg G, Darvasi A, Pinkas Kramarski R, Navon R, Peirce TR, et al (2006) The involvement of ErbB4 with schizophrenia: association and expression studies. Am J Med Genet B Neuropsychiatr Genet 141:142-148.

Straub RE, Weinberger DR (2006) Schizophrenia genes - Famine to feast. Biol Psychiatry 60:81-83.

Sullivan PF, Kendler KS, Neale MC (2003) Schizophrenia as a complex trait-Evidence from a meta-analysis of twin studies. Arch Gen Psychiat 60:1187-1192.

Talbot K, Eidem W, Tinsley CL, Benson MA, Thompson EW, et al (2004) Dysbindin-1 is reduced in intrinsic, glutamatergic terminals of the hippocampal formation in schizophrenia. J Clin Invest 113:1353-1363.

Talkowsi ME, HStelman H, Bassett AS, Brzustowicz LM, Chen X, et al (2006) Evaluation of a suscpetibility gene for schizophrenia: genotype based meta-analysis of RGS4 polymorphisms from thirteen independetn samples. Biol Psychiatry 60:152-162.

Theuns J, Remacle J, Killick R, Corsmit E, Vennekens Kl, et al (2003) Alzheimer-associated C allele of the promoter polymorphism -22C>T causes a critical neuron-specific decrease of presenilin 1 expression. Hum Mol Genet 12:869-877.

Tokuhiro S, Yamada R, Chang X, Suzuki A, Kochi Y, et al (2003) An intronic SNP in a RUNX1 binding site of SLC22A4, encoding an organic cation transporter, is associated with rheumatoid arthritis. Nat Genet 35:341-348.

Tunbridge EM, Weinberger DR, Harrison PJ (2006) Catechol-o-methyltransferase, cognition and psychosis: Val[158]Met and beyond. Biol Psychiatry 60:141-151.

Tunbridge EM, Weickert CS, Kleinman JE, Herman MM, Chen J, et al (2007) Catechol-o-methyltransferase enzyme activity and protein expression in human prefrontal cortex across the postnatal lifespan. Cereb Cortex (in press).

Valentonyte R, Hampe J, Huse K, Rosenstiel P, Albrecht M, et al (2005) Sarcoidosis is associated with a truncating splice site mutation in the gene BTNL2. Nat Genet 37:357-364.

Walss-Bass C, Liu W, Lew DF, Villegas R, Montero P, et al (2007) A Novel Missense Mutation in the Transmembrane Domain of Neuregulin 1 is Associated with Schizophrenia. Biol Psychiatry (in press).

Weickert CS, Straub RE, McClintock BW, Matsumoto M, Hashimoto R, et al (2004) Human dysbindin (DTNBP1) gene expression in normal brain and in schizophrenic prefrontal cortex and midbrain. Arch Gen Psychiatry 61:544-555.

Wilkins JF, Haig D (2003) What good is genomic imprinting: the function of parent-specific expression. Nat Rev Genet 4:1-10.

Xu B, Wratten N, Charych EI, Buyske S, Firestein BL, et al (2005) Increased expression in dorsolateral prefrontal cortex of CAPON in schizophrenia and bipolar disorder. PLoS Med 2:e263.

Yang J, Si T, Ling Y, Ruan Y, Han Y, Wang X et al (2003) Association study of the human FZD3 locus with schizophrenia. Biol Psychiatry 54:1298-1301.

Yin TCT, Chan JCK (1988) Neural mechanisms underlie interaural time sensitivity to tones and noise. In: W.E. Gall, G.M. Edelman and W.M. Cowans (Eds.), *Auditory Function: Neurobiological Bases of Hearing*. John Wiley, New York, pp. 385-430.

Chapter 2

Towards a Molecular Classification of Illness: Effects of Schizophrenia Susceptibility Loci on Clinical Symptoms and Cognitive Function

Anil K. Malhotra, Pamela DeRosse & Katherine E Burdick

Schizophrenia is a complex neuropsychiatric disorder characterized by a myriad of symptoms that include hallucinations, delusions, thought disorganization (or positive symptoms), apathy, avolition, and asociality (or negative symptoms), and cognitive impairment. No single symptom, or symptom cluster, is pathognomonic, and the current classification of the disorder into the Diagnostic and Statistical Manual IV (DSM-IV) (First et al. 1997) subtypes of paranoid, disorganized, catatonic and undifferentiated is solely based upon clinical phenomena.

The molecular basis of the heterogeneity of schizophrenia has not been elucidated. Although a number of strong candidate genes have recently been identified that predispose to illness, relatively limited work has been conducted to parse out the specific clinical phenomena associated with risk genotypes. It seems likely that schizophrenia susceptibility genes may confer overlapping and non-overlapping effects on different aspects of the clinical syndrome, and delineation of the role of specific genes on these clinical domains could lead to more refined approaches, as well as suggest novel treatment targets for the domains of illness.

Therefore, more specific examination of the role of schizophrenia risk genes is underway in well-characterized clinical populations. Although the data to date is limited, early reports from several independent research groups does provide some intriguing suggestions to support gene-based clinical profiles. In particular, the gene coding for dystrobrevin binding protein 1 (*DTNBP1*), or dysbindin, has been linked to a constellation of symptoms including general cognitive dysfunction and severity of negative symptoms (Fig. 2.1), whereas other schizophrenia susceptibility loci, such as disrupted in schizophrenia 1 (*DISC1*) may be more closely related to relatively specific cognitive deficits, and a distinct phenomenological profile.

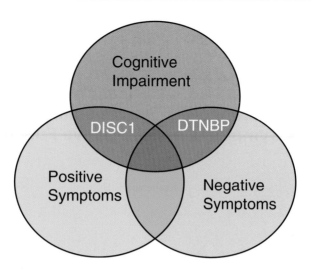

Fig. 2.1. Molecular classification of schizophrenia? Early attempts to dissect the clinical heterogeneity of schizophrenia suggest that risk genes such as *DTNBP1* may be associated with general cognitive dysfunction and greater severity of negative symptoms, whereas other genes including *DISC1* may be related to a different profile of cognitive symptomatology and more severe positive symptoms.

2.1 *DTNBP1* in Schizophrenia

Dysbindin-1 is a 40–50 kDa protein that binds to β-dystrobrevin, a component of the dystrophin glycoprotein complex in the brain. Dysbindin is broadly expressed throughout the brain, including regions in the frontal cortex, temporal cortex, hippocampus, caudate, putamen, nucleus accumbens, amygdala, thalamus, and midbrain (Weickert et al. 2004).

Genome scan and candidate gene analyses have linked *DTNBP1* with risk for schizophrenia (Kendler 2004) including our group's report (Funke et al. 2004) of an association between a 6–locus *DTNBP1* haplotype and schizophrenia, providing the first evidence of association in the U.S. population. Other support for a role for dysbindin in schizophrenia is derived from post-mortem studies indicating reductions in *DTNBP1* gene and protein expression in the dorsolateral prefrontal cortex and the hippocampal formation of patients with schizophrenia. These data linking dysbindin with schizophrenia have primarily utilized the broad diagnostic category of schizophrenia (and schizoaffective disorder), without examination of specific clinical manifestations of illness. More recent data, however, suggest that *DTNBP1* genotype may be associated with a form of the disorder that is characterized by prominent impairment in general cognitive ability and negative symptoms.

2.2 *DTNBP1* and Neurocognition

The first data stem from linkage studies that suggest that *DTNBP1* may be associated with cognitive function. Posthuma and colleagues (2005) completed a genome-wide linkage scan with two samples (475 Australian and 159 Dutch sibling pairs) in which Full Scale (FSIQ), Verbal (VIQ) and Performance IQ were assessed with the Multi-dimensional Aptitude Battery. Model free-multipoint linkage analysis revealed a LOD score of 3.20 for FSIQ and 2.33 for VIQ with a region in chromosome 6 (6p25.3-6p22.3) that encompasses *DTNBP1*. Similarly, Hallmayer and colleagues (2005) have reported significant linkage (maximum LOD of 3.07) to chromosome 6p24 in families comprised of "cognitive deficit" patients, characterized by poor performance on neurocognitive tests, soft neurological signs, and non-right handedness. This empirically-derived phenotype was relatively common, and accounted for up to 50% of a sample of 138 affected individuals from 112 Western Australian families. These linkage analyses provide substantial evidence that a gene in the region near *DTNBP1* influences cognition.

Candidate gene studies have also assessed the relationship between *DTNBP1* and cognition. Straub and colleagues reported that a three-locus haplotype originally observed to increase risk for schizophrenia in their Irish pedigrees was associated with a 5-point decrement in Full Scale IQ in schizophrenia (Straub et al. 2005). These data are consistent with Williams et al. (2004) who reported that a three-locus *DTNBP1* haplotype was associated with educational attainment, although this study did not formally test neurocognitive function.

These data linking *DTNBP1* with IQ and educational achievement suggest that dysbindin genotype may have a broad effect on cognition, as opposed to effects on a specific domain of cognition. The psychometric definition of general cognitive ability (*g*) is the product of an unrotated principal component analysis, which accounts for approximately 40% of the variance in performance on diverse cognitive measures (Carroll 1997). Because of the prior data suggesting a general role for dysbindin on cognition, our group (Burdick et al. 2006) has recently focused on g as a phenotype in studies of dysbindin and cognition. In a cohort of patients with schizophrenia and schizoaffective disorder (n=213) and in healthy volunteers (n=126), we have observed that the dysbindin 6-locus haplotype that increased risk for schizophrenia in our initial study (Funke et al. 2004) was associated with nearly a half standard deviation decrement in g in each group. No differences were observed between risk haplotype carriers and non-carriers on any demographic variable (age, sex, years of education) or clinical characteristic of illness (age of onset, GAF score) in the patient group. Similar to Hallmayer and colleagues (2005), we found that the risk haplotype was twice as frequent in a "cognitive deficit" (33.3%) group of patients who performed more than one standard deviation below the normative mean, as compared with "cognitively spared" patients (15.3%) who did not meet this threshold. Additionally, in a subsequent study (Burdick et al. 2007), we assessed cognitive decline (change in intellect over the disease course) in 183 Caucasian SZ patients using a proxy measure of premorbid intelligence (IQ) with which current general cognitive ability (*g*) was compared. We found that schizophrenia patients who carried the risk haplotype (n=35) demonstrated a significantly greater decline in IQ (residual mean change=13.5±13.6) as compared with non-carriers (n=148) (residual mean change=8.7±12.4). In the group of deteriorating patients, the risk haplotype had a frequency of 24%, as compared with a frequency of only 15% in the non-deteriorating group. These data suggest that *DTNBP1* also influences the severity of intellectual decline in schizophrenia. These data provide direct support for the hypothesis that *DTNBP1* genotype may influence cognition.

2.3 *DTNBP1* and negative symptoms of schizophrenia

Cognitive impairment may represent a trait-like characteristic of schizophrenia. Clinical symptomatology, however, is also critical in the phenomenology of the disorder, and much less attention has been paid to dissecting the genetic contribution to the heterogeneity in clinical symptom profiles. Negative symptoms are a particularly attractive phenotype for genetic investigation, as they tend to be more stable across the duration of illness, including perhaps presenting prior to the onset of illness in many cases, and are relatively unaffected by antipsychotic drug treatment. Moreover, heritability studies of clinical symptomatology have indicated that affected relative pairs with schizophrenia may share similar severity of negative symptoms more often than expected by chance (Kendler et al. 1997), indicating a substantial genetic component.

Recent data suggest that negative symptoms may be influenced by *DTNBP1* genotype. Fanous and colleagues (2005) assessed 755 subjects with psychotic illness in the Irish Study of High Density Schizophrenia Families with the Operational Criteria Checklist for Psychiatric Illness (McGuffin et al. 1991), with factor analyses revealing that an 8-locus schizophrenia risk haplotype was over-transmitted to subjects with greater negative symptom factor scores. Other factors for hallucinations, delusions, depressive and manic symptoms were not related to the high-risk haplotype, and no other haplotype within the sample was associated with negative symptoms. These data, indicating a specific relationship between a schizophrenia risk haplotype and negative symptoms, are consistent with our recent study (DeRosse et al. 2006) in which a schizophrenia risk haplotype was associated with significantly (p = 0. 001) greater negative symptoms in a cohort of 181 Caucasian patients. Risk haplotype carriers displayed significantly higher ratings than non-carriers on each of the three 3 negative symptom items assessed by the Structured Diagnostic Interview for DSM-IV; avolition (p = 0.039), alogia (p = 0.014) and flattened affect (p = 0.016). The relationship was not secondary to the previously described effect of the risk haplotype on neurocognition, as neurocognitive function was included as a covariate in these analyses. Finally, Tosato and colleagues (2007), utilizing a longitudinal design in a cohort of 81 patients with schizophrenia, recently reported an association between a

2-locus risk haplotype within *DTNBP1* and persistence of negative symptoms over a 6-year follow up period. Specifically, carriers of the risk haplotype had significantly higher ratings on the emotional withdrawal and conceptual disorganization items of the Brief Psychiatric Rating Scale (BPRS) than non-carriers ($p < 0.05$ and $p < 0.01$ respectively).

Taken together, these independent studies suggest that *DTNBP1* risk haplotypes may be associated with a greater severity of negative symptoms in patients with schizophrenia, and with cognitive dysfunction in both schizophrenia patients and healthy individuals. Recent data indicating a negative correlation between neuronal dystrobrevin and the glutamate transporter in schizophrenia (Talbot et al. 2004) suggest that the dysbindin genotype influences glutamatergic function. Since glutamatergic antagonists such as ketamine can produce negative symptoms in healthy individuals and exacerbate negative symptoms in schizophrenia, and produce marked cognitive dysfunction in healthy controls and schizophrenia patients (Malhotra et al. 1997) dysbindin genotype may exert its effects on negative symptoms through a glutamate receptor-related mechanism. Further studies elucidating the relation between the glutamatergic system and dysbindin genotype are warranted.

2.4 *DISC1* in schizophrenia

Although *DTNBP1* is a strong, perhaps the strongest, candidate gene for schizophrenia, several other candidate genes have been implicated in influencing susceptibility to the disorder. Of these, the gene disrupted in schizophrenia 1 (*DISC1*) has now been shown to be linked with schizophrenia, as well as other related psychiatric phenotypes, across a number of studies.

DISC1 is a component of the dynein protein complex and plays a central role in normal microtubular dynamics (Kamiya et al. 2005). *DISC1* is widely expressed in brain including the hippocampus, cerebral cortex, olfactory bulb and cerebellum (Morris et al. 2003). Due to its role in neurogenesis, *DISC1* may contribute to abnormalities in the neurodevelopmental process thereby contributing to the dysfunction believed to occur in schizophrenia.

The gene encoding *DISC1* is located at the breakpoint of balanced translocation between chromosomes 1 and 11, t(1;11)(q42;q14). This translocation, which may result in truncation of the *DISC1* protein, co-segregates with major psychiatric disorders including schizophrenia and schizoaffective disorder (St Clair et al. 1990). Genome scans and candidate gene analyses have linked *DISC1* with risk for psychiatric disorders including our group's report of multiple haplotypes within *DISC1* associated with schizophrenia and schizoaffective disorder (Hodgkinson et al. 2004).

Further evidence of *DISC1* association to schizophrenia is derived from postmortem studies that indicate altered protein expression of *DISC1* in orbitofrontal cortex in patients with schizophrenia relative to controls (Sawamura et al. 2005). These data, linking *DISC1* with schizophrenia and schizoaffective disorder, have primarily utilized broad diagnostic categories without examination of specific clinical manifestations of illness. More recent data, however, suggest that *DISC1* genotype may be associated with specific impairments in cognitive function, as well as confer an effect on the clinical profile of illness.

2.5 *DISC1* and Neurocognition

Data emerging from studies that have focused on cognitive function as an intermediate phenotype have consistently demonstrated an effect of *DISC1* on measures associated with working memory and verbal memory. Early evidence for a link between *DISC1* and cognition was provided from the original Scottish pedigree, in which translocation carriers, regardless of affection status, demonstrated impaired early information processing, as measure by P300 amplitude (Blackwood et al. 2001). Subsequent data linking *DISC1* to cognitive function were provided by Gasperoni and colleagues (2003) who reported evidence for linkage at a microsatellite marker on chromosome 1, previously associated with risk for schizophrenia (Eklund et al. 2001), to performance on tasks of spatial and visual working memory. Follow up association analyses of this region revealed a significant effect of a haplotype within *DISC1* (HEP3) on risk for schizophrenia (Hennah et al. 2003) that was later shown to influence performance on neurocognitive tasks of visual working memory in both patients with schizophrenia and healthy individuals (Hennah et al. 2005).

Moreover, Callicott and colleagues (2004) measured hippocampal structure and function in a cohort of patients with schizophrenia, their unaffected siblings, and healthy controls and found an association between a risk locus in *DISC1* (Ser704Cys), reduced hippocampal volume, and abnormal engagement of the hippocampus during cognitive measures of working memory and verbal memory. The effect on the neuroimaging parameters was most robust in the healthy control sample, with additional but weaker relationships reported between the *DISC1* Ser allele and impaired performance on memory and executive function tasks in the schizophrenia group. Convergent data were derived from a study conducted by our group, in which we assessed cognitive function in 250 patients with schizophrenia and tested for an association with the five *DISC1* SNPS that were linked with schizophrenia susceptibility in our cohort (Hodgkinson et al. 2004). We reported a relationship between two of these risk SNPs and impaired performance on measures of verbal working memory and rapid visual processing in patients with schizophrenia (Burdick et al. 2005). These results are consistent with a population-based twin cohort study by Cannon et al. (2005), in which a risk haplotype within *DISC1* was associated with worse performance on measures of verbal learning and memory and reduced hippocampal volume in families of patients with schizophrenia. Further, a second *DISC1* risk haplotype in these families was associated with impaired spatial working memory and decreased prefrontal gray matter. Finally, Thomson et al. (2005) described a complex relationship between *DISC1* Ser704Cys and cognitive ageing in healthy subjects. Specifically, Cys homozygosity was associated with a declining cognitive profile at age 79 in females and an improving cognitive profile at age 79 in males. These data are partially consistent with Callicott et al. (2005); however, allelic heterogeneity makes these results somewhat more difficult to interpret without taking haplotype structure and possible gene-gene interactions into account.

2.6 *DISC1* and Positive Symptoms of Schizophrenia

It should be noted that the data linking *DISC1* with neurocognitive phenotypes appears to suggest a relatively specific effect on aspects of memory function, as compared to the more generalized effect observed with *DTNBP1* genotype. Similarly, studies of the effects of *DISC1* on the clinical profile of illness have differed from that observed with *DTNBP1*, with, instead of an effect on negative symptoms, early data suggest that *DISC1* genotype may preferentially influence severity of positive symptoms; specifically psychotic delusions.

Hennah and colleagues (2003) assessed 458 Finnish schizophrenia families and identified a single common, under-transmitted haplotype (HEP3) that significantly increased risk for the development of the disease. Additional analyses using more refined phenotypes that included factor scores derived from a cross-sectional clinical instrument (Operational Criteria Checklist of Psychotic Illness: McGuffin et al. 1991), however, indicated that the risk haplotype was more significantly associated with a specific clinical profile. Specifically, HEP3 was significantly associated with factor scores that included delusions and hallucinations (p = 0.0003), a factor score that included manic symptoms (p = 0.0006), a factor score that included depressive symptoms (p = 0.0009) and a factor score that included negative symptoms (p = .0018). The strongest of these findings indicated association of *DISC1* HEP3 to the delusions and hallucination factor and within this factor, the strongest association to HEP3 was persecutory delusions (p < 0.00001). Convergent data for these findings are derived from work by our group using a lifetime measure of psychosis (Structured Clinical Interview for the DSM-IV: First, et al 1998). Specifically, DeRosse and colleagues (2007) found that a 5-locus haplotype encompassing the Ser704Cys locus was significantly associated with the lifetime severity of delusions in a cohort of 199 patients with schizophrenia (p = 0.001). Consistent with the work of Hennah et al. (2003) the strongest of finding in this cohort indicated association of *DISC1* to persecutory delusions (p = 0.01). This relationship was not secondary to the previously described effect of *DISC1* on neurocognition, as neurocognitive function was included as a covariate in these analyses.

A potential criticism of studies utilizing positive symptoms as an intermediate phenotype is that positive symptoms are not a particularly stable trait of the disease, and may therefore not be under strong genetic control. Specifically, hallucinations demonstrate considerable intra-individual variation, and may be most sensitive to antipsychotic drug treatment. Moreover, in a study of 256 sibling pairs concordant for schizophrenia assessed across a range of clinical symptoms (hallucinations, delusions, catatonia, etc) Kendler et al. (1997) reported that, although sibling pairs significantly resembled each other for 8 of the 9 symptom domains assessed, there was not a significant correlation between pairs for hallucinations, again suggesting a minimal role for a genetic contribution to this symptom domain. It should be noted, however, that the studies linking *DISC1* with positive symptoms, have reported the strongest associations with ratings reflecting persecutory delusions. Delusions may represent a more stable symptom than hallucinations, supported by at least one longitudinal study indicating consistent delusional beliefs in schizophrenia patients for up to eight years (Harrow et al. 1995). Therefore, delusions may resemble a more trait-like phenomenon than other positive symptoms and be a more robust phenotype for examination in clinical genetic studies of schizophrenia symptomatology.

Although the specific mechanism is unknown, recent data suggests that the *DISC1* protein mediates GABAergic function through its interaction with activating transcription factors (ATF) 4 and 5. ATF4/5 interact directly with GABA-B receptors and although the physiological consequences of this interaction are not clear, it is likely that the interaction regulates GABA-B receptor responses, signal transduction or both. Interaction of *DISC1* with ATF4/5 is lost upon *DISC1* truncation and ATF4/5 regulation of GABA-B receptor function may also be lost (Morris et al. 2004). Data indicating a relation between *DISC1* and hippocampal volume (Calicott et al. 2005) and between hippocampal volume and positive symptomatology (Jacobsen et al. 1998; Strasser et al. 2005) suggests that *DISC1* may exert its influence on positive symptoms through neural circuitry within the hippocampus and GABAergic neurons are located in all regions of the hippocampus. Thus, the activity of GABAergic neurons in the hippocampus may have significant effects on neuronal activity within the prefrontal cortex and other areas associated with positive symptoms (Keverne 1999).

2.7 Conclusions

The data to date suggest differential effects for *DISC1* and *DTNBP1* on neurocognitive function and clinical symptomatology, and there are less data for other putative schizophrenia risk genes. For example, catechol-o-methyltransferase (*COMT*) has been commonly implicated in cognitive function (Egan et al. 2001; Malhotra et al. 2002), however, there are only limited reports linking *COMT* genotype with clinical domains of illness (DeRosse et al. 2006). Nevertheless, the available data do begin to indicate that certain risk genes may have unique effects on the clinical characteristics of illness, with *DTNBP1* favoring a presentation of broad cognitive decline, and more severe negative symptoms, akin to prior hypotheses of "poor outcome" schizophrenia. In contrast, *DISC1* has relatively specific effects on aspects of cognitive function, particularly within the working memory domain, and may perhaps be more closely related to positive symptoms such as delusions. These data are limited, however, and the advent of whole genome association approaches to schizophrenia (Lencz et al. in press), and the more careful attention being paid to collection of phenotypic information other than simple diagnostic categorizations, should provide an opportunity to more comprehensively examine the role of genetic variation and clinical domains of illness. If successful, these data may represent the first steps towards the molecular classification of schizophrenia.

References

Blackwood DHR, Fordyce MT, Walker D, St Clair DJ, Porteus J, Muir WJ (2001) Schizophrenia and affective disorders-Cosegregation with a translocation at chromosome 1q42 that directly disrupts brain-expressed genes: Clinical and P300 findings in a family. Am J Hum Gen 69:428-433.

Burdick KE, Hodgkinson CA, Szeszko PR, Lencz T, Ekholm JM, Kane JM, Goldman D, Malhotra AK (2005) *DISC1* and neurocognitive function in schizophrenia. Neuroreport 16:1399-1402.

Burdick KE, Lencz T, Funke B, Finn CT, Szeszko PR, Kane JM, Kucherlapati R, Malhotra AK (2006) Genetic variation in *DTNBP1* influences general cognitive ability. Hum Mol Genet 15:1563-1568.

Burdick KE, Goldberg TE, Funke B, Bates JA, Lencz T, Kucherlapati R, Malhotra AK (2007) *DTNBP1* genotype influences cognitive decline in schizophrenia. Schizophr Res 89:169-172.

Callicott JH, Straub RE, Pezawas L, Egan MF, Mattay VS, Hariri AR, Verchinski BA, Meyer-Lindenberg A, Balkissoon R, Kolachana B, Goldberg TE, Weinberger DR (2004) Variation in *DISC1* affects hippocampal structure and function and increases risk for schizophrenia. Proc Natl.Acad Sci USA 102: 8627-8632.

Cannon TD, Hennah W, van Erp TG, Thompson PM, Lonnqvist J, Huttunen M, Gasperoni T, Tuulio-Henriksson A, Pirkola T, Toga AW, Kaprio J, Mazziotta J, Peltonen L (2005) Association of *DISC1*/TRAX haplotypes with schizophrenia, reduced prefrontal gray matter, and impaired short- and long-term memory. Arch Gen Psychiatry 62:1205-1213.

Carroll, JB (1997) Psychometrics, intelligence, and public policy. Intelligence 24:25-52.

DeRosse P, Funke B, Burdick KE, Lencz T, Ekholm JM, Kane JM, Kucherlapati R, Malhotra AK (2006) Dysbindin genotype and negative symptoms in schizophrenia. Am J Psychiatry 163:532-534.

DeRosse P, Funke B, Burdick KE, Lencz T, Goldberg TE, Kane JM, Kucherlapati R, Malhotra AK (2006) *COMT* genotype and manic symptoms in schizophrenia. Schizophr Res 87:28-31.

DeRosse, P, Hodgkinson, CA, Lencz, T, Burdick, KE, Kane, JM, Goldman, D, Malhotra, AK (2006) Disrupted in Schizophrenia 1 Genotype and Positive Symptoms in Schizophrenia. Biol. Psychiatry, Oct: Epub ahead of print.

Egan MF, Goldberg TE, Kolachana BS, Callicott JH, Mazzanti CM, Straub RE, Goldman D, Weinberger DR (2001) Effect of *COMT* Val108/158 Met genotype on frontal lobe function and risk for schizophrenia. Proc Natl Acad Sci USA 98:6917-6922.

Ekelund J, Hovatta I, Parker A, Paunio T, Varilo T, Martin R, Suhonen J, Ellonen P, Chan G, Sinsheimer JS, Sobel E, Juvonen H, Arajarvi R, Partonen T, Suvisaari J, Lonnqvist J, Meyer J, Peltonen L (2001) Chromosome 1 loci in Finnish schizophrenia families. Hum Mol Genet 10:1611-1617.

Fanous AH, van den Oord EJ, Riley BP, Aggen SH, Neale MC, O'Neill FA, Walsh D, Kendler KS (2005) Relationship between a high-risk haplotype in the *DTNBP1* (dysbindin) gene and clinical features of schizophrenia. Am J Psychiatry 162:1824-1832.

First MB, Spitzer RL, Gibbon M, Williams JBW (1997) Structured Clinical Interview for DSM-IV Axis I Disorders (SCID). *American Psychiatric Press*; New York, NY.

Funke B, Finn CT, Plocik AM, Lake S, DeRosse P, Kane JM, Kucherlapati R, Malhotra AK (2004) Association of the *DTNBP1* locus with schizophrenia in a U.S. population. Am J Hum Genet 75:891-898.

Gasperoni TL, Ekelund J, Huttunen M, Palmer CG, Tuulio-Henriksson A, Lonnqvist J, Kaprio J, Peltonen L, Cannon TD (2003) Genetic linkage and association between chromosome 1q and working memory function in schizophrenia. Am J Med Genet B Neuropsychiatr Genet 116:8-16.

Hallmayer JF, Kalaydjieva L, Badcock J, Dragovic M, Howell S, Michie PT, Rock D, Vile D, Williams R, Corder EH, Hollingsworth K, Jablensky A (2005) Genetic Evidence for a Distinct Subtype of Schizophrenia Characterized by Pervasive Cognitive Deficit. Am J Hum Genet 77:468-476.

Harrow M, MacDonald AW 3rd, Sands JR, Silverstein ML (1995) Vulnerability to delusions over time in schizophrenia and affective disorders. Schizophr Bull 21:95-109.

Hennah W, Varilo T, Kestila M, Paunio T, Arajarvi R, Haukka J, Parker A, Martin R, Levitzky S, Partonen T, Meyer J, Lonnqvist J, Peltonen L, Ekelund J (2003) Haplotype transmission analysis provides evidence of association for *DISC1* to schizophrenia and suggests sex-dependent effects. Hum Mol Genet 12:3151-3159.

Hennah W, Tuulio-Henriksson A, Paunio T (2005) A haplotype within the *DISC1* gene is associated with visual memory functions in families with a high density of schizophrenia. Mol Psychiatry 10:1097-1103.

Hodgkinson CA, Goldman D, Jaeger J, Persaud S, Kane JM, Lipsky RH, Malhotra AK (2004) Disrupted in schizophrenia 1 (*DISC1*): Association with schizophrenia, schizoaffective disorder, and bipolar disorder. Am. J. Hum. Genet 75:862-872.

Jacobsen LK, Giedd JN, Castellanos FX, Vaituzis AC, Hamburger SD, Kumra S, et al. (1998) Progressive Reduction in Temporal Lobe Structures in Childhood-Onset Schizophrenia. Am J Psychiatry 155:678-685.

Kamiya A, Kubo K, Tomoda T, Takaki M, Youn R, Ozeki Y, Sawamura N, Park U, Kudo C, Okawa M, Ross CA, Hatten ME, Nakajima K, Sawa A (2005) A schizophrenia-associated mutation of *DISC1* perturbs cerebral cortex development. Nat Cell Biol 7:1167-1178.

Keverne EB (1999) GABA-ergic Neurons and the Neurobiology of Schizophrenia and other Psychoses. Brain Res Bull 48:467-473.

Kendler KS, Karkowski-Shuman L, O'Neill FA, Straub RE, MacLean CJ, Walsh D (1997) Resemblance of psychotic symptoms and syndromes in affected sibling pairs from the Irish Study of High-Density Schizophrenia Families: evidence for possible etiologic heterogeneity. Am J Psychiatry 154:191-198.

Kendler KS (2004) Schizophrenia genetics and dysbindin: a corner turned? Am J Psychiatry 161:1533-1536.

Lencz T, Morgan TV, Athanasiou M, Dain B, Reed CR, Kane JM, Kucherlapati R, Malhotra AK (In press) Converging Evidence for a Pseudoautosomal Cytokine Receptor Gene Locus in Schizophrenia. Mol Psychiatry.

Malhotra AK, Pinals DA, Adler CM, Elman I, Clifton A, Pickar D, Breier A (1997) Ketamine-induced exacerbation of psychotic symptoms and cognitive impairment in neuroleptic-free schizophrenics. Neuropsychopharmacology 17:141-150.

Malhotra AK, Kestler LJ, Mazzanti C, Bates JA, Goldberg T, Goldman D (2002) A functional polymorphism in the *COMT* gene and performance on a test of prefrontal cognition. Am J Psychiatry 159:652-654.

McGuffin P, Farmer A, Harvey I (1991) A polydiagnostic application of operational criteria in studies of psychotic illness. Development and reliability of the OPCRIT system. Arch Gen Psychiatry 48:764-770.

Millar JK, Wilson-Annan JC, Anderson S, Christie S, Taylor MS, Semple CA, Devon RS, Clair DM, Muir WJ, Blackwood DH, Porteous DJ (2000) Disruption of two novel genes by a translocation co-segregating with schizophrenia. Hum Mol Genet 9:1415-1423.

Morris JA, Kandpal G, Ma L, Austin CP (2003) *DISC1* (Disrupted in Schizophrenia 1) is a Centrosome-Associated Protein that Interacts with AMP1A, MIPT3, ATF 4/5 and NUDEL: Regulation and Loss of Interaction with Mutation. Hum Mol Gen 12:1591-1608.

Posthuma D, Luciano M, Geus EJ, Wright MJ, Slagboom PE, Montgomery GW, Boomsma DI, Martin NG (2005) A genomewide scan for intelligence identifies quantitative trait Loci on 2q and 6p. Am J Hum Genet 77:318-326.

Sawamura N, Sawamura-Yamamoto T, Ozeki Y, Ross CA, Sawa A (2005) A form of *DISC1* enriched in nucleus: altered subcellular distribution in orbitofrontal cortex in psychosis and substance/alcohol abuse. Proc Natl Acad Sci USA 102:1187-1192.

St Clair D, Blackwood D, Muir W, Carothers A, Walker M, Spowart G, Gosden C, Evans HJ (1990) Association within a family of a balanced autosomal translocation with major mental illness. Lancet 336:13-16.

Strasser HC, Lilyestrom J, Ashby ER, Honeycutt NA, Schretlen DJ, Pulver AE, et al. (2005) Hippocampal And Ventricular Volumes In Psychotic And Nonpsychotic Bipolar Patients Compared With Schizophrenia Patients And Community Control Subjects: A Pilot Study. Biol Psychiatry 57:633-639.

Straub RE, Mayhew MB, Vakkalanka RK, Kolachana B, Goldberg TE, Egan MF, Weinberger DR (2005) Detection of dysbindin (*DTNBP1*) effects on clinical and cognitive phenotypes can be highly dependent on variation in *COMT*. Am J Med Gen 138B:135.

Talbot K, Eidem WL, Tinsley CL, Benson MA, Thompson EW, Smith RJ, Hahn CG, Siegel SJ, Trojanowski JQ, Gur RE, Blake DJ, Arnold SE (2004) Dysbindin-1 is reduced in intrinsic, glutamatergic terminals of the hippocampal formation in schizophrenia. *J. Clin. Invest* 113:1353-1363.

Thomson PA, Harris SE, Starr JM, Whalley LJ, Porteous DJ, Deary IJ (2005) Association between genotype at an exonic SNP in *DISC1* and normal cognitive aging. Neurosci Lett 389:41-45.

Tosato S, Ruggeri M, Bonetto C, Bertani M, Marrella G, Lasalvia A, Cristofalo D, Aprili G, Tansella M, Dazzan P, Diforti M, Murray RM, Collier DA (2007) Association study of dysbindin gene with clinical and outcome measures in a representative cohort of Italian schizophrenic patients. Am J Med Genet B Neuropsychiatr Genet, Feb 8; Epub ahead of print.

Weickert CS, Straub RE, McClintock BW, Matsumoto M, Hashimoto R, Hyde TM, Herman MM, Weinberger DR, Kleinman JE (2004) Human dysbindin (*DTNBP1*) gene expression in normal brain and in schizophrenic prefrontal cortex and midbrain. Arch Gen Psychiatry 61:544-555.

Williams NM, Preece A, Morris DW, Spurlock G, Bray NJ, Stephens M, Norton N, Williams H, Clement M, Dwyer S, Curran C, Wilkinson J, Moskvina V, Waddington JL, Gill M, Corvin AP, Zammit S, Kirov G, Owen MJ, O'Donovan MC (2004) Identification in 2 independent samples of a novel schizophrenia risk haplotype of the dystrobrevin binding protein gene (*DTNBP1*). Arch Gen Psychiatry 61:336-344.

Chapter 3

Homer: A Genetic Factor in Schizophrenia?

Karen K. Szumlinski and Todd E. Kippin

3.1 Introduction

The hypothesis that abnormal limbic-corticostriatal glutamate transmission underlies the pathophysiology of schizophrenia emerged from decades of neurological, brain imaging, pharmacological, genetic and biochemical research in affected individuals, as well as behavioral, molecular and neurochemical data derived from preclinical animal models of schizophrenia and related disorders (e.g., drug-induced psychosis) (Coyle 2006; Coyle and Tsai 2004; Krystal et al. 2003, 2006; MacDonald and Chafee 2006; Meador-Woodruff and Healy 2000; Moghaddam 1994; Pietraszek 2002; Tsai and Coyle 2002). In schizophrenia, pathology within the glutamatergic system is suggested by 1) reduced cortical volumes; 2) reduced glutamatergic somatic or neuropil size; 3) reduced dendritic spines; 4) a disarray of pyramidal cell orientation; 5) altered expression of subtypes of glutamate receptors; and 6) reduced expression of certain synaptic proteins within cortical areas (Coyle and Tsai 2004; Kristiansen et al. 2007; Krystal et al. 2003; Meador-Woodruff and Healy 2000; Pietraszek 2002; Tsai and Coyle 2002). The traditional glutamate hypothesis of schizophrenia holds that glutamatergic abnormalities culminate in reduced frontal cortical activity and/or activation (i.e., hypofrontality) (Andreasen et al. 1997; Davidson and Heinrichs 2003; Hazlett and Bushbaum 2001). However, other glutamate theories propose that glutamatergic hyperactivity (more specifically, a disinhibition of glutamate transmission) might underlie the psychotic behavior and cognitive pathology observed in schizophrenia (Krystal et al. 2003; Moghaddam et al. 1997). Although it is highly unlikely that alterations in the function of any single gene can account for all of the morphological, neurochemical and cellular abnormalities associated with schizophrenia, the cellular, biochemical and behavioral genetic data suggest a potential role for the glutamate-associated *Homer* gene products in the abnormalities in the structural and functional integrity of glutamatergic neurons within the brains of patients with schizophrenia.

3.2 The Molecular Features of Homer Proteins

The Homer family of proteins is the product of three independent mammalian genes (*Homer1-3*), one *Xenopus* gene and one *Drosophila* gene (Brakeman et al. 1997; Foa et al. 2005; Kato et al. 1998; Xiao et al. 1998). In humans, *Homer1*, *Homer2* and *Homer3* are localized to chromosomes 5, 15 and 19, respectively (Xiao et al. 1998) and *Homer* transcripts have been identified in many different tissues including: brain, retina, liver, kidney, spleen, testis, thymus, placenta, intestine, as well as cardiac, skeletal and smooth muscle (Shiraishi et al. 2004; Soloviev et al. 2000; Xiao et al. 1998). First described in the late 1990's, the original family of Homer proteins consisted of Homer1a/b/c, Homer2a/b and 3 (Brakeman et al. 1997; Kato et al. 1997, 1998; Xiao et al. 1998). Since that time, 21 *Homer* mRNAs have been isolated from rat, mouse and human brain (see Table 1); however, the proteins for some of these mRNAs have yet to be detected in mammalian brain tissue (Berke et al. 1998; Saito et al. 2002; Soloviev et al. 2000).

Table 3.1. Homer Isoforms

Homer isoform	Molecular mass (kDa)	Reference
Homer1a	~30	Brakeman et al. 1997
Ania-3	~30	Berke et al. 1998
Homer1b	47	Kato et al. 1998
Homer1c	47	Kato et al. 1998
Homer1d	48	Saito et al. 2002
Homer1e	27	Klugmann et al. 2005
Homer1g	24	Klugmann et al. 2005
Homer1h	28	Klugmann et al. 2005
Ania-3	33	Berke et al. 1998
Homer2a	47	Xiao et al. 1998
Homer2b	47	Xiao et al. 1998
Homer2c	29	Soloviev et al. 1998
Homer2d	29	Soloviev et al. 1998
Homer3a$_{00}$	48	Soloviev et al. 1998
Homer3a$_{01}$	48	Soloviev et al. 1998

Homer3a$_{10}$	48	Soloviev et al. 1998
Homer3a$_{11}$	48	Soloviev et al. 1998
Homer3b$_{00}$	45	Soloviev et al. 1998
Homer3b$_{01}$	45	Soloviev et al. 1998
Homer3b$_{10}$	45	Soloviev et al. 1998
Homer3b$_{11}$	45	Soloviev et al. 1998
Homer3c	16	Soloviev et al. 1998
Homer3d	14	Soloviev et al. 1998

The mammalian genes encoding the Homer family of proteins have open reading frames that spread over 10 exons and can give rise to both constitutively expressed, as well as immediate early gene (IEG) products (Bottai et al. 2002; Brakeman et al. 1997; Kato et al. 1998; Klugmann et al. 2005). Exon 1 encodes the 5' untranslated region (UTR) and contains the translational initiation codon ATG. Up-stream of the initiation codon lie several multiple start site elements downstream motifs [MED-1; GCTCC(G/C)] indicating that *Homer* genes may contain multiple transcription initiation sites. Also located upstream of the putative start site are a number of transcription factor binding sites, including: specific promoter 1, activator protein 1, GATA, octamer recognition site, enhancer box element and CRE (cyclic adenosine monophosphate response element; Bottai et al. 2002). Thus, the transcription of *Homer* genes can be influenced by multiple factors, including mitogen-activated protein kinase (MAPK), CRE binding protein (CREB) and serum response element, enzymes implicated in corticoaccumbens glutamate dysfunction (Ahn et al. 2005; Culm et al. 2004; Kawanishi et al. 1999; Kozlovsky et al. 2001; Svenningsson et al. 2003).

With the exception of the recently characterized Homer1g (Klugmann et al. 2005), exons 2-5 encode an Enabled/vasodilator-stimulated phosphoprotein (Vasp) homology 1 (EVH1) domain (Beneken et al. 2000; Gertler et al. 1996), which is similar in sequence to other Ena/Vasp proteins that regulate cytoskeleton dynamics (Reinhard et al. 2001). The EVH1 domain exhibits a RxxxxxGLGF sequence that is common to most post-synaptic density 95(PSD-95)/Drosophila discs large tumor suppressor gene (Dlg)/Zona occludens-1 (ZO-1) (PDZ) domains that mediate protein-protein interactions and are involved in ion channel and receptor targeting to the plasma membrane (Kim and Sheng 2004). The EVH1 domain exhibits a high degree of similarity across Homer isoforms and is essential for Homer interactions with a proline-rich sequence (PPSPF) displayed by proteins regulating neuronal development, synaptic architecture, as well as the integration of intracellular calcium signaling (Beneken et al., 2000; Ebihara et al. 2003; Foa et al. 2001; Gray et al. 2003; Kammermeier and Worley 2007; Naisbitt et al. 1999; Nakamura et al. 2004; Okabe et al. 2001; Sala et al. 2001, 2003, 2005; Shiraishi et al. 2004; Usui et al. 2003; Xiao et al. 1998). Of note, many of these Homer-interacting proteins (Ango et al. 2002; Brakeman et al. 1997; Kato et al. 1998; Minakami et al. 2000; Naisbitt et al. 1999; Rong et al. 2003; Shiraishi et al. 1999, 2003a; Soloviev et al. 2000; Tu et al. 1998; Usui et al. 2003) have been implicated in the pathophysiology of schizophrenia, including the mGluR1a and mGluR5 subtypes of Group 1 metabotropic glutamate receptors (mGluRs) (e.g., Brody et al. 2003, 2004; Moghaddam 2004; Shipe et al. 2005), the NMDA (*N*-methyl-D-aspartate) glutamate receptor scaffolding protein Shank (Castner and Williams 2007; Coyle 2006; Coyle and Tsai 2004; Kristiansen et al. 2007; Li and He 2007; MacDonald and Chafee 2006; Stone et al. 2007), the inositol-1,4,5-triphosphate (IP3) receptor (Gu et al. 2005; Kaiya et al. 1992), F-actin (Ishizuka et al. 2006), dynamin3 (Brebner et al. 2005), phosphoinositide 3 kinase enhancer-long (PIKE-L) (Kalkman 2006) and syntaxin 13 (Barr et al. 2004; Gabriel et al. 1997); but also include ryanodine receptors 1 and 2 (Feng et al. 2002; Hwang et al. 2003; Westhoff et al. 2003), the transient receptor potential canonical channels (Kim et al. 2006; Yuan et al. 2003), and most recently, the plasma membrane calcium ATPase (Sgambato-Faure et al. 2006).

In contrast to the high (~80%) sequence homology within exons 2-5 of the 3 *Homer* genes, exons 6-10 exhibit low homology (~20-30%) (Soloviev et al. 2000; Xiao et al. 1998). Exons 6-10 encode the carboxyl tail of the majority of Homer proteins that consists of a coiled-coil (CC) domain, 2 leucine zipper motifs and encode also the 3' UTR (Kato et al. 1998; Soloviev et al. 2000; Xiao et al. 1998). Homer proteins multimerize through CC/leucine zipper motif interactions (Kato et al. 1998; Hayashi et al. 2006; Soloviev et al. 2000; Sun et al. 1998; Tu et al. 1998; Xiao et al. 1998) and a recent elucidation of the quaternary structure of Homers indicate that these proteins form tetramers with each monomer oriented in parallel (Hayashi et al. 2006). The tetrameric structure of Homer oligomers confers slower turn-over rates and greater efficiency of localization to dendritic spines (Hayashi et al. 2006). Alternative transcript splicing in regions downstream from exon 5 has been reported for all 3 *Homer* genes and can result in premature termination of transcription prior to the sequences encoding the CC and leucine zipper motifs (Bottai et al. 2002; Brakeman et al. 1997; Soloviev et al. 2000; Xiao et al. 1998). This premature termination of transcription results in truncated or "short" Homer isoforms that lack the motifs necessary to multimerize.

In mammalian brain, transcripts encoding truncated *Homer* gene products have been identified for all 3 *Homer* genes (see Table 1). However, in neurons, the most characterized is the bimodal expression of *Homer1* constitutive (Homer1b-g) and IEG (Homer1a and Ania-3) products. Within the *Homer1* gene lies transcriptional stop codons in intron 5 and sequence comparison between rat cDNA for *Homer1a* (Brakeman et al. 1997; Kato et al. 1997) and the mouse *Homer1* gene established that *Homer1a* mRNA terminates ~4.4 kb into intron 5 (Bottai et al. 2002). The intronic sequence extends exon 5 by 33 bases (11 codons), followed by a translational stop codon, and a ~4.4 kb 3' UTR, colinear with the 5' portion of intron 5. Ania-3, the other known IEG Homer1 isoform (Berke et al. 1998), is generated by alternative transcript splicing from exon 5 into an intron 5 sequence that lies ~5.7 kb downstream of the poly adenylation site for *Homer1a* mRNA (Bottai et al. 2002). The precise sequence contribution to the premature transcript termination for Homer1a and Ania-3 is not known. Examination of the 30 kb intron 5 sequence of *Homer1* failed to reveal obvious contributing sequences and no conserved sequences have been reported 5' or 3' of the poly(A) sites for these IEG transcripts. Moreover, the human intron 5 of *Homer1* did not provide any indication of highly conserved sequence islands that might function in the activity-dependent termination of transcripts within this intron (Bottai et al. 2002). However, the mRNA for IEG *Homer1* isoforms does contain several AUUUA repeats at their 3' UTRs that may be responsible for destabilizing interactions with ribosomal translational machinery. While not involved in premature termination of transcription, this sequence likely contributes to the characteristic fast decay of mRNA transduction exhibited by *Homer1a* and *Ania-3* (Bottai et al. 2002).

Finally, Homer proteins contain a PEST (proline, glutamate, serine, threonine) sequence in their C-terminal regions that targets these proteins for degradation by ubiquitin-mediated systems (Soloviev 2000). The presence of this PEST sequence in the primary structure of IEG Homer1 isoforms might explain their high turnover rates (Kato et al. 1998) and account for incongruity in mRNA versus protein expression following synaptic activity. However, the C-terminal regions of certain isoforms of the constitutively expressed Homer1 and Homer2 (Homer2b, Homer2d) proteins also contain the PEST sequence, which may provide a means to differentially regulate the expression of these proteins (Soloviev 2000; Soloviev et al. 2000). In vitro mutational approaches confirmed that an 11-residue tail unique to Homer1a is necessary for its ubiquitin-mediated degradation (Ageta et al. 2001). Moreover, tetramerization of CC-Homer isoforms masks the PEST sequence and thus contributes to their considerably longer turnover rates (Hayaishi et al. 2006).

3.3. Functional aspects of Homer proteins related to glutamatergic neurotransmission

Originally hypothesized to serve as a protein scaffold that facilitated intracellular signaling through Group1 mGluRs and calcium-related interactions between these receptors and ionotropic NMDA receptors (e.g., Brakeman et al. 1997; Kato et al. 1998; Tu et al. 1998, 1999; Xiao et al. 1998), Homer proteins play critical roles in regulating many aspects of the functional architecture of glutamatergic synapses. As discussed above, Homers interact via their EVH1 domains with a wide variety of proteins and thus, function not only to scaffold receptors and ion channels on the plasma membrane to the cytoskeleton and intracellular signaling complexes, but also to regulate the function of plasma membrane ion channels and intracellular messenger systems that impact cellular signaling and cell excitability (for detailed reviews: de Bartolomeis and Iasevoli 2003; Duncan et al. 2005; Fagni et al. 2002; Szumlinski et al. 2006c; Xiao et al. 2000). Moreover, Homer protein interactions with other cytosolic scaffolding proteins, in particular Shank, are necessary for morphological aspects of glutamate synapses (Sala et al. 2001, 2003). Given the putative role for functional and morphological abnormalities within the central glutamate systems in schizophrenia (e.g., Coyle 2006; Coyle and Tsai 2004; Krystal et al. 2003, 2006;

Lewis and Moghaddam 2006; Stone et al. 2007; Tamminga 2006; Tsai and Coyle 2002), this subsection will review the functional aspects of Homer proteins as they relate to the architecture of glutamatergic synapses and in vivo glutamate transmission.

3.3.1. Homers regulate the morphology of excitatory synapses

Patients with schizophrenia exhibit a number of morphological abnormalities within cortical areas that include a decreased number of dendritic spines and reduced expression of synaptic proteins (Eastwood et al. 1995; Glantz and Lewis 1997, 2000, 2001; Rosoklija et al. 2000), a disarray of neuronal orientation (Kovelman and Scheibel 1984) and alterations in glutamate receptor gene expression and ligand binding (c.f., Krystal et al. 2003; Meador-Woodruff and Healy 2000). The results of postmortem studies are paralleled by in vivo structural neuroimaging findings that collectively indicate a disturbance in cortical connectivity and in dysregulation of activity within cortical networks (e.g., Buchsbaum et al. 1998; Hoffman et al. 1991; Koenig et al. 2001; Kubicki et al. 2002; Lawrie et al. 2002; Lim et al. 1996, 1999; Meyer-Lindenberg et al. 2001; Tauscher et al. 1998; Weinberger et al. 1986, 1992; Wible et al. 1995; Winterer et al. 2001), which may account for abnormalities (decreases or pathological increases) in task-related cortical activity (Barch et al. 2001; Fletcher et al. 1998; Haig et al. 2000; Kwon et al. 1999; Lee et al. 2001; Mathalon et al. 2000; McCarley et al. 2002; Weinberger et al. 1986). As abnormalities in task-related brain activation are considered to contribute to the disabling cognitive impairments observed in patients with schizophrenia (Bell and Bryson 2001; Bryson et al. 1998; Gold et al. 2002; Goldberg et al. 2001; Lysaker et al. 1996), thus, likely molecular candidates in the pathophysiology of schizophrenia are those regulating the morphology and development of cortical neurons.

Dendritic spines are tiny membrane protuberances that are the postsynaptic contact site for the vast majority (>90%) of excitatory synapses in the brain. Dendritic spines are highly motile structures that rapidly undergo shape changes, which contribute to alterations in synaptic strength, in part by enhancing connective opportunities (Alvarez and Sabatini 2007; Schubert and Dotti 2007; Sheng and Hoogenraad 2006). Alterations in spine size, emergence of new spines and the perforation of the postsynaptic density are correlated with a long-lasting enhancement of synaptic efficacy (Alvarez and Sabatini 2007; Engert and Bonhoeffer 1999; Maletic-Savatic et al. 1999; Schubert and Dotti 2007; Sheng and Hoogenraad 2006; Toni et al. 1999). In addition, spines possess an electron-dense fibrous structure referred to as the postsynaptic density (PSD), which is a tridimensional complex consisting of clusters of neurotransmitter receptors, as well as associated scaffolding and signaling proteins (Garner et al. 2000; Kennedy 2000; Kim and Sheng 2004; Okabe 2007; Scannevin and Huganir 2000; Sheng and Hoogenraad 2006). Spines also contain high concentrations of cytoskeletal associated proteins that include the Homer family of proteins, which cross-link actin filaments with the plasma membrane, glutamate receptors and ion channels, other components of intracellular signaling, and the smooth endoplasmic reticulum present in some spines (Ehlers 2002; Fagni et al. 2000).

Indeed, Homers are critical for regulating dendritic morphology and axonal pathfinding and their expression is dysregulated within limbo-corticostriatal structures in post-mortem brains of aged patients with schizophrenia (Meador-Woodruff et al. 2005) and by environmental factors that influence the manifestation of psychotic symptoms (see below). The co-transfection of Homer1b with Shank1b in developing hippocampal neuronal cultures induces the maturation of spines and causes an enlargement in dendritic spine size, compared to the transfection of either protein alone (Sala et al. 2001, 2003, 2005). Moreover, co-transfection of Homer1b with Shank1b also recruits IP3 receptors and endoplasmic cisternae to the PSD (Sala et al. 2001, 2005). The maturation and enlargement of spines requires the physical interaction of Homer and Shank as co-transfection of Homer1b with a mutant form of Shank that is incapable of binding Homer or the co-transfection of Shank1b with the truncated Homer1a did not alter spine size or shape (Sala et al. 2001, 2005). Within the PSD, CC-Homer proteins colocalize with F-actin/Shank/PSD-95/GKAP (guanylate kinase associated protein)/NMDA receptor clusters (Shiraishi et al. 2003; Usui et al. 2003). Corresponding to the maturation and enlargement of dendritic spines by Homer-Shank interactions (e.g., Sala et al. 2001) is the recruitment of the NR2B subunit of the NMDA receptor, as well as, a number of other scaffolding and signaling molecules associated with glutamate receptors, including GKAP, PSD-95, IP3 receptors, F-actin and various proteins related to the endoplasmic reticulum cisterna (Sala et al. 2001, 2003, 2005; Shiraishi et al. 2003, 2004; Usui et al. 2003). In mature neurons, the distribution and clustering of CC-Homers within dendritic spines is regulated by neuronal activity. High potassium-stimulated neuronal depolarization and subsequent fast, transient calcium entry induces the translocation of CC-Homers to spines and a marked increase in the number of CC-Homer-NMDA receptor clusters (Okabe et al. 2001). Moreover, calcium entry via voltage-gated calcium channels also stimulates CC-Homer-NMDA clustering (Okabe et al. 2001). In contrast, the application of glutamate and stimulation of calcium entry through NMDA receptors causes Homer de-clustering and reduces the amount of CC-Homer-NMDA clusters within the spine (Okabe et al. 2001). Although activation of protein kinase C (PKC) does not affect the intracellular distribution of CC-Homers, it induces the recruitment of Homer1a from the soma to the dendritic spine (Kato et al. 2001). Moreover, the administration of high potassium, phorbol esters, brain-derived neurotrophic factor (BDNF) and NT3 all promote the accumulation of Homer1a at synaptic sites in hippocampal neurons and these effects are dependent upon the activation of the extracellular signal-regulated kinase (ERK) (Kato et al. 2001, 2003). Thus, neuronal activation, particularly that leading to the activation of the MAPK-ERK signaling pathway would be expected to weaken synaptic efficacy due to disruption in CC-Homer interactions. Consistent with this, the over-expression of Homer1a in hippocampal cultures reduces the sizes of CC-Homer clusters and decreases the size of F-actin clusters (Inoue et al. 2004; Sala et al. 2003).

The formation of Homer-containing multi-protein clusters is developmentally regulated (Ebihara et al. 2003; Shiraishi et al. 2003). Image analysis of developing hippocampal neuronal cultures revealed the localization of Homer2/NR2B/PSD-95 clusters along the soma and proximal dendrites at 7 days in vitro (DIV) and these clusters migrated towards the dendritic head and developing spines where they co-localized with F-actin by 21 DIV (Shiraishi et al. 2003). The size of Homer1c/PSD-95 clusters increases progressively between 11-17 DIV and newly formed dendrites rapidly accumulates Homer1c, the density of which can be enhanced by activation of cAMP-dependent protein kinase (PKA) (Ebihara et al. 2003), furthering the evidence that neuronal activity regulates the translocation of Homer proteins to dendritic spines. While not yet assessed in developing cortical neurons, both IEG and CC-Homers also play a critical role in axonal pathfinding in optical tectal neurons (Foa et al. 2001). Transfection of tectal neurons with Homer1a, Homer1c or a mutant of Homer1c that interferes with the EVH1 domain induces aberrant axonal projections and irregular dendritic arborization, while transfection with a mutant of Homer1a prevented alterations in neuronal morphology and the alterations in axonal trajectories. Thus, there appears to be an optimal level of functional CC-Homer required in the growth cone for normal axonal pathfinding (Foa et al. 2001) and it has been suggested that abnormalities in Homer expression may contribute to the reported abnormalities in not only dendritic morphology, but also neuronal orientation, observed within cortical structures of patients with schizophrenia (e.g., Harrison 2004; Krystal et al. 2003; Selemon and Rajkowska 2003). In support of this hypothesis, preliminary postmortem data indicate an upregulation in *Homer2* transcript levels in the frontal cortex of aged patients with schizophrenia (Meador-Woodruff et al. 2005).

3.3.2. Homers regulate Group 1 mGluR expression and function

Group 1 mGluRs have been highly implicated in synaptic plasticity, neuronal development and neuroprotection (e.g., Bortollotto et al. 2005; Hannan et al. 2001; Harney et al. 2006; Liu et al. 2005; Rong et al. 2003) and growing histochemical, pharmacological and behavioral genetics evidence supports a role for abnormalities in Group1 mGluRs, particularly within cortical structures, in the pathophysiology of schizophrenia (e.g., Brody and Geyer 2004; Brody et al. 2003; Devon et al. 2001; Gupta et al. 2005; Moghaddam 2004; Ohnuma et al. 1998; Ritzen et al. 2005). Group 1 mGluRs are Gq-coupled receptors that are localized to the lateral edge of the PSD (Coutinho et al. 2001). Their activation leads to the generation of diacylglycerol, stimulation of phospholipase C, and the subsequent release of intracellular calcium from sarcoplasmic stores via activation of IP3 receptors (Conn and Pin 1997). The mGluR-mediated rise in intracellular calcium leads to the activation of a number of intracellular effectors that include adenylyl cyclase, tyrosine-kinase, MAPK, PKC), PI3K and glycogen synthase kinase 3beta (Calbresi et al. 2001; Heuss et al. 1999; Liu et al. 2005; Olive et al. 2005; Rong et al. 2003) and abnormalities in these signaling molecules have all been implicated either directly or indirectly in the cognitive or sensorimotor impairment associated with schizophrenia (e.g., Ahn et al. 2005; Culm et al. 2004; Kawanishi et al. 1999; Kozlovsky et al. 2001; Lovestone et al. 2007; Svenningsson et al. 2003). Group1 mGluR stimulation also results in the depolarization of post-synaptic neurons (Charpak et al. 1990; Guerineau et al. 1995), influences NMDA receptor activity within the limbic-corticostriatal circuit (Awad et al. 2000; Holohean et al. 1999; Homayoun and Moghaddam 2006; Pisani et al. 1997; Pisani et al. 2001; Prisco et al. 2002; Salt and Binns 2000), and can increase calcium-mediated potassium channel conductance within dopamine neurons located within this circuit (Fiorillo and Williams 1998; Paladini et al. 2001). Thus, molecules mediating or regulating the intracellular consequences of Group1 mGluR stimulation can have widespread impact upon the activity of the limbic-corticostriatal system implicated in the clinical manifestation of schizophrenia.

Table 3.2: Effects of Homer1 and Homer2 deletion upon measures of glutamate neurotransmission

Brain Region	Measure	Homer1 KO	Homer2 KO	Reference
Nucleus Accumbens	Basal glutamate content	↓	↓	Szumlinski et al. 2004, 2005a, 2005b
	Cocaine-stimulated glutamate release	↑	↑	Szumlinski et al. 2004
	Alcohol-sensitized glutamate release	ND	↓	Szumlinski et al. 2005b
	Cystine-stimulated glutamate release	ND	↓	Szumlinski et al. 2004
	Group1 mGluR agonist-stimulated glutamate release	ND	↓	Szumlinski et al 2004
	mGluR1, mGluR5 total protein content	—	↓ for GluR1	Szumlinski et al 2004 Williams et al 2007
	mGluR1,mGluR5 plasma membrane content	ND	—	Szumlinski et al. 2005b
	NR2a, NR2b Total protein content	—	—	Szumlinski et al. 2004 Williams et al. 2007
	NR2a, NR2b plasma membrane content	ND	↓ NR2a	Szumlinski et al. 2005b
	Xc total protein expression	ND	↓	Szumlinski et al. 2004
PFC	Basal glutamate content	↑	—	Lominac et al. 2005

				Szumlinski et al. 2005b
	K⁺-stimulated release	↓	ND	Szumlinski 2006
	Cocaine-stimulated glutamate release	↓	—	Lominac et al 2005 Szumlinski et al 2005a
	mGluR1,mGluR5, NR2a, NR2b total protein content	—	ND	Williams et al. 2007
Hippocampus	mGluR1,mGluR5, NR2a, NR2b total protein content	—	ND	Williams et al. 2007
	Whole-cell NMDA current	—	↓	Smothers et al. 2003, 2004
	Whole-cell AMPA current	—	↓	Smothers et al. 2003

↑ increase relative to WT, ↓ decrease relative to WT, — no change relative to WT, ND not yet determined.

Both the mGluR1a and the mGluR5 subtypes of Group1 mGluRs and Homer proteins co-immunoprecipitate in brain tissue (Brakeman et al. 1997; Kato et al. 1998; Xiao et al. 1998). Site-directed mutational analyses demonstrate that the physical interaction between Homer proteins and Group1 mGluRs occurs via the EVH1 domain and the PPSPF motif located on the C-terminal region of these receptors (Xiao et al. 1998). Cellular transfection studies have demonstrated an active role for both IEG and constitutively expressed Homer isoforms in regulating the subcellular localization of Group1 mGluRs and the clustering of these receptors to the PSD; however, discrepancies exist concerning the precise roles for IEG and constitutively expressed Homer proteins in mGluR trafficking, owing presumably to differences in experimental preparation. In human embryonic kidney-293 (HEK-293) cells, HeLa cells and cerebellar Purkinje neurons, cellular transfection with Homer1a increases or permits the cell surface expression of mGluR1a or mGluR5 (Abe et al. 2003; Ciruela et al. 1999; Minami et al. 2003; Roche et al. 1999) and induces the translocation of mGluR5 to dendrites and axons of cerebellar granule cells (Ango et al. 2000). It should be noted that an increase in mGluR1a surface expression has also been observed in HEK-273 cells following co-transfection with Homer1c, an effect attributed to a reduction in the rate of receptor turnover owing presumably to a Homer-mGluR interaction that masks the PEST sequence responsible for mGluR degradation (Ciruela et al. 2000; Minami et al. 2003). Alternatively, CC-Homers may reduce cell surface expression of these receptors via their retention within the endoplasmic reticulum. Co-transfection of mGluR5 with the CC-Homer isoform Homer1b in HeLa cells or cerebellar granule cells retains mGluR5 in the endoplasmic reticulum (Ango et al. 2002; Coutinho et al. 2001; Roche et al. 1999) and this retention requires Homer binding to the PPSPF motif on mGluR5 (Roche et al. 1999).

However, the in vivo data to date does not support a major role for Homer in regulating Group1 mGluR expression. While mice with null mutations of *Homer2* exhibit an approximately 25% reduction in the total protein expression of mGluR1a within the ventral striatum (Szumlinski et al. 2004), this reduction does not reflect changes in the content of this receptor subtype (or mGluR5) in the plasma membrane fraction of accumbens tissue homogenates (Szumlinski et al. 2005b). Moreover, recent immunoblotting studies in our laboratory have failed to detect changes in total Group1 mGluR protein expression in *Homer1* knock-out (KO) mice (see Table 2) (Williams et al. 2007), results consistent with immunocytochemical data from Jaubert et al. (2007) indicating no obvious effect of *Homer1* deletion upon the density or pattern of mGluR5 staining within the hippocampus. Moreover, the over-expression of Homer proteins does not significantly alter Group1 mGluR expression in vivo. While transgenic mice over-expressing Homer1a throughout the limbic-corticostriatal system fail to exhibit the co-immunoprecipitation of Group1 mGluRs with either Homer1b/c or Homer2a/b proteins, these mice do not differ from wild-type (WT) animals regarding Group1 mGluR content (Xiao et al. 1997). While it may be argued that developmental compensations in other Homer proteins may have masked the effects of gene deletion upon Group1 mGluR expression, very recently, a study of conditional transgenic mouse lines over-expressing Homer1a in striatum also reported no effect of Homer1a over-expression upon Group1 mGluR expression within this region (Tappe and Kuner 2006). Moreover, in rats, adeno-associated viral (AAV) transfection of hippocampal neurons with cDNA for *Homer1a*, *Homer1c* or the novel Homer1 isoform *Homer1g* all failed to alter significantly total Group1 mGluR expression or the pattern of receptor expression (Klugmann et al. 2005). As the majority of in vivo investigations employed measures of total protein expression, which cannot provide information regarding receptor trafficking, it is clear that further study is required in order to elucidate the relative roles for different Homer isoforms in regulating the trafficking and turn-over of Group1 mGluR in vivo.

In addition to regulating the subcellular compartmentalization of Group1 mGluRs, cellular transfection with CC-Homer isoforms facilitates receptor clustering on the plasma membrane (Ango et al. 2002; Ciruela et al. 2000; Coutinho et al. 2001; Kammermeier 2006; Kammermeier et al. 2000; Roche et al. 1999; Serge et al. 2002; Tadakoro et al. 1999) and induces the translocation of mGluR5 from the soma to dendrites in cerebellar granule cells (Ango et al. 2000). At least in COS-7 cells, the induction of mGluR clustering by constitutively expressed Homer isoforms requires the leucine-zipper motif (Tadakoro et al. 1999). As IEG Homer isoforms are incapable of this clustering function, their activity-dependent induction disrupts mGluR-containing receptor clusters, enabling their translocation and the functional rearrangement of the PSD, effects which are predicted to alter synaptic efficacy (Ango et al. 2000, 2002; Kammermeier and Worley 2007; Sala et al. 2001, 2003). Indeed, considerable in vitro evidence indicates that IEG and CC-Homer proteins actively regulate Group1 mGluR function and where examined, exert opposing effects on the ability of these receptors to modulate ion channel conductance and intracellular effectors (de Bartolomeis and Iasevoli 2003; Duncan et al. 2005; Fagni et al. 2002;

Xiao et al. 2000). In support of an inhibitory role for CC-Homer proteins in regulating aspects of Group1 mGluR signaling and as evidence for an opposing role for Homer1a in this regard, co-transfection of *Homer2b* with *mGluR1α* in superior cervical ganglion neurons decreases the modulation of N-type calcium and M-type potassium channels by mGluR1α, an effect that is reversed by co-transfection with *Homer1a* (Kammermeier et al. 2000). Depolarization-mediated activation of Purkinje cells induces Homer1a expression concomitant with enhanced intracellular calcium responses and inward currents, an effect that depends upon MAPK (Minami et al. 2003). In cerebellar granule cells, anti-sense-mediated knock-down of *Homer3* augments constitutive mGluR1 activity, as well as calcium-dependent big potassium (BK) channel activity and inositol phosphate accumulation (Ango et al. 2002). The effect of *Homer3* knock-down can be mimicked by co-transfection with *Homer1a* or co-transfection with an F1128R mutant form of mGluR5 that does not bind Homers (Ango et al. 2000, 2002). In a related study, co-transfection of cerebellar granule cells with *Homer1b* and *mGluR5* completely prevented the augmentation in BK channel activity produced by receptor agonism (Ango et al. 2002). In contrast to the data supporting a facilitatory role for Homer1a in regulating constitutive Group1 mGluR activity and potassium conductance, a recent study by Kammermeier and Worley (2007) showed that transfection of autaptic hippocampal neurons with *Homer1a*, but not a mutant isoform incapable of binding to Group1 mGluRs, reduces the capacity of Group1 mGluR agonists to inhibit glutamate-driven excitatory postsynaptic potentials and to elevate intracellular calcium. Moreover, in hippocampal cultures, the over-expression of *Homer1a* (or a mutant form of Shank that cannot bind Homers) interferes with the capacity of Group1 mGluR agonists to activate BK channel activity and to elevate IP3 levels (Sala et al. 2005). Such data are consistent with earlier indications of a dominant negative role for IEG Homer isoforms in Group1 mGluR function, in particular agonist-stimulated calcium release from internal stores (Tu et al. 1998; Xiao et al. 1998) and are consistent with the in vivo observation that *Homer2* deletion reduces agonist-stimulated glutamate release within the nucleus accumbens (Szumlinski et al. 2004). It should be noted that in many of the studies to date, effects of altering Homer expression upon Group1 mGluR function occur in the absence of significant changes in total or cell surface receptor expression (Ango et al. 2002; Kammermeier and Worley 2007; Roche et al. 1999; Sala et al. 2005; Szumlinski et al. 2004; Szumlinski et al. 2005b). Thus, Homers appear to regulate these aspects of Group1 mGluR transmission independently.

In sum, through interactions between their EVH1 domains at the C-terminus of Group1 mGluRs, both IEG and CC-Homer isoforms influence the intracellular retention, trafficking and clustering of Group1 mGluRs, at least in vitro. Given the important role for these receptors in regulating neuronal excitability within the limbic-corticostriatal circuit (e.g., Awad et al. 2000; Fiorillo and Williams 1998; Holohean et al. 1999; Homayoun and Moghaddam 2006; Paladini et al. 2001; Pisani et al. 1997, 2001, 2002; Salt and Binns 2000), factors affecting the balance between IEG and CC-Homer protein expression would be predicted to alter the subcellular localization of these receptor subtypes and impact receptor function within the limbic-corticostriatal circuit.

3.3.3. Homers regulate ionotropic glutamate receptor function and expression

Decades of behavioral pharmacological, genetic and histological evidence support ionotropic glutamate receptor (iGluR) hypofunction in the manifestation of psychotic symptoms and cognitive dysfunction (c.f., Abi-Saab et al. 1998; Coyle 2006; Kristiansen et al. 2007; Krystal et al. 2000; Krystal et al. 2003; Meador-Woodruff and Healy 2000; Nabeshima et al. 2006; Pietraszek 2003; Tsai and Coyle 2002). Ionotropic glutamate receptors (iGluRs) are divided into 3 subtypes based on pharmacology [NMDA, AMPA (α-amino-3-hydroxy-5-methyl-4-isoxazolepropionic acid) and kainate] and of these, the NMDA receptor has been most implicated in the etiology of schizophrenia and related psychotic disorders (c.f., Coyle 2006; Kristiansen et al. 2007; Krystal et al. 2003; Meador-Woodruff and Healy 2000; Pietraszek 2003; Tsai and Coyle 2002). The NMDA receptor is a tetramer containing multiple subunits, including the obligatory NR1, NR2A-D and NR3A-B, which affect the biophysical and pharmacologic characteristics of the receptor (Lynch and Guttmann 2001; Matsuda et al. 2003; Stephenson 2001). Activation of the NMDA receptor by glutamate requires the co-agonist glycine (Javitt and Zukin 1989; Johnson and Ascher 1987) and leads to an influx of calcium, sodium and potassium, which in turn depolarizes the neuronal membrane and can lead to the activation of various calcium-dependent kinases, including PI3K, MAPK/ERK and calcium-calmodulin kinase II (e.g., Hong et al. 2004; Mao et al. 2004; Perkinton et al. 1999; Wang et al. 2004; Wang et al. 2007).

In contrast to the direct interaction between Homers and Group1 mGluRs, Homers bind indirectly to NMDA receptors via an EVH1-mediated interaction with a proline-rich domain on Shank (Tu et al. 1999), which in turn links to NMDA receptors through a GKAP/PSD-95 scaffold (Naisbitt et al. 1999; Tu et al. 1999). Homers co-immunoprecipitate with Shank in forebrain and cerebellum (Tu et al. 1999) and co-localize with Shank at the PSD (Naisbitt et al. 1999; Sala et al. 2001; Sala et al. 2003; Shiraishi et al. 2003; Tu et al. 1999). During neuronal development, CC-Homer isoforms (incl. Homer 1b/c, Homer 2a/b, and Homer 3) localize closely with NR2B-containing NMDA receptor complexes throughout dendritic and synaptic differentiation, while they localize with clusters of the AMPA subtype of iGluRs only during the later stages of synaptic development (Shiraishi et al. 2003). As discussed above, the majority of studies revealed that the co-transfection of CC-Homers with Shank clusters mGluR5 in heterologous cells, as well as in neuronal cultures (Sala et al. 2001; Sala et al. 2003; Tu et al. 1999), induces the co-clustering of Homer with GKAP/PSD-95 (Tu et al. 1999) and augments the cluster size of NMDA, as well as AMPA, receptors within the PSD (Sala et al. 2003). Although, in one study by Hennou et al. (2003), virus-mediated transfection of dissociated hippocampal neurons with Homer1c did not change the size of either NMDA or AMPA receptor clusters. Over-expression of Homer1a in hippocampal cultures decreases the size of PSD-95 clusters, the number of NMDA receptor clusters, as well as the surface expression of AMPA receptors, indicating that IEG Homers exert a dominant negative effect upon iGluR trafficking and localization within the PSD (Sala et al. 2003). However, an enhancement in hippocampal AMPA, but not NMDA, receptor clusters by Homer1a over-expression has also been reported (Hennou et al. 2003). While not yet assessed in hippocampus, *Homer2* deletion reduces the content of NR2A and to a lesser extent NR2B, within the plasma membrane fraction of ventral striatal tissue homogenates, without altering the total content of either protein in this region (Szumlinski et al. 2004, 2005b). In contrast, a more recent examination of the effects of *Homer1* deletion upon NR2 subunit expression within the limbic-corticostriatal circuit revealed a reduction in total NR2A expression within the dorsal striatum, but not in other regions examined (see Table 2; Williams et al. 2007). Whether or not *Homer1* deletion affects the plasma membrane content and localization of NMDA receptor within the PSD remains to be determined.

Although discrepancies in findings exist, both IEG and CC-Homers also appear to be involved in regulating iGluR function. Homer1a over-expression in hippocampal neurons diminishes AMPA and NMDA receptor postsynaptic currents, an effect that requires the interaction between the EVH1 domain on Homer1a and the proline-rich motif on Shank (Sala et al. 2003). However, in another study, Homer1a over-expression was reported to augment, not reduce, AMPA-mediated currents in dissociated hippocampal neurons, while Homer1c over-expression was without effect upon the ion conductance of either iGluR (Hennou et al. 2003). Consistent with an important role for CC-Homers in the development of the PSD and the localization of NMDA receptor therein (Shiraishi et al. 2003), electrophysiological recordings on embryonic hippocampal neurons derived from embryonic *Homer2* KO mice revealed a shift downwards in the dose-effect curve for NMDA-mediated whole-cell currents at 14 DIV, but not at 7 DIV (Smothers et al. 2003). Consistent with earlier reports in transfected neurons (Hennou et al. 2003; Sala et al. 2003), *Homer2* deletion also reduces AMPA/kainate-mediated currents in 14 DIV embryonic hippocampal neuronal cultures, although the magnitude of this effect was less than that observed for the NMDA receptor (Smothers et al. 2003). Despite these effects of *Homer2* deletion upon iGluR conductance, a follow-up study failed to detect effects of gene deletion upon either the amplitude or frequency of spontaneous excitatory postsynaptic potentials, indicating that elimination of *Homer2* gene products does not produce gross abnormalities in glutamate-mediated synaptic signaling (Smothers et al. 2005). In contrast to the data from *Homer2* KO mice (Smothers et al. 2003), *Homer1* deletion does not alter whole-cell currents mediated by either NMDA or AMPA receptor stimulation in 14 DIV embryonic hippocampal cultures (Smothers et al. 2004). These latter data are inconsistent with the reported effects of *Homer1* manipulation in vitro (Hennou et al. 2003; Sala et al. 2003), which may reflect either compensatory developmental alterations in receptor function or the masking of an effect by elimination of both IEG and CC-Homer1 isoforms.

Despite significant physiological evidence supporting the cross-talk between Group1 mGluRs and iGluRs (Awad et al. 2000; Holohean et al. 1999; Homayoun and Moghaddam 2006; Pisani et al. 1997; Pisani et al. 2001; Prisco et al. 2002; Salt and Binns 2000; Yang et al. 2004) and the co-localization of Homer with Group1 mGluRs/Shank/GKAP/PSD-95/NMDA receptors (Brakeman et al. 1997; Kato et al. 1997; Kato et al. 1998; Naisbitt et al. 1999; Shiraishi et al. 2003; Tu et al. 1998, 1999; Xiao et al. 1998), only a few studies have investigated the putative role for Homers in regulating the cross-talk between Group1 mGluRs and iGluRs (Tu et al. 1999). In cultured striatal neurons, the synergistic increase in ERK phosphorylation and the activation of Elk-1 and CREB induced by the co-application of NMDA and the Group1 mGluR agonist DHPG [(S)-3,5-dihydroxyphenylglycine] was blocked by the administration of either a Tat peptide disrupting NMDA receptor-PSD-95 binding or a small interference RNA against Homer1b/c (Yang et al. 2004). Interestingly, the capacity of NMDA and Group1 mGluR agonists to synergistically elevate ERK phosphorylation levels was independent of both calcium entry through the NMDA receptor and calcium release from internal stores (Yang et al. 2004). Thus, despite playing an important role in juxtaposing Group1 mGluRs, NMDA receptors and IP3 receptors to promote glutamate-driven intracellular calcium signaling (e.g., Fagni et al. 2002; Sala et al. 2005; Tu et al. 1999; Xiao et al. 2002), Homer proteins can apparently facilitate receptor cross-talk and activate intracellular signaling pathways also through calcium-independent mechanisms (Yang et al. 2004).

In sum, through binding to Shank via their EVH1 domains, both IEG and CC-Homer isoforms influence the cell surface clustering and function of iGluRs, at least in vitro. Given the putative role for iGluR hypofunction, particularly that of the NMDA receptor, in the pathophysiology of schizophrenia (c.f., Coyle and Tsai 2004; Krystal et al. 2003; MacDonald and Chafee 2006; Kristiansen et al. 2007; Meador-Woodruff and Healy 2000; Pietraszek 2002; Tsai and Coyle 2002), factors affecting the balance between IEG and CC-Homer protein expression would be predicted to affect iGluR function, iGluR expression and cross-talk with Group1 mGluRs and thereby disrupt glutamate neurotransmission necessary for normal cognitive, motivational, emotional and sensorimotor processing.

3.3.4. Homers regulate extracellular glutamate levels in vivo.

In addition to their roles in regulating the subcellular localization and signaling through glutamate receptors (de Bartolomeis and Iasevoli 2003; Duncan et al. 2005; Fagni et al. 2002; Xiao et al. 2000), neurochemical phenotyping of *Homer* mutant mice has revealed an important role for these proteins in regulating extracellular levels of glutamate within the limbo-corticostriatal pathway (see Table 2) (c.f., Szumlinski et al. 2006c). *Homer* KO mice exhibit a number of abnormalities in corticofugal glutamate transmission, some of which depend upon the deleted gene. Both *Homer1* and *Homer2* KO mice exhibit an approximately 50% reduction in basal extracellular levels of glutamate within the ventral striatum (Szumlinski et al. 2004, 2005a, 2005b). Importantly, restoration of ventral striatal Homer2 levels via the local infusion of an AAV carrying *Homer2b* cDNA normalizes glutamate levels in *Homer2* KO mice, supporting an active role for this CC-Homer isoform in regulating basal glutamate content in this region (Szumlinski et al. 2004, 2005b). Interestingly, *Homer2* deletion does not affect basal glutamate content within the prefrontal cortex (PFC), while *Homer1* deletion elevates PFC glutamate content by approximately 50% (Lominac et al. 2005; Szumlinski et al. 2004, 2005a). AAV-mediated transfection of PFC neurons with *Homer1c* cDNA, but not *Homer1a* cDNA, restores PFC basal glutamate levels to WT controls, implicating CC-Homer1 isoforms in regulating extracellular glutamate levels within this region (Lominac et al. 2005).

At the present time, it is not entirely clear how deletion of *Homer* genes lead to such pronounced changes in basal glutamate content within the corticoaccumbens pathway. Ventral striatal extracellular glutamate levels are regulated primarily through a sodium-independent cystine-glutamate anti-porter, system Xc- (Baker et al. 2002). *Homer2* deletion attenuates ventral striatal cystine-glutamate anti-porter function, a finding attributed to reduced protein expression of the Xc- catalytic subunit (Szumlinski et al. 2004). As the cystine-glutamate anti-porter does not contain the PPSPF domain necessary for Homer binding, it is likely that Homer regulation of anti-porter function is indirect. Stimulation of presynaptically localized mGluR1 receptors leads to elevations in extracellular glutamate levels within both the ventral striatum and in the PFC (Melendez et al. 2005; Swanson et al. 2001). While *Homer2* deletion reduces the total mGluR1 protein expression within the ventral striatum and reduces agonist-stimulated glutamate release within this region (Szumlinski et al. 2004), recent immunoblotting studies conducted on *Homer1* WT and KO mice have failed to detect genotypic differences in Group1 mGluR expression (Table 2) (Williams et al. 2007). Thus, the effects of *Homer1* deletion upon basal glutamate content can be dissociated from effects upon Group1 mGluR function/expression. Homers regulate the plasma membrane trafficking of NMDA receptors via interactions with a trimeric Shank-GKAP-PSD95 complex that binds to NR2 subunits of the NMDA receptor (Naisbitt et al. 1999; Szumlinski et al. 2005b; Tu et al. 1999). While neither deletion of *Homer1* nor *Homer2* alters total NMDA receptor subunit expression within the ventral striatum or PFC (Table 3) (Szumlinski et al. 2004; Williams et al. 2007), *Homer2* deletion reduces the ventral striatum plasma membrane expression of NR2a and NR2b NMDA receptor subunits in vivo (Szumlinski et al. 2005b). Through the Shank-GKAP-PSD95 complex, NR2 subunits colocalize with nitric oxidase synthase-1 (Brenman et al. 1996; Burette et al. 2002; Christopherson et al. 1999), an enzyme responsible for the synthesis of the retrograde messenger nitric oxide (Bredt and Snyder 1994a; Moncada 1992). Thus, one testable hypothesis to account for the effects of Homer deletion upon basal and stimulated corticoaccumbens glutamate transmission relates to alterations in NMDA-mediated nitric oxide retrograde signaling (Bredt and Snyder 1994b; Moncada et al. 1991). While the precise molecular mechanisms involved in Homer regulation of corticostriatal glutamate transmission remain elusive, it is clear from the phenotypic characterization of *Homer* mutant mice that this family of proteins is necessary for the normal in vivo regulation of corticostriatal glutamate transmission.

3.4. Homers are dynamically regulated within the corticolimbic-striatal circuit

3.4.1. Anatomical localization of Homers in brain.

With the exception of the inducible, IEG, Homer1 isoforms, CC-Homer proteins are expressed in similar quantities in brain, but differ somewhat in their regional distribution (Shiraishi et al. 2004; Soloviev et al. 2000; Sun et al. 1998; Xiao et al. 1998). Of particular relevance to schizophrenia, Homer transcripts or proteins are present in many of the structures within the limbic-corticostriatal circuit – a neural circuit exhibiting pathology in schizophrenia and related psychotic disorders (Krystal et al. 2003). CC-Homer proteins are found throughout the cerebral cortex, including the frontal, parietal and somatosensory cortices, where they are localized to the superficial and deep layers (Xiao et al. 1998) and when induced, the expression of *Homer1a* mRNA is also present in these cortical layers (de Bartolomeis et al. 2002; Polese et al. 2002; Xiao et al. 1998). Homers appear to be differentially distributed within hippocampus; Homer1b/c is localized to the CA1, CA2, CA3, dentate gyrus and subiculum, where Homer3a/b labeling is high in CA3, but intermediate in CA2 regions (Shiraishi et al. 2004). In contrast to CC-Homer1 isoforms, the induction of *Homer1a* reveals mRNA expression within the CA1, CA2 and CA3 regions, but not in the dentate gyrus (Sun et al. 1998). Homer2a/b labeling is intense in CA1 and CA2, with intermediate labeling in the subiculum (Shiraishi et al. 2004). All three Homer isoforms, including the inducible forms, are located within the dorsal and ventral aspects of the striatum (Ary et al. 2007b; de Bartolomeis et al. 2002; Polese et al. 2002; Shiraishi et al. 2004; Sun et al. 1998; Swanson et al. 2001; Xiao et al. 1998; Zhang et al. 2007), but only Homer1 and Homer2 isoforms have been localized within thalamus (Shiraishi et al. 2004; Xiao et al. 1998), and immunoblotting has detected Homer1a, Homer1b/c and Homer2a/b within the amygdala (Ary et al. 2007a). In addition to these limbic-corticostriatal structures, all three Homer isoforms are located within cerebellum, with labeling of Homer1b/c and Homer2a/b within the internal granule layer and lighter labeling within the molecular layer (Shiraishi et al. 2004). In contrast, Homer3 expression within cerebellum is higher than that for Homer1 with labeling within the molecular and Purkinje cell layers (Shiraishi et al. 2004). Finally, while the expression of Homer3 within the olfactory bulbs is developmentally down-regulated, the expression of both Homer1 and Homer2 isoforms persist throughout development in this structure (Shiraishi et al. 2004).

Within neurons, Homer proteins are localized predominantly in the soma and apical dendrites (Brakeman et al. 1997; Kaja et al. 2003; Sun et al. 1998; Xiao et al. 1998). At a subcellular level, Homer proteins are enriched within the PSD fraction, but are also present in the crude nuclear, synaptosomal and microsomal fractions (Sun et al. 1998; Xiao et al. 1998). Homer2 differs from the other Homer isoforms as it is also localized to the soluble fraction and the synaptic vesicle fraction (Xiao et al. 1998), implicating these isoforms in vesicular trafficking of receptor proteins.

While the etiology of schizophrenia and related psychotic disorders is unknown, the manifestation of psychotic symptoms clearly involves a combination of genetic and environmental factors. Indeed, the expression of both IEG and constitutively expressed Homer isoforms can be regulated within the limbo-corticostriatal circuit by factors that either exacerbate or precipitate psychosis (incl. stressors and psychotomimetic drugs) or by antipsychotic treatments. First characterized as a gene whose mRNA expression is upregulated within 1-3 hours following the application of supra-physiological electrical stimulation of the hippocampus (Brakeman et al. 1997; Kato et al. 1998), the capacity of synaptic activation to upregulate both *Homer1a* and *Ania-3* mRNA expression has been demonstrated to occur throughout the limbo-corticostriatal circuit following various experimental manipulations of relevance to either schizophrenia or animal models of psychosis (de Bartolomeis and Iasevoli 2003; Szumlinski et al. 2006c). More recently, a number of studies have extended the data for IEG Homer mRNA to the protein level and revealed a regulation of both IEG and CC-Homer expression throughout the limbo-corticostriatal circuit by psychotomimetic and anti-psychotic manipulations (see Table 3).

Table 3.3: Regulation of Homer mRNA or protein by factors impacting the manifestation of psychosis.

Homer isoform	Region	Treatment	Effect	Reference
Stressors				
Homer1a	Forebrain	PTZ, CHX	mRNA ↑ at 3 hrs blocked by MK-801	Kato et al. 1997
	Parietal cortex & Hippocampus	Exposure to a novel environment	mRNA ↑ between 25-25 min	Vazdarjanova et al. 2002
Homer1a	Hippocampus	Footshock	mRNA ↑ at 3 & 24 h	Igaz et al. 2004
	PFC, Amygdala & Hippocampus	Repeated maternal stress	Protein ↑ at 3 weeks	Ary et al. 2007a
Homer1b/c	PFC	Repeated maternal stress	Protein ↑ at 3 weeks	Ary et al. 2007a
Homer2a/b	Amygdala & Hippocampus	Repeated maternal stress	Protein ↑ at 3 weeks	Ary et al. 2007a
Acute Psychotomimetic				

Drug Exposure				
Homer1a	Neocortex	Acute methamphetamine or cocaine	mRNA ↑ at 1 hr	Fujiyama et al. 2003
	Neocortex or PFC	Acute cocaine injection	mRNA ↑ at 1-2 hrs	Fujiyama et al. 2003; Ghasemzadeh et al. 2006
	Ventral striatum	Acute cocaine injection	mRNA ↑ at 2 hrs	Ghasemzadeh et al. 2006
	Dorsal striatum	Acute cocaine injection	mRNA ↑ at 2 hrs	Brakeman et al. 1997
	Dorsal striatum	Acute cocaine injection	protein ↑ at 2 hrs	Zhang et al. 2007
	VTA	Acute cocaine injection	mRNA ↑ at 2 hrs	Ghasemzadeh et al. 2006
	PFC, auditory and granular retrosplenial cortices	Acute phencyclidine	mRNA ↑ at 2 & 24 hrs	Cochran et al. 2002
	PFC	Acute lysergic acid diethylamine (LSD)	mRNA ↑ at 1.5 hrs	Nichols and Sanders-Bush 2002
	Dorsal striatum	Quinpirole (D2/D3 receptor agonist)	mRNA ↑ <1 hr	Berke et al. 1998
Ania-3	PFC	Acute LSD injection	mRNA ↑ at 1.5-3 hrs	Nichols and Sanders-Bush 2002 Nichols et al. 2003
Homer1b/c	Ventral striatum	Cocaine injection	Protein ↑ at 1 hr	Fourgeaud et al. 2004
	Cerebellum	Cocaine	Protein ↑ at 15 hr	Dietrich et al. 2007
Homer3a/b	Cerebellum	Cocaine	Protein ↑ at 15 hr	Dietrich et al. 2007
Repeated Psychotomimetic Drug Exposure				
Homer1a	Neocortex	Repeated methamphetamine + methamphetamine or cocaine challenge	mRNA ↑ at 1 hr	Fujiyama et al. 2003
	Accumbens shell	Repeated cocaine	Protein ↓ at 3 weeks, but not at 24 hr	Ary et al. 2007b Swanson et al. 2001
	Hippocampus	Repeated cocaine	Protein ↑ at 3 weeks	Ary et al. 2007b
	Accumbens core, PFC, Striatum	Repeated cocaine	No protein change at 24 hr or 3 weeks	Ary et al. 2007b Swanson et al. 2001

	Cerebellum	Repeated cocaine	Protein ↑ at 15 hr	Dietrich et al. 2007
	PFC	1-hr cocaine self-administration	No protein change at 24 h or 2 weeks	Ary et al. 2006 Obara et al. 2007
Homer1a	PFC	6-hr cocaine self-administration	Protein ↑ at 24 hr, but not at 2 weeks	Ary et al. 2006 Obara et al. 2007
	Accumbens core	1-hr cocaine self-administration	Protein ↓ at 24 hr, but not at 2 weeks	Obara et al. 2007
	Accumbens core	6-hr cocaine self-administration	Protein ↓ at 24 hr, but not at 2 weeks	Obara et al. 2007
	Ventral striatum	Methamphetamine injections	No protein change at 24 hr Protein ↑ at 3 weeks	Lominac and Szumlinski 2006
Homer2a/b	PFC	Cocaine injections	Protein ↑ at 3 weeks	Ary et al. 2007b
	Hippocampus	Cocaine injections	Protein ↑ at 3 weeks	Ary et al. 2007b
	Ventral striatum	Cocaine injections	Protein ↓ at 3 weeks	Ary et al. 2007b
	PFC	1-hr cocaine self-administration	No protein change at 24 hr or 2 weeks	Ary et al. 2006
	PFC	6-hr cocaine self-administration	No protein change at 24 hr or 2 weeks	Ary et al. 2006
	Accumbens core	1-hr cocaine self-administration	No protein change at 24 hr Protein ↓ at 2 weeks	Obara et al. 2007
	Accumbens core	6-hr cocaine self-administration	No protein change at 24 hr Protein ↓ at 2 weeks	Obara et al. 2007
	Ventral striatum	Methamphetamine injections	No protein change at 24 hr Protein ↑ at 3 weeks	Lominac and Szumlinski 2006
Homer3a/b	Cerebellum	Cocaine injections	Protein ↑ at 15 hrs	Dietrich et al. 2007
Antipsychotic treatment				
Homer1a	Frontal & Parietal Cortex	Acute haloperidol or acute quetiapine	No mRNA change at 1.5 hrs	Ambesi-Impiombato et al. 2007
	Ventral striatum	Acute haloperidol	mRNA ↑ at 1.5 or 3 hrs	de Bartolomeis et al. 2002 Polese et al. 2002

	Ventral striatum	Acute olanzapine	mRNA ↑ at 3 hrs	Polese et al. 2002
	Ventral striatum	Acute clozapine	mRNA ↑ at 1.5 hrs	de Bartolomeis et al. 2002
	Ventral striatum	Acute quetiapine	No mRNA change at 1.5 hrs	Ambesi-Impiombato et al. 2007
	Dorsal Striatum	Acute haloperidol	mRNA ↑ at 1.5 or 3 hrs	de Bartolomeis et al. 2002 Polese et al. 2002 Ambesi-Impiombato et al. 2007
	Dorsal Striatum	Acute olanzapine, clozapine or quetiapine	No mRNA change at 1.5 or 3 hrs	de Bartolomeis et al. 2002 Polese et al. 2002 Ambesi-Impiombato et al. 2007
	Frontal & Parietal Cortex	Repeated haloperidol or quetiapine	No mRNA change at 1.5 hrs	Ambesi-Impiombato et al. 2007
Homer1a	Frontal Cortex	Repeated olanzapine	mRNA ↑ at 1.5 hrs	Fatemi et al. 2006
	Ventral Striatum	Repeated haloperidol or quetiapine	No mRNA change at 1.5 hrs	Ambesi-Impiombato et al. 2007
	Ventral Striatum	Repeated haloperidol or olanzapine	No protein change at 24 hrs	Fig. 3.1
	Dorsal Striatum	Repeated haloperidol	mRNA ↑ at 1.5 hrs	Ambesi-Impiombato et al. 2007
Ania-3	Frontal or Parietal Cortex	Acute haloperidol or quetiapine	No mRNA change at 1.5 hrs	Ambesi-Impiombato et al. 2007
	Ventral striatum shell subregion	Acute haloperidol or quetiapine	mRNA ↑ at 1.5 hrs	Ambesi-Impiombato et al. 2007
	Dorsal striatum	Acute haloperidol	mRNA ↑ at 1.5 hrs	Ambesi-Impiombato et al. 2007
	Dorsal striatum	Acute quetiapine	No mRNA change at 1.5 hrs	Ambesi-Impiombato et al. 2007

	Frontal or parietal cortex	Repeated haloperidol or quetiapine	No mRNA change at 1.5 hrs	Ambesi-Impiombato et al. 2007
	Ventral striatum	Repeated haloperidol or quetiapine	No mRNA change at 1.5 hrs	Ambesi-Impiombato et al. 2007
	Dorsal striatum	Repeated haloperidol	mRNA ↑ at 1.5 hrs	Ambesi-Impiombato et al. 2007
	Dorsal striatum	Repeated quetiapine	No mRNA change at 1.5 hrs	Ambesi-Impiombato et al. 2007
Homer1b/c	Ventral striatum	Repeated haloperidol	No protein change at 24 hrs	Fig. 3.1
	Ventral striatum	Repeated olanzapine	Protein ↓ at 24 hrs	Fig. 3.1
Homer2a/b	Ventral striatum	Repeated haloperidol	Protein ↓ at 24 hrs	Fig. 3.1
	Ventral striatum	Repeated olanzapine	Protein ↓ at 24 hrs	Fig. 3.1
Repeated Nicotine Administration				
Homer1a	NAC	3, 7 or 14 days of nicotine injections	mRNA ↓ after 3 days of Rx no mRNA change after 7 or 14 days of Rx no protein change	Kane et al. 2005
	Amygdala or VTA		No mRNA or protein change	
Homer1b/c	Ventral striatum		mRNA ↑ after 14, but not 3 or 7 days Rx no protein change	
	Amygdala		no mRNA change with any Rx Protein ↑ after 3, but not 7 or 14 days Rx	
	VTA		Protein ↑ after 3, but not 7 or 14 days of Rx; no mRNA change with any Rx	
Homer2a/b	Ventral striatum		No change in mRNA or protein	

Amygdala	mRNA ↑ after 3, but not 7 or 14 days Rx no protein change with any Rx
VTA	no mRNA change with any Rx Protein ↑ after 3, but not 7 or 14 days of Rx

↑ increase relative to WT, ↓ decrease relative to WT, — no change relative to WT, ND not yet determined.

3.4.2. Limbic-corticostriatal regulation of Homer proteins by stressors

Stress is highly implicated in the etiology of schizophrenia and is considered a major pre-disposing or precipitating factor in the etiology of this disease (Castine et al. 1998; Koenig 2006; Muller et al. 2004; Phillips et al. 2006; Seedat et al. 2003). Evidence that stressors can affect the expression of Homers in brain was first described in the seminal report by Kato et al. (1997), in which a modest rise in *Homer1a* mRNA expression was detected within hippocampal homogenates at 3 hrs following the systemic administration of the pharmacological stressors and convulsants pentylenetetrazol (PTZ) and cycloheximide (CHX). Similarly, exposure to a mild stressor such as placement into a novel environment also augments Homer1a mRNA within hippocampus, as well as within parietal cortex (Vazdarjanova et al. 2002). Intriguingly, Igaz et al. (2004) reported no change in Homer1a protein expression within the hippocampus at 3 hours following delivery of a mild footshock in a one-trial aversive learning paradigm; however, Homer1a protein expression was found to be elevated in animals euthanized 24 hours following shock delivery. This footshock-induced change in the pattern of Homer1a expression is very atypical for an IEG and is distinct from that produced by other forms of synaptic activity (e.g., Brakeman et al. 1997; Kato et al. 1997; Vazdarjanova et al. 2002), raising the possibility that IEG Homers may regulate both the consolidation and recall of memories concerning stressful or aversive events (Igaz et al. 2004).

The ability of stressors to regulate Homer proteins in the limbic-corticostriatal circuits has recently been extended to periods of brain development. The prenatal stress model of psychosis induces profound cognitive, social and sensorimotor deficits in adult offspring akin to those observed in schizophrenia and has been utilized for approximately a decade to understand the contribution of early exposure to stressors upon the developmental trajectory that underlies abnormal brain functioning (e.g., Koenig 2006; Weinstock 2001), as well as inducing abnormalities in glutamate neurotransmission within the limbic-corticostriatal circuit (Kippin et al. 2007). In weanling offspring of rat dams subjected to repeated restraint stress during the last 7 days of gestation, marked increases in Homer1a protein expression were detected within several corticolimbic structures, including the prefrontal cortex, amygdala and hippocampus, while Homer1a levels were reduced within the dorsal and ventral aspects of the striatum (Table 3). The rise in Homer1a protein expression was accompanied by elevations in Homer2a/b levels with hippocampus and the amygdala, while Homer1b/c levels were increased in both the amygdala and prefrontal cortex (Ary et al. 2007a). As the changes in protein expression were observed at 3 weeks following the last maternal restraint stress, these data indicate that repeated prenatal stress can elicit persistent changes not only in CC-, but also, IEG-Homer expression, which are likely to impact synaptic development, morphology, and plasticity that contribute to behavioral abnormalities in later life.

3.4.3. Limbic-corticostriatal regulation of Homer proteins by psychotomimetic drugs

Psychomotor stimulant (e.g., cocaine and the amphetamines), hallucinogenic (lysergic acid diethylamide or LSD) and dissociative anesthetic (e.g., phencyclidine or PCP, ketamine) drugs all induce behavioral or cognitive symptoms of schizophrenia in non-affected individuals and can precipitate psychotic episodes in those who are affected with the disorder (Curran et al. 2004; Dore and Sweeting 2006; Krystal et al. 2003; Ross 2006; Schuckit 2006). For decades, the phenomenon of drug-induced psychosis has been modeled in laboratory animals using either acute or repeated drug administration regimens and has yielded considerable insight into the role of various receptors and neurotransmitter systems in the pathogenesis of psychotic disorders. Consistent with the data for stress presented above, Homer1a or Ania-3 mRNA expression is increased within limbo-corticostriatal structures by a number of psychotomimetic drugs, that include cocaine and methamphetamine (Brakeman et al. 1997; Fourgeaud et al. 2004; Fujiyama et al. 2003), LSD (Nichols et al. 2003; Nichols and Sanders-Bush, 2003), and PCP (Cochran et al. 2002) (see Table 3). Most recently, a report by Zhang and colleagues extended earlier data for cocaine-induced changes in Homer1a mRNA to the protein level by demonstrating a dose-dependent increase in Homer1a within both the dorsal and ventral aspects of the striatum following acute cocaine administration. In contrast to stressors (Igaz et al. 2004; Ary et al. 2007a), the increase in Homer1a protein was transient, peaking at 2-3 hrs post-injection and returning to baseline values at 12 hrs (Zhang et al. 2007). The cocaine-mediated increase in Homer1a expression likely relates to this drug's ability to block dopamine reuptake as the selective dopamine reuptake inhibitor GBR 12909 can also induce increases in Homer1a or Ania-3 expression throughout the striatum, as well as in frontal and parietal cortices (Ambesi-Impiombato et al. 2007). Indeed, an earlier report indicated that Homer1a mRNA could be induced within striatum by the administration of the D_2/D_3 dopamine receptor agonist quinpirole (Berke et al. 1998);

however, the cocaine-induced rise in Homer1a protein reported by Zhang et al. (2007) could be prevented by D_1, but not by D_2, receptor antagonists and involved the activation of PKA and Ca^{2+}/calmodulin-dependent protein kinases, which in turn stimulated gene transcription by CREB (Zhang et al. 2007).

While the majority of studies on how psychotomimetic drugs affect Homer expression have employed acute drug treatment regimens, a few studies have examined the consequences of repeated drug treatment upon Homer levels within brain. The repeated administration of psychomotor stimulant drugs induces a progressive increase in their motor hyper-activating effects. This phenomenon is referred to as behavioral sensitization and is considered a valid animal model of drug-induced psychosis (c.f., Pierce and Kalivas 1997; Robinson and Becker 1983; Segal et al. 1981; Vandershuren and Kalivas 2000). In a study of methamphetamine-induced behavioral sensitization, Fujiyama et al. (2003) reported no change in the magnitude of the methamphetamine-induced increase in Homer1a mRNA with repeated drug injections, despite the development of sensitized behavior. Related to this, a tolerance appears to develop for the capacity of cocaine to elicit a rise in ventral striatal Homer1a mRNA levels with daily injections (Ghasemzadeh et al. 2006). Thus, at least within these animal models of psychosis, the precise relationship between drug-induced elevations in Homer1a expression and behavior are unclear. However, accumulating proteomic data supports a role for drug-induced changes in CC-Homer expression in the development of drug-sensitized behavior (see Table 3). For example, 3 weeks withdrawal from a repeated cocaine regimen that induces behavioral sensitization down-regulates by approximately 25% the expression of Homer1b/c and Homer2a/b protein, as well as Group1 mGluRs within the medial aspect of the ventral striatum, but not within the more lateral ventral striatum or the dorsal striatum (Swanson et al. 2001; Ary et al. 2007b). The effects of withdrawal from repeated cocaine appear to be regionally selective as cocaine-sensitized rats exhibit an increase in both Homer1b/c and Homer2a/b levels within hippocampus (Ary et al. 2007b), suggesting that enduring alterations in CC-Homer expression might contribute to the enhancement in long-term potentiation observed in this brain region following cocaine self-administration (del Olmo et al. 2006) and contribute to cue- and cocaine-induced relapse (Rogers and See 2007). In contrast to the lack of change in Homer1b/c levels, Homer2a/b levels are approximately doubled within the PFC of cocaine-treated animals, concomitant with increases in the expression of mGluR1 and the activated form of PI3K (Ary et al. 2007b). This finding supports earlier data for an important role for PI3K activation in mediating the expression of cocaine-sensitized behavior (Izzo et al. 2002) and implicate this neuroadaptation in mediating behavioral disinhibition and impairments in working memory associated with repeated stimulant exposure (e.g., Hester and Garavan 2004; Javanovski et al. 2005; Kubler et al. 2005; Verdejo-Garcia and Perez-Garcia 2007).

Interestingly, recent studies examining the short- and long-term consequences of repeated methamphetamine administration upon Group1 mGluR/Homer/PI3K expression have revealed a pattern of changes within the ventral striatum that is markedly distinct from that observed following cocaine (Lominac et al. 2007; see Table 3).

As observed following repeated cocaine (Swanson et al. 2001), methamphetamine-induced changes in protein expression were only apparent within the ventral striatum following 3 weeks, but not 24 hrs, withdrawal; however, methamphetamine elicited an increase in the expression of Homer1a, Homer1b/c and Homer2a/b, concomitant with increases in both mGluR1 and mGluR5, NR2A, as well as phosphorylated PI3K (Lominac et al. 2007). The precise reason for the apparently opposite effects of repeated cocaine and methamphetamine upon the ventral striatal expression of members of the mGluR/Homer/PI3K signaling complex is currently unclear. However, cocaine elicits a sensitized glutamate response within this region following withdrawal from repeated treatment (Pierce et al. 1996; Szumlinski et al. 2006), while the capacity of methamphetamine to elevate extracellular glutamate within the ventral striatum exhibits a mild tolerance (Lominac et al. 2007; Zhang et al. 2003). Thus, it appears that the changes in Homer protein expression, at least within the ventral striatum, may reflect a compensatory change in response to alterations in glutamate transmission produced by repeated psychotomimetic administration. While this hypothesis requires further investigation, these data nonetheless support a relationship between the long-term behavioral-sensitizing and proteomic effects of repeated psychotomimetic drug administration and the functional implications of this relationship will be discussed in greater detail below.

3.4.4. Limbic-corticostriatal regulation of Homer proteins by anti-psychotic drugs

In addition to their regulation by environmental and pharmacological factors that mimic or precipitate the manifestation of psychotic symptoms, Homer expression is regulated by pharmacological agents that ameliorate symptoms of schizophrenia (see Table 3). Typical and atypical antipsychotic drugs are the primary pharmacotherapeutic agents employed in the treatment of schizophrenia and can differently regulate neuronal gene expression in several preclinical paradigms. Consistent with other reports for the IEG c-fos (Hiroi and Graybiel 1996; MacGibbon et al. 1994; Semba et al. 1999), IEG Homer mRNA are differentially regulated by these two major classes of antipsychotic treatments (see Table 3) (Ambesi-Impiombato et al. 2007; de Bartolomeis et al. 2002; Fatemi et al. 2006; Polese et al. 2002). In a series of in situ hybridization studies conducted by de Bartolomeis and colleagues, neither acute nor repeated administration of typical or atypical antipsychotics altered Homer1a mRNA within prefrontal or parietal cortices (Ambesi-Impiombato et al. 2007; de Bartolomeis et al. 2002). However, repeated (21 days) olanzapine treatment has been reported to elevate Homer1a levels within the frontal cortex when assessed by cDNA microarray, a finding confirmed by real-time polymerase chain reaction (RT-PCR) (Fatemi et al. 2006). The acute and repeated administration of the prototypical D_1/D_2 dopamine receptor antagonist and neuroleptic haloperidol elevates Homer1a or Ania-3 expression within both the dorsal and ventral aspects of the striatum (Ambesi-Impiombato et al. 2007; de Bartolomeis et al. 2002; Polese et al. 2002). In contrast, acute administration of the atypical antipsychotics olanzapine (de Bartolomeis et al. 2002) or clozapine (Polese et al. 2002) increased *Homer1a* mRNA in the ventral, but not the dorsal, aspect of the striatum. Moreover, the co-administration of D-cycloserine, a partial agonist at the strychnine-insensitive glycine site of the NMDA receptor complex (Goff et al. 1999) prevented the antipsychotic-induced rise in *Homer1a* mRNA expression in both brain regions, suggesting a role for this site on the NMDA receptor in mediating this effect (Polese et al. 2002).

More recent studies have extended these data to a novel fast-dissociating D_2 antagonist quetiapine (Kapur et al. 2000) and demonstrated that the acute administration of higher quetiapine doses increases *Ania-3*, but not *Homer1a*, expression within the ventral striatum, but not within dorsal striatum and frontal or parietal cortices. Moreover, while repeated haloperidol administration induced robust increases in both *Homer1a* and *Ania-3* mRNA throughout the striatum, the repeated administration of quetiapine (21 days) was without effect upon mRNA expression in any of the brain regions examined (Ambesi-Impiombato et al. 2007). Given the marked differences in the capacity of slow-dissociating typical and fast-dissociating atypical antipsychotics to induce extrapyramidal side effects, these data for IEG Homers pose a potential role for the Homer family of proteins in regulating not only the therapeutic, but also the motor side-effect profile of these agents. In support of this assertion, recent behavioral phenotyping of transgenic mice that over-express *Homer1a* within the extrastriosomal matrix of the striatum exhibit fine motor impairments and fail to exhibit motor learning (Tappe and Kuner, 2006).

Only preliminary data exist regarding the effects of anti-psychotic treatment upon Homer protein expression and to date, the effects of repeated haloperidol versus olanzapine upon Homer1a, Homer1b/c and Homer2a/b have been characterized within the ventral striatum only (Fig. 3.1). At 24 hrs following the last of 10 daily injections of either 1 mg/kg haloperidol or 2 mg/kg olanzapine, Homer1a expression is unchanged, indicating that the antipsychotic-induced rise in striatal *Homer1a* expression (Ambesi-Impiombato et al. 2007; de Bartolomeis et al. 2002; Polese et al. 2002) does not translate into persistent effects upon protein expression. In contrast to their lack of an effect upon Homer1a, both typical and atypical antipsychotics differentially down-regulate the expression of Homer1b/c and Homer2a/b within the ventral striatum (Fig. 3.1). When compared to saline-treated controls, Homer1b/c levels were significantly reduced in mice treated repeatedly with haloperidol, but not with olanzapine. In contrast, both antipsychotic drugs lowered Homer2a/b expression but the magnitude of the effect was greater in mice treated with olanzapine (Fig. 3.1). It remains to be determined whether or not typical and atypical antipsychotics differentially regulate Homer protein expression within the dorsal aspects of the striatum, or within frontal cortical structures or produce specific actions to oppose changes observed in models of psychoses, for instance reversal of the effects of prenatal stress upon limbo-corticostriatal Homer protein levels. Moreover, while the effects of repeated antipsychotic administration upon Homer expression are more or less opposite to those observed in the prenatal stress and methamphetamine sensitization models of psychosis (Ary et al. 2007a; Lominac et al. 2007), they do approximate those observed in cocaine-sensitized animals (Ary et al. 2007b; Swanson et al. 2001; see Table 3). At the present time, these discrepancies in findings remain unresolved. Nevertheless, these data support the notion that pharmacotherapies for the treatment of schizophrenia exert different, perhaps regionally selective, effects upon the expression of Homer proteins, which likely impact corticolimbic glutamate neurotransmission and the behaviors regulated by this neurotransmission.

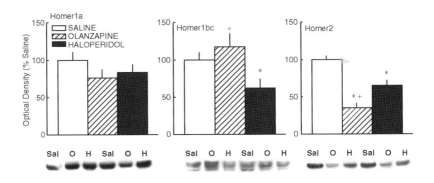

Fig. 3.1. Effects of 14 daily intraperitoneal injections of 1.0 mg/kg haloperidol (H), 2.0 mg/kg olanzapine (O) or saline (S) upon the expression of Homer proteins within the ventral striatum at 24 hrs following the last injection. Sample sizes ranged from 8-10 animals/data point. *$p<0.05$ vs. saline, +$p<0.05$ olanzapine vs. haloperidol.

Cigarette smoking is notoriously prevalent amongst patients with schizophrenia and while a number of theories exist to account for this co-morbidity, one prevalent theory proposes that nicotine intake ameliorates cognitive deficits or periods of high agitation (e.g., Dalack et al. 1998; Kumari and Postma 2005; Lohr and Flynn 1992). In agreement with the data for anti-psychotic drugs above, a study conducted by Kane and colleagues (2005) examined the regulation of various Homer isoforms at both the mRNA and protein levels, following subchronic (3 days) and chronic (7 or 14 days) nicotine treatment regimens (Table 3). While subchronic nicotine increased Homer2a/b expression within the amygdala, this result did not translate to the protein level. Similarly, within the ventral striatum, Homer2a/b expression was also increased by 14 days of nicotine administration, but this was not accompanied by a change in protein expression. In contrast, subchronic nicotine administration failed to alter Homer1b/c but produced a small, albeit significant, rise in protein expression within this brain region. Within the ventral tegmental area (VTA), subchronic nicotine also increased the protein expression of both Homer1b/c and Homer2a/b (Kane et al. 2005). Thus, consistent with other reports (e.g., Kato et al. 1998), there is a disconnect between the effects of nicotine upon the mRNA and protein expression for the different Homer isoforms but these data nonetheless support a potential role for nicotine-induced regulation of Homer expression in this drug's prophylactic effect.

In sum, both IEG and CC-Homer mRNA or protein expression is regulated throughout the limbic-corticostriatal circuit by environmental factors influencing the manifestation of psychotic symptoms. The precise relevance of these Homer changes for behavior is not fully characterized (see below). However, given the roles for Homer proteins in cell morphology, glutamate receptor trafficking and subcellular localization, as well as the maintenance and regulation of extracellular levels of glutamate within the limbic-corticostriatal circuit (see subsections above), these molecular/proteomic data implicate altered Homer expression in the biobehavioral abnormalities exhibited by patients with schizophrenia. Consistent with this possibility, Homer2 transcript levels are elevated in post-mortem brains derived from aged patients with schizophrenia (Meador-Woodruff et al. 2005) and linkage analysis identified a single nucleotide polymorphism (SNP) in intron 4 of *Homer1* that was associated significantly with schizophrenia in one of two cohorts of patients derived from a large, heterogeneous population (Norton et al. 2003). This polymorphism could impact both IEG and CC-Homer1 transcription (Bottai et al. 2002), implicating both IEG and CC-Homer1 isoforms in the psychopathology of schizophrenia and related disorders.

3.5. Functional aspects of Homer proteins related to behavior

Extensive behavioral phenotyping of *Homer* mutant mice has demonstrated important roles for several Homer proteins in regulating cognitive, sensorimotor, emotional and motivational processing, as well as in determining behavioral sensitivity to psychotomimetic drugs (Jaubert et al. 2007; Lominac et al. 2005; Szumlinski et al. 2004, 2005a, 2005b). In particular, *Homer1* KO mice exhibit endophenotypes that are observed in other animal models of psychosis, including the prenatal stress, ventral hippocampal-lesion and stimulant-sensitization models, and are consistent with those observed in patients with schizophrenia or related disorders (Table 4). Importantly for the validation of this KO as a potential genetic model of schizophrenia, the studies to date indicate that some of the behavioral abnormalities can be reversed by pretreatment with anti-psychotic administration (see below). This final section reviews the preclinical data implicating Homer proteins in regulating behaviors associated with limbo-corticostriatal function.

3.5.1. Homer1 isoforms in cognitive, attentional, motivational and emotional processing

Behavioral screens of adolescent and adult *Homer1* and *Homer2* KO mice revealed striking differences in the "undrugged" phenotype of these two mutant strains

(Szumlinski et al. 2006c), supporting an important role for *Homer1* polymorphisms in the clinical presentation of psychosis (Norton et al. 2003). As outlined in Table 4, *Homer1* KO mice exhibit deficits in sensorimotor and cognitive processing that are not observed in *Homer2* KO mice. *Homer1* KO mice exhibit pronounced learning deficits during the acquisition of both Morris water and radial arm mazes, indicating poor reference and working memory in these mice (Jaubert et al. 2007; Lominac et al. 2005; Szumlinski et al. 2004, 2005a). Normal working memory, at least as assessed by a water version of the radial arm maze, requires Homer1c in the PFC as AAV-mediated restoration of Homer1c, but not Homer1a, reverses radial arm maze deficits in *Homer1* KO mice (Lominac et al. 2005). The learning deficits produced by *Homer1* deletion may stem from abnormalities in attentional processing as *Homer1* KO mice are poor at locating platforms upon maze rotation (Jaubert et al. 2007). Moreover, when tested as adolescents or young adults, *Homer1* KO mice show deficits in pre-pulse inhibition (PPI) of acoustic startle (Lominac et al. 2005; Szumlinski et al. 2005a), although this deficit may be age-dependent as another study failed to detect genotypic differences when mice were tested as older adults (Jaubert et al. 2007). The deficits in PPI exhibited by *Homer1* KO mice can also be reversed by restoration of Homer1c to the PFC (Lominac et al. 2005), or by systemic administration of the typical antipsychotic haloperidol (Szumlinski et al. 2005a). Whether or not antipsychotic drugs may exert their "therapeutic" effects via increasing PFC levels of CC-Homers remains to be determined. However, the majority of evidence indicates little regulation of PFC IEG Homer expression by antipsychotic drug treatment (see Table 3), posing the regulation of CC-Homers by antipsychotic treatments as a potential mechanism contributing to their therapeutic effects (Fig. 3.1). AAV "rescue" studies have yet to be performed in hippocampus of *Homer1* KO mice to assay the relative role for IEG versus CC-Homer isoforms in regulating cognitive function; however, when over-expressed in the hippocampus of rats, Homer1g enhances, while Homer1a impairs, Morris water maze performance (Klugmann et al. 2005). Thus, IEG and CC-Homer1 isoforms both play active, but distinct, roles in attention, learning and memory; the induction of IEG Homer1 isoforms impairs, while the over-expression facilitates, attentional and cognitive processing.

Table 3.4: Effects of Homer1 and Homer2 deletion upon behavioral measures relevant to animal models of psychosis.

Measure	Homer1	Homer2	Reference
Cognitive Function			
Morris water maze performance	WT>KO	WT=KO	Jaubert et al. 2006
			Szumlinski et al. 2004
Radial arm maze performance	WT>KO	WT=KO	Lominac et al. 2005
			Szumlinski et al. 2005a
Sensorimotor Function			

Pre-pulse inhibition of acoustic startle	WT>KO (5 week-old) WT=KO (10-11 week-old)	WT=KO	Lominac et al. 2005 Szumlinski et al. 2005a Jaubert et al. 2006
Emotional Reactivity			
Floating in Porsolt Swim test	WT<KO	WT=KO	Lominac et al. 2005
Locomotor activity in novel environment	WT<KO	WT=KO	Lominac et al. 2005 Szumlinski et al. 2004, 2005a
Exploration of a novel object	WT>KO	WT=KO	Szumlinski et al. 2005a
Habituation of locomotor activity	WT>KO	WT=KO	Lominac et al. 2005 Szumlinski et al. 2004, 2005a
Motor Activity			
Spontaneous locomotor activity	WT<KO	WT=KO	Lominac et al. 2005 Szumlinski et al. 2004, 2005a, 2005b
Cocaine-induced locomotion	WT<KO	WT<KO	Lominac et al. 2005 Szumlinski et al. 2004, 2005a
Methamphetamine-induced locomotion	WT<KO	WT<KO	Lominac et al. 2005
NMDA antagonist-induced locomotion	WT<KO	WT<KO	Szumlinski et al. 2004, 2005a, 2005b, Fig. 3.2

In addition to abnormal cognitive and attentional processing, deletion of Homer1, but not of Homer2, enhances the expression of certain anxiety- and depression-related behaviors in mice. Compared to WT controls, Homer1 KO mice exhibit (1) increased reactivity to handling when experimentally naïve; (2) decreased latencies to interact with novel objects and reduced time spent in contact with novel objects; (3) increased locomotor hyperactivity in response to novel environments and reduced habituation to a familiar environment; and (4) decreased latencies to float and increased incidences of floating behavior in a forced swim test (Lominac et al. 2005; Szumlinski et al. 2004; 2005a). Moreover, increased aggression has also been observed in Homer1 heterozygous mice, relative to WT and KO animals (Jaubert et al. 2007). The "hyper-emotional" behavioral phenotype produced by Homer1 deletion in mice appears to result from an inability to induce Homer1a in the PFC as AAV-mediated restoration of Homer1a to the PFC of KO mice reverses genotypic differences in a forced swim test and in locomotor hyperactivity in response to familiar environments (Lominac et al. 2005). In contrast, PFC restoration of Homer1c in Homer1 KO mice enhances genotypic differences in both paradigms (Lominac et al. 2005), a finding consistent with a reduction in elevated-plus maze exploration by rats over-expressing another CC-Homer1 isoform, Homer1g, within hippocampus

(Klugmann et al. 2005). Collectively, these data pose active and distinct roles for IEG and CC Homer1 isoforms in behavioral responding to repeated stressors; whereas the induction of IEG Homer1 isoforms within cortical structures facilitates the ability to cope with stress, the over-expression of CC Homer1 isoforms leads to behavioral debilitation. Consistent, in part, with this suggestion, prenatally stressed weanling rats exhibit enduring elevations in PFC levels of both Homer1a and Homer1b/c (Ary et al. 2007a). When combined with the behavioral genetic data, we suggest that a balance in PFC levels of IEG and CC-Homer isoforms is critical for maintaining normal executive functioning and that disruption in this balance, be it by environmental factors (i.e., prenatal stress or drug experience) or genetic factors (i.e., polymorphisms in the gene or epigenetic regulation of gene expression) reduces executive processing and behavioral inhibition.

Patients with schizophrenia exhibit abnormal processing of primary rewards, a behavioral deficit attributed to alterations in ventral striatal activation (Garmezy 1952; Juckel et al. 2006a, 2006b; Layne and Wallace 1982). In conjunction with abnormalities in emotional, cognitive and attentional processing, Homer1 KO mice also exhibit deficits in primary reward processing and certain reward-related learning measures relative to WT and to Homer2 KO mice (Szumlinski et al. 2004, 2005a). While deletion of neither Homer1 nor Homer2 affects the development of a food-conditioned place-preference when tested under preprandial (high hunger/low thirst) conditions (Szumlinski et al. 2005a), Homer1 KO mice exhibit a gross delay in the acquisition of an instrumental lever-pressing response for a palatable sucrose reinforcer even when food-restricted (Szumlinski et al. 2004). While the instrumental learning impairment exhibited by Homer1 KO mice could reflect deficits in procedural memory, impairments in attention and reference memory, or a combination thereof (Lominac et al. 2005; Szumlinski et al. 2005a), these mice also exhibited lowered lever-pressing for sucrose when tested on a progressive ratio schedule of reinforcement once the lever-press response was acquired (Szumlinski et al. 2005a). Moreover, Homer1 KO mice emit fewer lever-presses for sucrose than WT mice under both high and low hunger conditions, indicating that Homer1 deletion produces abnormalities in the motivational drive for natural primary reinforcers. However, as will be discussed below, the blunted processing of natural rewards exhibited by Homer1 KO mice does not generalize to drugs of abuse and that consistent with the clinical features of schizophrenia (see below), Homer KO mice exhibit heightened reward-related behaviors vis-à-vis psychomotor stimulant drugs that relates to abnormalities in limbo-corticostriatal glutamate transmission.

3.5.2. Homer proteins and psychotomimetic sensitivity

Psychomotor stimulant drugs

Patients with schizophrenia or related psychotic disorders exhibit an increased pro-pensity to develop dependence upon psychomotor stimulant drugs, when compared to the average population (Alterman et al. 1982; Barbee et al. 1989; Batel 2000; Gerber and Tonegawa, 2004; Green et al. 2007; Meuser et al. 1992; Schneier and Siris 1987; Spring et al. 2003; Winklebauer et al. 2006). Homer expression is dynamically regulated in limbo-corticostriatal structures by the acute or repeated administration of psychomotor stimulant drugs, and several of these alterations are paralleled in the brains of prenatally stressed animals (Table 3). These data pose an imbalance in Homer expression in dysregulated limbic-corticostriatal glutamate transmission pro-posed as a common neuropathology in schizophrenia and psychomotor stimulant de-pendence disorders (e.g., Chambers et al. 2001; Coyle 2006; Gerber and Tonegawa 2004; Krystal et al. 2006; Moghaddam 2002; Szumlinski et al. 2006c). Indeed, stud-ies have established clearly an important role for CC-Homer proteins in regulating sensitivity to the behavioral or neurochemical effects of the psychomotor stimulants methamphetamine and cocaine. While only Homer1 KO mice exhibit deficits in cognitive, emotional, motivational and sensorimotor processing (see Table 4), both Homer1 and Homer2 KO mice exhibit enhanced cocaine-induced place-conditioning, increased locomotor hyperactivity in response to methamphetamine and cocaine, and Homer2 KO mice exhibit a shorter latency to acquire a lever-press response for intravenous cocaine, than do WT mice (Szumlinski et al. 2004, 2005a). Similarly, a reduction in ventral striatal Homer1 levels, produced by infusion of anti-sense oligonucleotides against Homer1, elicits a sensitized motor response in cocaine-naïve rats given an acute cocaine injection (Ghasemzadeh et al. 2003). In cocaine-naïve Homer1 and Homer2 KO mice, the increased behavioral sensitivity to cocaine are accompanied by a reduction in ventral striatal glutamate content and an enhanced capacity of cocaine to elevate ventral striatal glutamate levels (Szumlinski et al. 2004, 2005a), as well as reduced function and expression of both Group1 mGluRs and the cystine-glutamate transporter within this region (Szumlinski et al. 2004). Such behavioral and neurochemical alterations are akin to those observed in rodents with a history of repeated cocaine exposure (e.g., Baker et al. 2003; Pierce et al. 1996; Swanson et al. 2001; Szumlinski et al. 2006a). In contrast to Homer1 and Homer2 gene products, Homer3 isoforms are localized primarily to hippocampus and cerebellum (Shiraishi et al. 2004; Xiao et al. 1998). While both acute and re-peated cocaine elevates cerebellar Homer3a/b expression (Dietrich et al. 2007), Homer3 KO mice do not exhibit a cocaine "presensitized" behavioral phenotype (Szumlinski et al. 2004). Thus, extensive evidence from transgenic KO studies in-dicate that a reduction in CC-Homer1 and CC-Homer2 expression can elicit a "pre-sensitized" cocaine phenotype – an endophenotype associated with psychotic disor-ders (c.f., Kaiser et al. 2005; Yui et al. 1999).

Further support for active roles of specific Homer proteins in responsiveness to psychomotor stimulants has been obtained in studies employing several methods of over-expression. Transgenic mice over-expressing Homer1a within the striosome patches of (primarily the dorsal aspects of) the striatum exhibit increased locomotor sensitivity in response to an acute challenge injection of amphetamine (Tappe and Kuner 2006), whereas AAV-mediated over-expression of Homer1a within the ventral striatum or PFC does not alter the acute or the sensitized locomotor responses to cocaine (Lominac et al. 2005; Szumlinski et al. 2004, 2006a). However, AAV-mediated restoration of Homer2b to the ventral striatum of Homer2 KO mice reversed genotypic differences in cocaine-conditioned reward and motor activity (Szumlinski et al. 2004) and either AAV- or TAT-mediated over-expression of Homer1c or Homer2b in the ventral striatum of repeated cocaine-treated rats prevents the expression of cocaine-induced behavioral and glutamate sensitization when assessed at 3 weeks withdrawal from repeated cocaine treatment (Szumlinski et al. 2006a). To date the effects of manipulating corticostriatal Homer expression upon behavioral sensitivity to other psychomotor stimulants (such as amphetamine or methamphetamine) remains to be determined. As repeated cocaine administration can produce a time-dependent down-regulation in CC-Homer1 and –Homer2 isoforms within the ventral striatum (Ary et al. 2006, 2007b; Obara et al. 2007; Swanson et al. 2001), this collection of data, at least for cocaine, supports a role for ventral striatal CC-Homer1 and –Homer2 isoforms in regulating sensitivity to both the rewarding and psychomotor-activating properties of stimulant drugs.

Although withdrawal from repeated cocaine does not alter PFC basal glutamate content (Baker et al. 2003), it sensitizes the capacity of a challenge injection to elevate glutamate levels in this region (Williams and Steketee 2004). Despite the pronounced abnormalities in ventral striatal glutamate, Homer2 KO mice do not exhibit alterations in basal or cocaine-stimulated glutamate release within the PFC (Szumlinski et al. 2004). In contrast, the cocaine "presensitized" phenotype of Homer1 KO mice is accompanied by an approximately 50% elevation in basal levels of glutamate and a blunting of the cocaine-induced rise in extracellular glutamate within the PFC (Lominac et al. 2005; Szumlinski et al. 2005a). Infusion of AAV-Homer1c to the PFC of Homer1 KO mice reverses the elevated basal glutamate content, while infusion of AAV-Homer1a has no effect (Lominac et al. 2005). Additionally, infusion of AAV-Homer1c, but not -Homer1a, reverses the genotypic difference in cocaine-induced elevations of glutamate in the PFC (Lominac et al. 2005), indicating that cocaine-stimulated glutamate release within the PFC requires Homer1c. However, an intra-PFC infusion of AAV-Homer1a blunts the capacity of cocaine to elevate PFC glutamate in WT mice and reduces glutamate levels below baseline in KO animals, indicating an active and inhibitory role for this IEG isoform in cocaine-stimulated glutamate release (Lominac et al. 2005). When combined, these behavioral and neurochemical data demonstrate that a balance in corticostriatal IEG and CC-Homer isoforms is important in regulating sensitivity to both the rewarding and psychomotor-activating properties of stimulant drugs. Thus, idiopathic, stress- or stimulant-induced imbalances between IEG and CC-Homer expression within the limbo-corticostriatal pathway are likely contributors to glutamate dysfunction hypothesized to be central to psychotic, as well as, addictive disorders (e.g., Chambers et al. 2001; Coyle 2006; Gerber and Tonegawa 2004; Moghaddam 2002; Krystal et al. 2006; Szumlinski et al. 2006c) and might underlie, not only the increased sensitivity of psychotic patients to psychomotor stimulant drugs, but also the increased rate of co-morbidity between schizophrenia and stimulant addiction. Consistent with this possibility, a SNP in intron 1 of Homer1 is associated significantly with cocaine addiction in an African-American population (Dahl et al. 2005). While not yet assessed in dually diagnosed populations, this raises the possibility that interacting SNPs, either on Homer1 (Dahl et al. 2005; Norton et al. 2003) or between those on Homer1 and other Homer-associated genes may increase vulnerability to develop disease co-morbidity.

NMDA receptor antagonists

Predictive of an animal model of psychosis (e.g., Geyer et al. 2001; Lindsley et al. 2006; Krystal et al. 2003; Nabeshima et al. 2006), both Homer1 KO and Homer2 KO mice also exhibit increased behavioral sensitivity to dissociative anesthetics or NMDA receptor antagonists (Table 4). Deletion of either gene shifts the dose-response functions for locomotor hyperactivity induced by systemic PCP or MK-801 to the left of WT controls (Fig. 3.2; Szumlinski et al. 2004, 2005a, 2005b). In the case of the locomotor hyperactivity induced by systemic PCP administration, pretreatment with olanzapine reverses the genotypic differences observed between Homer1 WT and KO mice (Fig. 3.2), further validating the Homer1 KO mouse as a genetic model of schizophrenia. As observed for cocaine (Szumlinski et al. 2004, 2006a), the ventral striatum may be an important substrate involved in the increased behavioral responsiveness to NMDA receptor blockade as Homer2 KO mice also exhibit increased locomotor hyperactivity upon the local infusion of the competitive NMDA receptor antagonist CPP [4-(3-phosphonopropyl)piperazine-2-carboxylic acid] (Szumlinski et al. 2005b).

Alcohol is a drug of abuse with known NMDA antagonist properties (e.g., Lovinger 1993; Woodward 2000). As observed with psychomotor stimulant and nicotine dependence (see above), there exists a higher rate of alcohol use disorders in schizophrenia, relative to the general population (Bartels et al. 1993; Coyle 2006; Drake et al. 1989, 2000; Green et al. 2007; Krystal et al. 2006; Schukit 2006; Wiklbauer et al. 2006) and alcohol use negatively impacts the course and outcome of schizophrenia (e.g., Cuffel and Chase 1994; Drake et al. 1989; Xie et al. 2005). Patients with schizophrenia exhibit a shift to the left in the dose-response function for alcohol-induced euphoria and stimulation (D'Souza et al. 2006) and alcohol-dependent patients with schizophrenia tend to drink less frequently than do non-psychotic alcohol-dependent individuals (Krystal et al. 2001), possibly owing to an increase in the subjective effects of this drug. Both *Homer1* and *Homer2* KO mice exhibit increased sensitivity to the sedative effects of a moderate dose of alcohol (Smothers et al.

2004; Szumlinski et al. 2005b). This increased sensitivity to alcohol-induced sedation is associated with a reduced capacity of alcohol to sensitize ventral striatal glutamate release and both the alcohol behavioral and neurochemical phenotype of *Homer2* KO mice can be reversed by an intra-ventral striatal infusion of AAV-Homer2b (Szumlinski et al. 2005b). Conversely, Homer2b over-expression within the ventral striatum promotes the development of tolerance to alcohol's sedative effects and augments the capacity of alcohol to sensitize ventral striatal glutamate levels (Szumlinski et al. 2007). Interestingly, repeated alcohol administration, be it self-administered or experimenter-administered, up-regulates ventral striatal Homer2a/b protein expression, concomitant with elevations in Group1 mGluRs, NR2 subunits and the activation of PI3K (Szumlinski 2006; Szumlinski et al. 2006b, 2007). The alcohol-induced increase in ventral striatal Homer2a/b expression can persist for months following chronic alcohol consumption (Szumlinski et al. 2007), posing a persistent up-regulation in mGluR/Homer/PI3K, as well as NMDA receptor, signaling in the development of neuroplasticity induced by chronic NMDA receptor blockade. It remains to be determined whether or not similar alterations in mGluR/Homer/NMDA/PI3K signaling occur following the chronic administration of more selective NMDA receptor antagonists; however, acute PCP administration elevates IEG Homer1 expression in several cortical structures, including the PFC (Cochran et al. 2002), implicating NMDA antagonist-induced imbalances in IEG versus CC-Homer expression also in the development of dissociative anesthetic-induced psychosis, alcoholism and the co-morbidity between schizophrenia and alcohol dependence disorder.

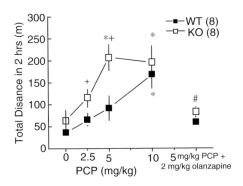

Fig. 3.2. Dose-response function for phencyclidine (PCP)-induced locomotion in Homer1 knock-out (KO) mice relative to wild-type (WT) controls and reversal of the genotypic difference by pretreatment (30 min earlier) with 2.0 mg/kg olanzapine. Sample sizes are indicated in parentheses. *p<0.05 vs. 0 mg/kg PCP, +p<0.05 vs. WT.

3.5.3. Homers, seizures, psychosis and mood

Epileptic seizures can elicit psychotic symptoms during the postictal and interictal state and considerable evidence supports comorbidity between epilepsy, psychosis and affective disorders (Boro and Haut 2003; Boylan 2002; Kanner 2000, 2004; Morrow et al. 2006; Sachdev 2007). Dissociative anesthetic drugs, in particular PCP and ketamine, increase risk of seizure (Baldridge and Bessen 1990; Garey 1979; Schneck and Rupreht 1989; Sioris and Krenzelok 1978) and the psychosis elicited by dissociative anesthetics and NMDA antagonists is associated with increased blood flow and metabolism within the thalamocortical seizure circuit of patients with psychosis caused by these psychotomimetics (Sharp and Hendren 2007). Convulsive seizures produced by electrophysiological or pharmacological stimulation induce the expression of IEG Homer isoforms in several cortical structures (Bottai et al. 2002; Brakeman et al. 1997; French et al. 2001; Kato et al. 1997, 1998; Morioka et al. 2001; Potschka et al. 2002). Transgenic mice with a mosaic over-expression of Homer1a within hippocampus exhibit lowered seizure susceptibility and retarded kindling upon successive corneal stimulation (Potschka et al. 2002). Consistent with these findings, AAV-mediated over-expression of Homer1a in hippocampus strongly suppresses the induction of self-sustaining limbic status epilepticus induced within hippocampus by continuous focal electrical stimulation (Klugmann et al. 2005), whereas hippocampal CC-Homer over-expression was without effect (Klugmann et al. 2005). Collectively, these data pose an inhibitory role for Homer1a induction in the propagation of seizures and implicate a failure to induce Homer1a upon excessive synaptic activity in epilepsy and peri-ictal psychosis. Related to this, electroconvulsive shock therapy produces a reduction in neuronal excitability with a concomitant rise in Homer1a in the neocortex of rats. Moreover, injecting an antibody against Homer1a prevents neuronal hyperpolarization induced by electroconvulsive shock therapy (Sakagami et al. 2005). Thus, the induction of Homer1a by such procedures may account, in part, for the clinical efficacy of this therapy for the treatment of psychotic and other disorders (e.g., Bowman and Coons 1002; Little et al. 2003; Russel et al. 2003).

3.6. Conclusions

Homer proteins are critical regulators of both presynaptic and postsynaptic facets of glutamatergic neurotransmission within the limbo-corticostriatal circuit. Both IEG and constitutively expressed members of the Homer family of postsynaptic proteins are dynamically regulated within the limbo-corticostriatal circuit by environmental factors that precipitate or aggravate psychotic symptoms and the ability of antipsychotic treatments to alter Homer expression has been implicated in both their therapeutic and side-effect profiles. Behavioral evidence, derived primarily from genetic studies employing gene mutant and targeted gene delivery approaches, supports important roles for members of the Homer protein family in cognitive, sensorimotor, emotional and motivational processing. Moreover, such studies have demonstrated active roles for Homer proteins in regulating sensitivity to psychotomimetic drugs, alcohol and supra-physiological electrical activity, thereby implicating Homers in idiopathic, drug-induced and peri-itcal psychoses, as well as the co-morbidity between schizophrenia and substance abuse disorders. While it is unlikely that a single high penetrant mutation in a *Homer* gene results in the complex phenotype of schizophrenia, mutational analyses in humans have linked certain SNPs in *Homer1* with schizophrenia and cocaine addiction and preliminary data supports an imbalance in Homer transcript levels within the limbo-corticostriatal circuit of postmortem brains of patients. Based on the collection of data to date, it is predicted that future investigations of clinical populations will likely reveal further abnormalities within the *Homer* gene family that may account for various endophenotypes exhibited by patients suffering from schizophrenia and related neuropsychiatric disorders.

Acknowledgments

Portions of the work described in this review were supported by funding from the National Alliance for Research on Schizophrenia and Affective Disorders (NARSAD) to KKS and TEK. Other portions of the work described in this review were supported by United States Public Health Service grants AA-013519 (pilot project, Integrative Neuroscience Initiative on Alcoholism West), AA-015351 and AA-016650 (Integrative Neuroscience Initiative on Alcoholism West) to KKS and DA-021161 to TEK.

Abbreviations

AAV adeno-associated viral vector
AMPA alpha-amino-3-hydroxy-5-methyl-4-isoxazolepropionic acid
BK calcium-dependent big potassium
CC coiled-coil
cDNA complementary deoxyribonucleic acid
CRE 3'-5'-cyclic adenosine monophosphate response element
CREB 3'-5'-cyclic adenosine monophosphate response element binding protein
C-terminus carboxyl terminus
DIV days in vitro
DNA deoxyribonucleic acid
ERK extracellular signal-regulated kinase
EVH1 Enabled/vasodilator-stimulated phosphoprotein homology 1
F-actin filamentous actin
GKAPguanylate kinase associated protein
HEK human embryonic kidney
IEG immediate early gene

iGluR ionotropic glutamate receptor
IP3 inositol triphosphate
KO knock-out
LSD lysergic acid diethylamide
MAPK mitogen-activated protein kinase
mGluR metabotropic glutamate receptor
mRNA messenger ribonucleic acid
NMDA N-methyl-D-aspartic acid
PCP phencyclidine
PDZ post-synaptic density 95
PFC prefrontal cortex
PIKE-L phosphoinositide 3-kinase enhancer-long
PI3K phosphoinositide 3-kinase
PKA cAMP-dependent protein kinase or protein kinase A
PKC protein kinase C
PPI pre-pulse inhibition
PSD postsynaptic density
SNP single nucleotide polymorphism
UTR untranslated region
VTA ventral tegmental area

References

Badner JA, Gershon ES (2002) Meta-analysis of whole-genome linkage scans of bipolar disorder and schizophrenia. Mol Psychiatry 7:405-411.

Bassett AS, Chow EWC (1999) 22q11 deletion syndrome: A genetic subtype of schizophrenia. Biol Psychiatry 46:882-891

Bertram L, Tanzi RE (2004) Alzheimer's disease: one disorder, too many genes? Hum Mol Genet 13 (R1):R135-141.

Bertram L, Hiltunen M, Parkinson M, Ingelsson M, Lange C, et al (2005) Family-Based Association between Alzheimer's Disease and Variants in UBQLN1. N Engl J Med 352:884-894.

Botstein D, Risch N (2003) Discovering genotypes underlying human phenotypes: past successes for mendelian disease, future prospects for complex disease. Nat Genet 33 Suppl: 228-237.

Bray NJ, Preece A, Williams NM, Moskvina V, Buckland PR, et al (2005) Haplotypes at the dystrobrevin binding protein 1 (DTNBP1) gene locus mediate risk for schizophrenia through reduced DTNBP1 expression. Hum Mol Genet 14:1947-1954.

Callicott JH, Straub RE, Pezawas L, Egan MF, Mattay VS, et al (2005) Variation in DISC1 affects hippocampal structure and function and increases risk for schizophrenia. Proc Natl Acad Sci USA 102:8627-8632.

Carter CJ (2006) Schizophrenia susceptibility genes converge on interlinked pathways related to glutamatergic transmission and long-term potentiation, oxidative stress and oligodendrocyte viability. Schizophr Res 86:1-14.

Cavalleri GL, Lynch JM, Depondt C, Burley M-W, Wood NW et al (2005) Failure to replicate previously reported genetic associations with sporadic temporal lobe epilepsy: where to from here? Brain 128:1832-1840.

Chumakov I, Blumenfeld M, Guerassimenko O, Cavarec L, Palicio M, et al (2002) Genetic and physiological data implicating the new human gene G72 and the gene for D-amino acid oxidase in schizophrenia. Proc Natl Acad Sci USA 99:13675-13680.

Duan J, Wainwright MS, Comeron JM, Saitou N, Sanders AR, et al (2003) Synonymous mutations in the human dopamine receptor D2 (DRD2) affect mRNA stability and synthesis of the receptor. Hum Mol Genet 12:205-216.

Eastwood SL, Burnet PWJ, Harrison PJ (2005a) Decreased hippocampal expression of the susceptibility gene PPP3CC and other calcineurin subunits in schizophrenia. Biol Psychiatry 57:702-710.

Eastwood SL, Salih T, Harrison PJ (2005b) Differential expression of calcineurin A subunit mRNA isoforms during rat hippocampal and cerebellar development. Eur J Neurosci 22:3017-3024.

Eastwood SL, Harrison PJ (2007) Decreased mRNA expression of Netrin-G1 and Netrin-G2 in the temporal lobe in schizophrenia and bipolar disorder [abstract]. Schizophr Res (in press).

Egan MF, Straub RE, Goldberg TE, Yakub I, Callicott JH, et al (2004). Variation in GRM3 affects cognition, prefrontal glutamate, and risk for schizophrenia. Proc Natl Acad Sci USA 101:12604-12609.

Emamian ES, Hall D, Birnbaum MJ, Karayiorgou M, Gogos JA (2004) Convergent evidence for impaired AKT1-GSK3β signaling in schizophrenia. Nat Genet 36:131-127.

Falls DL (2003) Neuregulins: functions, forms, and signaling strategies. Exp Cell Res 284:14-30.

Frankle WG, Lerma J, Laruelle M (2003) The synaptic hypothesis of schizophrenia. Neuron 39:205-216.

Fujii K, Maeda K, Hikida T, Mustafa A, Balkissoon R et al (2006) Serine racemase binds to PICK1: potential relevance to schizophrenia. Mol Psychiatry 11:150-157.

Goff DC, Coyle JT (2001) The emerging role of glutamate in the pathophysiology and treatment of schizophrenia. Am J Psychiatry 158:1367-1377.

Goltsov AY, Loseva JG, Andreeva TV, Grigorenko AP, Abramova LI, et al (2006) Polymorphism in the 5'-promoter region of serine racemase gene in schizophrenia. Mol Psychiatry 11:325-326.

Gurling HMD, Critchley H, Datta SR, McQuillin A, Blaveri E, Thirumalai S et al (2006) Genetic association and brain morphology studies and the chromosome 8p22 pericentriolar material 1 (PCM1) gene in susceptibility to schizophrenia. Arch Gen Psychiatry 63:844-854.

Hahn CG, Wang HY, Cho DS, Talbot K, Gur RE, et al (2006) Altered neuregulin 1-erbB4 signaling contributes to NMDA receptor hypofunction in schizophrenia. Nat Med 12:824-828.

Harrison PJ (2007) Schizophrenia susceptibility genes and neurodevelopment. Biol Psychiatry (in press).

Harrison PJ, Eastwood SL (2001) Neuropathological studies of synaptic connectivity in the hippocampal formation in schizophrenia. Hippocampus 11:508-519.

Harrison PJ, Law AJ (2006) Neuregulin 1 and schizophrenia: genetics, gene expression, and neurobiology. Biol Psychiatry 60:132-140.

Harrison PJ, Owen MJ (2003) Genes for schizophrenia? Recent findings and their pathophysiological implications. Lancet 361:417-419.

Harrison PJ, Weinberger DR (2005) Schizophrenia genes, gene expression, and neuropathology: on the matter of their convergence. Mol Psychiatry 10:40-68.

Harrison PJ, West VA (2006) Six degrees of separation: on the prior probability that schizophrenia susceptibility genes converge on synapses, glutamate, and NMDA receptors. Mol Psychiatry 11: 981-983.

Hashimoto R, Straub RE, Weickert CS, Hyde TM, Kleinman JE, Weinberger DR (2004) Expression analysis of neuregulin-1 in the dorsolateral prefrontal cortex in schizophrenia. Mol Psychiatry 9:299-307.

Jaenisch R, Bird A (2003) Epigenetic regulation of gene expression: how the genome integrates intrinsic and environmental signals. Nat Genet 33:245-254.

Kamiya A, Kubo K, Tomoda T, Takaki M, Youn R, et al (2005) A schizophrenia-associated mutation of DISC1 perturbs cerebral cortex development. Nat Cell Biol 7:1067-1078.

Kleinjan DA, van Heyningen V (2005) Long-range control of gene expression: emerging mechanisms and disruption in disease Am J Hum Genet 76:8-32.

Law AJ, Weickert CS, Hyde T, Kleinman J, Harrison PJ (2004) Neuregulin-1 (NRG1) mRNA and protein in the adult human brain. Neuroscience 127:125-136.

Law AJ, Lipska BK, Weickert CS, Hyde TM, Straub RE, et al (2006) Neuregulin 1 transcripts are differentially expressed in schizophrenia and regulated by 5' SNPs associated with the disease. Proc Natl Acad Sci USA 103:6747-6752.

Lemonde S, Turecki G, Bakish D, Du L, Hrdina PD, et al (2003) Impaired repression at a 5-hydroxytryptamine 1A receptor gene polymorphism associated with major depression and suicide. J Neurosci 23:8788-8799.

Lewis CM, Levinson DF, Wise LH, DeLisi LE, Straub RE, et al (2003) Genome scan meta-analysis of schizophrenia and bipolar disorder, part II: Schizophrenia. Am J Hum Genet 73:34-48.

Li D, Collier DA, He L (2006) Meta-analysis shows strong positive association of the neuregulin 1 gene with schizophrenia. Hum Mol Genet 15:1995-2002.

Lipska BK, Peters T, Hyde TM, Halim N, Horowitz C, et al (2006a) Expression of DISC1 binding partners is reduced in schizophrenia and associated with DISC1 SNPs. Hum Mol Genet 15:1245-1258.

Lipska BK, Mitkus S, Caruso M, Hyde TM, Chen J, et al (2006b) RGS4 mRNA expression in postmortem human cortex is associated with COMT Val[158]Met genotype and COMT enzyme activity. Hum Mol Genet 15:2804-2812.

Liu H, Abecasis GR, Heath SC, Knowles A, Demars S, et al (2002) Genetic variation in the 22q11 locus and susceptibility to schizophrenia. Proc Natl Acad Sci USA 99:16859-16864.

Maraganore DM, de Andrade M, Elbaz A, Farrer MJ, Ioannidis JP, et al (2006) Collaborative Analysis of {alpha}-Synuclein Gene Promoter Variability and Parkinson Disease. JAMA 296:661-670.

Matlin AJ, Clark F, Smith CWJ (2005) Understanding alternative splicing: towards a cellular code. Nat Rev Mol Cell Biol 6:386-398.

Mexal S, Frank M, Berger R, Adams CE, Ross RG, et al (2005) Differential modulation of gene expression in the NMDA postsynaptic density of schizophrenic and control smokers. Mol Brain Res 139:317-332.

Mirnics K, Levitt P, Lewis DA (2006) Critical appraisal of DNA microarraysin psychiatric genomics. Biol Psychiatry 60:163-176.

Moghaddam B (2003) Bringing order to the glutamate chaos in schizophrenia. Neuron 40:881-884.

Moises HW, Zoetga T, Gottesman II (2002) The glial growth factors deficiency and synaptic destabilization hypothesis of schizophrenia. BMC Psychiatry 2.

Mukai J, Liu H, Burt RA, Swor DE, Lai WS, et al (2004) Evidence that the gene encoding ZDHHC8 contributes to the risk of schizophrenia. Nat Genet 36:725-731.

Munafó MR, Thiselton DL, Clark TG, Flint J (2006) Association of the NRG1 gene and schizophrenia: a meta-analysis. Mol Psychiatry 11:539-546.

Nicodemus KK, Kolachana BS, Vakkalanka R, Straub RE, Giegling I, et al (2006) Evidence for statistical epistasis between Catechol-O-Methyltransferase (COMT) and polymorphisms in RGS4, G72 (DAOA), GRM3 and DISC1: Influence on risk of schizophrenia. Hum Genet in press.

Norton N, Moskvina V, Morris D, Bray NJ, Zammit S, et al (2006) Evidence that interaction between neuregulin 1 and its receptor erbB4 increases susceptibility to schizophrenia. Am J Med Genet B Neuropsychiatr Genet 141:96-101.

Olney JW, Farber NB (1995) Glutamate receptor dysfunction and schizophrenia. Arch Gen Psychiatry 52:998-1007.

Owen MJ, Craddock N, O'Donovan MC (2005) Schizophrenia: genes at last? Trends Genet 21:518-525.

Pastinen T, Hudson TJ (2004) Cis-acting regulatory variation in the human genome. Science 306:647-650.

Paterlini M, Zakharenko SS, Lai WS, Qin J, Zhang H, et al (2005) Transcriptional and behavioral interaction between 22q11.2 orthologs modulates schizophrenia-related phenotypes in mice. Nat Neurosci 8:1586-1594.

Peirce TR, Bray NJ, Williams NM, Norton N, Moskvina V, et al (2006) Convergent evidence for 2',3'-cyclic nucleotide 3'-phosphodiesterase as a possible susceptibility gene for schizophrenia. Arch Gen Psychiatry 63:18-24.

Porteous DJ, Thomson P, Brandon NJ, Millar JK (2006) The genetics and biology of DISC-1-An emerging role in psychosis and cognition. Biol Psychiatry 60:123-131.

Rademakers R, Melquist S, Cruts M, Theuns J, Del-Favero J, et al (2005) High-density SNP haplotyping suggests altered regulation of tau gene expression in progressive supranuclear palsy. Hum Mol Genet 14:3281-3292.

Rebbeck TR, Spitz M, Wu X (2004) Assessing the function of genetic variants in candidate gene association studies. Nat Rev Genet 5:589-597.

Rosand J, Bayley N, Rost N, de Bakker PI (2006) Many hypotheses but no replication for the association between *PDE4D* and stroke. Nat Genet 38:1091-1092.

Rowe S, Miller S, Sorscher E (2005) Cystic fibrosis. N Eng J Med 352:1992-2001.

Sartorius LJ, Nagappan G, Lipska BK, Lu B, Sei Y, et al (2006) Alternative splicing of human metabotropic glutamate receptor 3. J Neurochem 96:1139-1148.

Sawamura N, Yamamoto T, Ozeki Y, Ross C, Sawa A (2005) A form of DISC1 enriched in nucleus: altered subcellular distribution in orbitofrontal cortex in psychosis and substance/alcohol abuse. Proc Natl Acad Sci USA 102:1187-1192

Schadt EE, Lamb J, Yang X, Zuh J, Edwards S, et al (2005) An integrative genomics approach to infer causal associations between gene expression and disease. Nat Genet 37:710-717.

Schmidt Kastner R, van Os J, W M Steinbusch H, Schmitz C, Kamiya A, et al (2006) Gene regulation by hypoxia and the neurodevelopmental origin of schizophrenia. Schizophr Res 84:253-271.

Seitz A, Gourevitch D, Zhang X-M, Clark L, Chen P, et al (2005) Sense and antisense transcripts of the apolipoprotein E gene in normal and ApoE knockout mice, their expression after spinal cord injury and corresponding human transcripts. Hum Mol Genet 14:2661-2670.

Shen LX, Bassilion JP, Stanton VP (1999) Single-nucleotide polymorphisms can cause different structural folds of mRNA. Proc Natl Acad Sci USA 96:7871-7876.

Shifman S, Levit A, Chen ML, Chen CH, Bronstein M, et al (2006) A complete genetic association scan of the 22q11 deletion region and functional evidence reveal an association between DGCR2 and schizophrenia. Hum Genet 120:160-170.

Silberberg G, Darvasi A, Pinkas Kramarski R, Navon R, Peirce TR, et al (2006) The involvement of ErbB4 with schizophrenia: association and expression studies. Am J Med Genet B Neuropsychiatr Genet 141:142-148.

Straub RE, Weinberger DR (2006) Schizophrenia genes - Famine to feast. Biol Psychiatry 60:81-83.

Sullivan PF, Kendler KS, Neale MC (2003) Schizophrenia as a complex trait-Evidence from a meta-analysis of twin studies. Arch Gen Psychiat 60:1187-1192.

Talbot K, Eidem W, Tinsley CL, Benson MA, Thompson EW, et al (2004) Dysbindin-1 is reduced in intrinsic, glutamatergic terminals of the hippocampal formation in schizophrenia. J Clin Invest 113:1353-1363.

Talkowsi ME, HStelman H, Bassett AS, Brzustowicz LM, Chen X, et al (2006) Evaluation of a suscpetibility gene for schizophrenia: genotype based meta-analysis of RGS4 polymorphisms from thirteen independetn samples. Biol Psychiatry 60:152-162.

Theuns J, Remacle J, Killick R, Corsmit E, Vennekens Kl, et al (2003) Alzheimer-associated C allele of the promoter polymorphism -22C>T causes a critical neuron-specific decrease of presenilin 1 expression. Hum Mol Genet 12:869-877.

Tokuhiro S, Yamada R, Chang X, Suzuki A, Kochi Y, et al (2003) An intronic SNP in a RUNX1 binding site of SLC22A4, encoding an organic cation transporter, is associated with rheumatoid arthritis. Nat Genet 35:341-348.

Tunbridge EM, Weinberger DR, Harrison PJ (2006) Catechol-o-methyltransferase, cognition and psychosis: $Val^{158}Met$ and beyond. Biol Psychiatry 60: 141-151.

Tunbridge EM, Weickert CS, Kleinman JE, Herman MM, Chen J, et al (2007) Catechol-o-methyltransferase enzyme activity and protein expression in human prefrontal cortex across the postnatal lifespan. Cereb Cortex in press.

Valentonyte R, Hampe J, Huse K, Rosenstiel P, Albrecht M, et al (2005) Sarcoidosis is associated with a truncating splice site mutation in the gene BTNL2. Nat Genet 37:357-364.

Walss-Bass C, Liu W, Lew DF, Villegas R, Montero P, et al A (2007) Novel Missense Mutation in the Transmembrane Domain of Neuregulin 1 is Associated with Schizophrenia. Biol Psychiatry (in press).

Weickert CS, Straub RE, McClintock BW, Matsumoto M, Hashimoto R, et al (2004) Human dysbindin (DTNBP1) gene expression in normal brain and in schizophrenic prefrontal cortex and midbrain. Arch Gen Psychiatry 61:544-555.

Wilkins JF, Haig D (2003) What good is genomic imprinting: the function of parent-specific expression. Nat Rev Genet 4:1-10.

Xu B, Wratten N, Charych EI, Buyske S, Firestein BL, et al (2005) Increased expression in dorsolateral prefrontal cortex of CAPON in schizophrenia and bipolar disorder. PLoS Med 2:e263.

Yang J, Si T, Ling Y, Ruan Y, Han Y, Wang X et al (2003) Association study of the human FZD3 locus with schizophrenia. Biol Psychiatry 54:1298-1301.

Yin, T.C.T. and Chan, J.C.K. (1988) Neural mechanisms underlie interaural time sensitivity to tones and noise. In: W.E. Gall, G.M. Edelman and W.M. Cowans (Eds.), Auditory Function: Neurobiological Bases of Hearing. John Wiley, New York, pp. 385-430.

Chapter 4

Neuregulin-erbB Signaling and the Pathogenesis of Schizophrenia

Christine Roy and Gabriel Corfas

Schizophrenia is a devastating psychiatric disease, typically producing life-long disability. Initially considered a neurodegenerative disorder, schizophrenia is now viewed as a developmental disease due to the lack of neurodegenerative processes and because affected individuals exhibit cognitive and social impairment before the first psychotic episode (Lewis and Lieberman 2000). It has been proposed that schizophrenia may result from defects in early brain development that increase the risk of developing the disease as normal maturational events occur during adolescence or early adulthood. In addition, defects in late developmental events such as myelination or sexual maturation may also play a role. Genetic factors clearly account for a significant proportion of the underlying causes of schizophrenia. For example, studies of identical twins show that if one twin is affected, the likelihood that the other twin will also be affected (the concordance rate) is 41-65% whereas fraternal twins have concordance rates of up to 28% (Tsuang et al. 2001). This has led to heritability estimates as high as 85% (Tsuang et al. 2001). Numerous loci on different chromosomes have been linked to the disease, indicating that more than one gene underlies susceptibility to schizophrenia (Sklar 2002), but the identification of these genes has been difficult. Recent genetic studies have implicated the genes encoding the trophic factor neuregulin1 (*NRG1*) and one its receptors, erbB4, in this disease. While *NRG1* and erbB4 are not the only genes to predispose an individual to schizophrenia, the extensive knowledge of the biological roles of the NRG1/erbB pathway provides an opportunity to think about the molecular and cellular mechanisms of the disease.

4.1 Neuregulins and their Receptors

The NRGs constitute a family of factors that signal through receptor tyrosine kinases of the EGF (epidermal growth factor) receptor family, i.e. erbB2, erbB3 and erbB4 (reviewed in Buonanno and Fischbach 2001). Early studies demonstrated that NRG1 expression is highest in the brain, where it is expressed primarily by neurons (Corfas et al. 1995; Falls 2003; Marchionni et al. 1993). Other members of the NRG family, i.e. NRG2 and NRG3, are also expressed in the nervous system (Carraway et al. 1997; Carteron et al. 2006; Longart et al. 2004; Rimer et al. 2005), but their biological functions remain poorly understood.

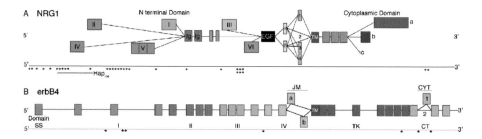

Fig. 4.1. Structure of the NRG1 and erbB4 genes depicting their isoforms and the location of markers associated with schizophrenia. A) Scheme of the human NRG1 gene showing the relative position of the known exons (the distance between exons is not drawn to scale). Alternative splicing yields at least 15 distinct isoforms that are primarily named based on the N-terminal domain contained (type I through IV) but are further distinguished based on the sequences present in their extracellular juxtamembrane domain (α vs. β; 1-4). Note that in the α2 and β2 isoforms the α/β domain connects directly to the transmembrane domain and that the α3 and β3 isoforms lack the cytoplasmic domain. A single exon encodes a transmembrane (TM) domain that is part of most isoforms, followed by several exons that can produce 3 different types of intracellular domains (a, b and c). Asterisks represent the location of markers found to be associated with schizophrenia, most of them lying within intronic regions. The at-risk haplotype defined by Stefansson and colleagues [121] is indicated by a horizontal line. Ig = immunoglobulin-like domain; EGF = epidermal growth factor-like domain. B) Scheme of the human erbB4 gene (HER4) showing the relative position of the known exons (the distance between exons is not drawn to scale). Alternative splicing in the extracellular juxtamembrane region (JM) generates the JMa and JMb isoforms, which differ in sensitivity to proteolytic cleavage. Removal of the CYT-1 exon the carboxy-terminal domain results in the CYT-2 isoform lacking a phosphatidyl inositol 3-kinase binding site. Exons of the same color together form specific domains within the protein. Domains II and IV are cysteine rich domains. (SS = signal sequence; TM = transmembrane domain; TK = tyrosine kinase domain; CT = carboxy-terminal domain. Asterisks represent the location of markers that have been found to be significantly associated with schizophrenia.

NRG1 was initially isolated independently by four research groups based on several biological activities: its ability to activate the erbB2 receptor tyrosine kinase (these proteins we originally called neu differentiation factor or heregulin) (Holmes et al. 1992; Peles et al. 1992), to induce acetylcholine receptor synthesis in skeletal muscle (originally named ARIA for acetylcholine receptor inducing activity) (Falls et al., 1993) and to induce Schwann cell proliferation (named GGF for glial growth factor) (Goodearl et al. 2001; Marchionni et al. 1993). It soon became clear that all of these proteins were derived from the same gene that gives rise to numerous isoforms via alternative splicing (Figure 4.1).

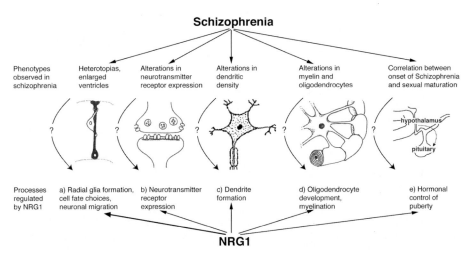

Fig. 4.2. Potential relationships between NRG1 biological functions and schizophrenia phenotypes. NRG1 has been implicated in a) cortical development by regulating radial glia morphology, neuronal migration and neural stem cell fate choices, b) synapse formation and function by regulating and the expression of neurotransmitter receptors and c) dendrite formation, d) myelination by regulating oligodendrocyte differentiation, and e) control of the onset of puberty through the induction of LHRH release in the hypothalamus. All these developmental processes have been proposed to be altered or involved in schizophrenia. Thus, defects in NRG1 function can potentially contribute to the disease by altering one or more of these processes. The arrows with question marks indicate the current hypothetical nature of these possible links between NRG1 function and schizophrenia.

NRG1 is a pleiotropic growth factor known to affect both neurons and glia. All the NRG1 receptors are present in the central nervous system but with distinct expression patterns, suggesting that different erbB receptor combinations may be at work in distinct cell types or brain regions. NRG1 is believed to signal primarily by being released as a soluble protein that binds and activates erbB receptors on its target cells, or by being presented on the cell surface as a transmembrane protein which activates the receptors in the target cells by cell-cell contact. Binding of NRG1 induces the formation of erbB receptor dimers, activating their tyrosine kinase activity that then regulates many aspects of cell physiology (Buonanno and Fischbach 2001; Falls 2003). However, it has been suggested that NRG1 also has the potential to signal within the NRG1-expressing cell through its intracellular domain that, upon cell depolarization, can be released and move to the cell nucleus where it may regulate gene expression (Bao et al. 2003).

NRG1 has been implicated in key neurodevelopmental processes in the central nervous system, including cortical development (Anton et al. 1997, 2004; Flames et al. 2004; Ghashghaei et al. 2006; Patten et al. 2003, 2006; Sardi et al. 2006; Schmid et al. 2003), dendritic development (Bermingham-McDonogh et al. 1996; Gerecke et al. 2004; Rieff and Corfas 2006; Zhang et al. 2004), regulation of neurotransmitter receptor expression (Chang and Fischbach 2006; Gu et al. 2005; Hahn et al. 2006; Kwon et al. 2005; Liu et al. 2001; Okada and Corfas 2003; Ozaki et al. 1997; Rieff et al 1999; Stefansson et al. 2002), oligodendrocyte development (Calaora et al. 2001; Cannella et al. 1999; Canoll et al. 1996; Fernandez et al. 2000; Park et al. 2001; Schmucker et al. 2003; Sussman et al. 2005; Vartanian et al. 1994, 1999) and hormonal control of puberty (Prevot et al. 2003). Interestingly, these developmental processes are thought to be involved, directly or indirectly, in schizophrenia (Fig. 4.2).

4.2 Altered Expression of NRG1 and erbB Receptors in Schizophrenia

The first indication that NRG1-erbB signaling may be altered in schizophrenia came from comparisons of gene expression in the prefrontal cortex of chronic patients and control subjects, which revealed that *erbB3* mRNA was significantly reduced in patients (Bahn 2002; Bray et al. 2003; Hakak et al. 2001; Mimmack et al. 2002; Tkachev et al. 2003). More recent studies have shown that the levels of *erbB4* are also altered in schizophrenia, though in contrast to *erbB3*, this receptor seems to be expressed at higher levels in patients (Law et al. 2007; Silberberg et al. 2006). Quantitative PCR analysis of *erbB4* expression in the dorsolateral prefrontal cortex (DLPFC) showed an almost 2 fold increase in the levels of a particular *erbB4* isoform (the one containing a phosphoinositide 3-kinase docking site (CYT-1)) in patients (Fig. 4.1) (Law et al. 2007; Silberberg et al. 2006). In addition, increased levels of expression of another alternative splice variant of *erbB4* (the JM-a juxtamembrane isoform) have been observed (Law et al. 2007; Silberberg et al. 2006). Furthermore, the extent of erbB4 activation by NRG1, measured by tyrosine phosphorylation and activation of downstream targets, appeared to be significantly increased in post-mortem tissue obtained from patients (Hahn et al., 2006). These observations suggest that complex alterations in the levels of expression and activity of the different erbB receptors may occur in schizophrenia patients.

Several studies have also reported alterations in the levels of *NRG1* expression in the brains of schizophrenia patients. Type I *NRG1* expression appears to be increased in the DLPFC of patients leading to alterations in the relative ratios of

NRG1 mRNA isoforms (Hashimoto et al. 2004). Similar alterations have been reported in the hippocampus, where patients were found to have 34% more type I *NRG1* mRNA than control subjects without changes in the levels of type II, III or IV (Law et al. 2004). These studies indicate that components of the NRG1-erbB signaling machinery are differentially altered in the brains of schizophrenia patients. This has given rise to the hypothesis that alterations in the relative levels of different ligand isoforms and/or receptors may be involved in the disease. While the alterations in the levels of expression of *NRG1* or the *erbB* receptors suggested a link between this signaling pathway and schizophrenia, these studies do not provide insights into whether the alterations in expression are involved in the pathogenesis of the disease or are a consequence of it.

4.3 Evidence for a Genetic Linkage between NRG1 and erbB4 and Schizophrenia

The hypothesis that defects in NRG1-erbB signaling directly contribute to schizophrenia is supported by genetic studies linking the *NRG1* and *erbB4* genes and the disease. A genome-wide scan that took advantage of the extensive pedigree information and relative genetic isolation of the Icelandic population identified the *NRG1* gene as a susceptibility locus for schizophrenia (Stefansson et al. 2002). This study identified an "at-risk haplotype" (HAP$_{ice}$) in the *NRG1* gene that is present at a frequency of 15.4% in the affected individuals, as compared to 7.5% in controls (relative risk ratio 2.1). This "at-risk haplotype" maps to the proximal promoter and first exon of type II *NRG1* (also known as *GGF2*) and type IV *NRG1* (Figure 1) (Stefansson et al. 2002; Steinthorsdottir et al. 2004). Since this initial report, research into the genetic association between *NRG1* and schizophrenia has flourished such that regions in or near HAP$_{ice}$ have been linked to the disease in patient populations in the United Kingdom, China, Portugal, Korea, Netherlands (Fig. 4.1) (Bakker et al. 2004; Corvin et al. 2004; Fukui et al. 2006; Kim et al. 2006; Li et al. 2004; Petryshen et al. 2005; Tang et al. 2004; Williams et al. 2003; Yang et al. 2003). Furthermore, other regions of *NRG1*, unrelated to HAP$_{ice}$, also appear to be associated with schizophrenia (Li et al. 2004; Liu et al, 2005; Petryshen et al. 2005; Yang et al. 2003). Most notably, the 3' end of *NRG1* was found to be associated with schizophrenia in a population of Chinese Han and Portuguese patients while two areas of *NRG1* upstream of HAP$_{ice}$ were linked with the disorder in Chinese Han and Taiwanese populations (Fig. 4.1). It is worth noting that studies of some Japanese, Danish, Chinese and Irish schizophrenia patients did not find an association between the *NRG1* gene and the disease (Hong et al. 2004; Ingason et al. 2006; Iwata et al. 2004; Thiselton et al. 2004). This highlights the fact that *NRG1* is not the only gene underlying susceptibility to schizophrenia and that *NRG1* is unlikely to be implicated in all cases of schizophrenia or in all populations; rather, this disease is very complex with multiple contributing genes. Nevertheless, the replication of the association of *NRG1* and schizophrenia in diverse populations by different groups provides very strong correlative genetic evidence for the association between the *NRG1* gene and schizophrenia.

These studies show that the *NRG1* gene may predispose an individual to schizophrenia or that *NRG1* is in linkage disequilibrium with a schizophrenia locus. Since regions nearby *NRG1* on chromosome 8p are relatively gene-poor, with no identified genes at least 500 kb upstream or downstream of the *NRG1* gene, it is unlikely that these results are due to linkage to a close gene. Another important concern is that few *NRG1* polymorphisms linked to schizophrenia produce changes in the amino acid sequence of the protein (Walss-Bass et al. 2006). Rather, most studies have found associations between schizophrenia and alleles that are predicted to produce silent changes or which occur in non-coding regions. It has been speculated that these polymorphisms should not alter the bioactivity of the NRG1 protein, but could instead affect *NRG1* gene expression. This is supported by the studies described above reporting increased expression of type I and type IV *NRG1* mRNA in patients (Hashimoto et al. 2004; Law et al. 2004). Interestingly, sequence analysis suggests that two of the markers in HAP_{ice} may result in changes in putative binding sites for transcription factors, and thus, could result in changes in gene expression (Law et al. 2004, 2006). However, these ideas are solely based on *in silico* data and need to be tested in more direct ways.

Recently, the functional consequences of the at-risk HAP_{ice} haplotype were examined by correlating the presence of specific at-risk marker alleles of *NRG1* with response to treatment (Kampman et al. 2004; Law et al. 2004). In a Finnish patient population, the presence of two T alleles at SNP8NRG221533 occurred significantly more often in patients who did not respond to conventional antipsychotic drugs (Kampman et al. 2004). This suggests that patients with alterations in *NRG1* may have different endophenotypes. This type of genetic information could be of use in the future to determine the most effective course of treatment.

The NRG1 receptor, erbB4, has also been linked to schizophrenia. Initially, a significant interaction between HAP_{ice} and *erbB4* was found in an Irish patient population such that subjects carrying a 'C' allele at marker IVS12-15 of *erbB4*, who also carried the *NRG1* HAP_{ice} haplotype, were more likely to receive a diagnosis of schizophrenia than those with the 'T' allele (Figure 1) (Norton et al. 2006). Evidence for a direct link between erbB4 and schizophrenia comes from two case control studies which showed an association between three SNP's encompassing the third exon of *erbB4* with the disease (Fig. 4.1) (Law et al. 2006; Silberberg et al. 2006). Two of these SNP's (rs839523 and rs7598440) were also found to be associated with schizophrenia in a study of 3 independent family-based samples, although the opposite alleles were implicated in the disease compared to those reported previously (Law et al. 2007; Silberberg et al. 2006). Thus, these studies warrant further replication. Another study found an additional triplet of SNPs near the 3' end of the *erbB4* gene that was also positively associated with schizophrenia (Nicodemus et al., 2006). Taken together with the alterations in the expression levels of different components of this pathway, these studies suggest that abnormal NRG1-erbB signaling may underlie certain aspects of schizophrenia.

Given the changes in levels of *NRG1* and *erbB4* in the brains of schizophrenia patients, it is intriguing that altering the levels of *nrg1* or *erbB4* in mice leads to behavioral abnormalities related to the features of the disease. Several genetically modified mice in which different parts of the *nrg1* gene have been deleted exhibit hyperactivity that was reduced by the anti-psychotic drug Clozapine (Gerlai et al. 2000; O'Tuathaigh et al. 2006; Stefansson et al. 2002). Furthermore, erbB4 heterozygous mice displayed similar phenotypes. In contrast, brain-specific erbB4 null mice were less active than their wild type counterparts and no alterations were observed in the activity levels of mice lacking one functional copy of erbB2, erbB3 or the Ig domain of nrg1 (Gerlai et al. 2000; Golub et al. 2004; Rimer et al. 2005; Thuret et al. 2004). Alterations in behaviors thought to be related to cognitive defects have also been found in some of these mouse models including alterations in prepulse inhibition, a measure of sensory gating, and latent inhibition, which reflects impairment in selective attention, both of which are abnormal in schizophrenia (Rimer et al. 2005; Stefansson et al. 2002). The variable alterations in behavior observed in the different nrg1 and erbB receptor mouse mutants suggest that changes in the location and expression levels of multiple ligands and/or receptors may be important to the manifestation of the disease.

Components of the NRG1 signaling pathway are robustly associated with schizophrenia, but how would altering this pathway increase susceptibility to the disease? Given that NRG1 and its receptors are pleiotropic, acting in diverse processes of brain development, changing the expression levels or function of NRGs by altering the specific isoforms expressed could affect the development of the brain in multiple ways. In the next pages, we will review the information regarding the possible links between schizophrenia and several developmental processes that are regulated by NRG1 signaling, the roles of NRG1, and the ways in which NRG1 may contribute to the disease through these processes.

4.4 Neuronal Migration and Cortical Development

The evidence for structural alterations in the brains of patients is subtle but compelling (reviewed in Arnold and Trojanowski, 1996). For example, imaging studies have shown that there is an enlargement of the lateral ventricles and a reduction in some associative cortical areas in affected individuals compared to their healthy monozygotic twins (Schlaepfer et al. 1994; Suddath et al. 1990). The cause of these alterations remains unclear, but it has been proposed that the initial steps of brain development, like neuronal migration and differentiation, could be altered. Numerous histological studies of postmortem tissues have demonstrated there are cytoarchitectural alterations in several areas of the brain implicated in schizophrenia. For example, interstitial neurons (remnants of the subplate) are abnormally distributed in the white matter of prefrontal and temporal cortices of patients (Akbarian et al. 1996; Roiux et al. 2003).

Early studies indicated that NRG1-erbB signaling regulates the morphology and function of radial glia, cells that play critical roles in cortical development such as supporting the movement of newborn neurons and acting as neural and glial precursors (Rakic 2003). NRG1 is expressed by newborn neurons and its receptors in the radial glial cells (Anton et al. 1997; Rio et al. 1997). Addition of NRG1 to cerebellar astrocyte cultures induces these cells to change their shape into a radial glial-like morphology in vitro (Patten et al. 2003, 2006) whereas loss of erbB2 leads to a reduction in the number of radial glial cells derived from neural progenitor cells in the telencephalon (Schmid et al. 2003). Functional erbB receptors are also required for neuronal migration in the cerebellum (Patten et al. 2003; Rio et al. 1997) and cerebral cortex (Anton et al. 1997). For example, cerebellar granule neurons in vitro fail to migrate along radial glia that do not respond to NRG1 (Rio et al. 1997). Similar results have been found in cortical cells (Anton et al. 1997). Thus, NRG1-erbB signaling appears to mediate critical interactions between radial glia and migrating neurons. This signaling pathway seems to function in non-radial glial-dependent migration as well. For example, loss of erbB4 has been shown to result in alterations in neuronal migration along the rostral migratory stream (RMS) (Anton et al. 2004) while infusion of NRG1 can halt the initiation and progression of migration in this migratory pathway (Ghashghaei et al. 2006). It has been also suggested that NRG1 may act as a short and long range attractant for the migration of GABAergic interneurons (Flames et al. 2004).

More recently, erbB4 was shown to play a key role in regulating the timing at which neural precursors generate the different cell types of the CNS, neurons vs. glia (Sardi et al. 2006), and that this depends on the function of the JMa isoform, the isoform that has been linked to schizophrenia (Law et al. 2007; Silberberg et al. 2006). In vitro studies showed that activation of erbB4-JMa inhibits neural precursors from generating astrocytes while leaving their neurogenic potential intact. Consistent with these findings, lack of erbB4 in mice results in precocious astrogenesis in the cerebral cortex.

These lines of evidence suggest that defects in NRG1-erbB signaling during brain development could lead to alterations in neuronal migration and cortical development. This could disrupt cortical connectivity and, subsequently, lead to behavioral defects. It is important to note that while several studies have shown defects in brain cytoarchitecture in schizophrenia, others do not reproduce these findings (Harrison 1999). It is possible that NRG1 is involved in the pathogenesis of the disease in only a portion of patients. Thus, it would be important to analyze the correlation between genotype and cortical defects in human cases.

4.5 Neuregulin and Neuronal Maturation

Once neurons reach their final destination they must form proper connections with other cells and there is mounting evidence that defects in neuronal connectivity exist in the brains of schizophrenia patients. For example, it has been reported that the length of dendrites and the incidence of basal dendrites on granule cells in the dentate gyrus of the hippocampus are significantly increased in patients (Cotter et al. 2000; Lauer et al. 2003; Senitz and Beckmann 2003). In contrast, the number and arborization of basal dendrites in layers III and V of the prefrontal cortex appear to be reduced in schizophrenia (Black et al. 2004; Broadbelt et al. 2002; Kalus et al. 2000). Similarly, neurons in other regions of the brain (spiny neurons of the olfactory tubercle and interneurons in the anterior cingulate cortex) are also affected in schizophrenia, displaying reduced dendritic length (Kalus et al. 2000: Markova et al. 2000). Together, these data suggest that dendritic outgrowth is abnormal in schizophrenia though this phenomenon is both region and cell-type specific.

In this respect, it is significant that erbB signaling has been shown to influence dendritic outgrowth in cerebellar granule cells. When erbB receptor function is blocked in vivo by expression of a dominant negative erbB receptor, the number of dendrites is significantly reduced in developing cerebellar granule cells (Rieff and Corfas 2006). Tissue culture experiments demonstrate that type I NRG1 induces neurite outgrowth in retinal cells (Bermingham-McDonogh et al. 1996), cerebellar granule cells (Rieff et al. 1999), mesencephalic neurons (Zhang et al. 2004) and hippocampal (Gerecke et al. 2004) neurons. Collectively, these results suggest that NRG1-erbB signaling may regulate dendritic development in several brain regions.

4.6 Neurotransmitter Receptor Expression and Signaling

A hypothesis favored by numerous investigators is that alterations in the expression levels or function of specific neurotransmitter receptors play a role in schizophrenia (reviewed in Blum and Mann 2002; Konradi and Heckers 2003). Defects in neurotransmitter expression could lead to synaptic dysfunction and, ultimately, abnormal information processing in the brain. The strongest candidates in this respect are the dopamine, serotonin, glutamate, GABA (γ-aminobutyric acid) and acetylcholine receptors. Of particular interest to this chapter are the last three, whose expression has been reported to be directly influenced by NRG1.

The link between glutamatergic, NMDA (N-methyl-D-aspartic acid) receptors and schizophrenia developed, at least in part, from the observation that PCP and ketamine, NMDA receptor blockers, induce a schizophrenia-like state in humans and relapse in schizophrenia patients. This gave rise to the glutamate dysfunction model of schizophrenia (reviewed in Konradi and Heckers 2003). Supporting this, Akbarian et al. (1996) found an increase in the relative abundance of NMDA receptor 2D subunits (NR2D) and a reduction in NR2C subunit mRNA levels in all cortical layers of the prefrontal cortex in patients compared to controls. Similarly, quantitative autoradiography of ^{3}H-MK-801 binding to NMDA receptors showed a 17% increase in binding in layers II/III of the anterior cingulate cortex (Zavitsanou and Huang 2002; Zavitsanou et al. 2002). Further support for the involvement of NMDA receptors in schizophrenia comes from studies showing that mice expressing NR1 at low levels (5-10% of normal) display behaviors related to schizophrenia (Mohn et al. 1999). From this and other data, it has been proposed that the changes in the pattern of expression of NMDA receptors in schizophrenia may compensate for impaired glutamatergic signaling (Konradi and Heckers 2003). As glutamatergic signaling is known to influence both GABAergic and dopaminergic neurotransmission, changes in NMDA receptors could produce a range of effects. This suggests that alterations in glutamate play a role in the etiology of schizophrenia although the mechanisms are not well understood.

The evidence that NRG1 influences glutamatergic neurotransmission is mounting. Ozaki et al. (1997) reported that treatment of cerebellar slices with NRG1 induces the upregulation of NR2C subunit mRNA, a subunit whose expression is reduced in the prefrontal cortex of schizophrenia patients (Ozaki et al. 1997). It has also been shown that NRG1 treatment can induce a reversible decrease in NMDA receptor current amplitude in dissociated prefrontal cortical neurons, which appears to be mediated, in part, by increasing NR1 subunit internalization (Gu et al. 2005). In the hippocampus, Kwon and colleagues (2005) showed that NRG1 application inhibits LTP at the Schaeffer collateral to CA1 synapse while erbB tyrosine kinase inhibitors potentiate these synapses and block LTP reversal that is normally induced by negative pulse stimulation. *In vivo* evidence that NRG1 regulates glutamate receptor expression comes from *nrg1* heterozygous mice, which have fewer functional NMDA receptors in the prefrontal cortex (Stefansson et al. 2002). In addition, studies in human post-mortem tissue indicate that when cells from the prefrontal cortex are stimulated with NMDA in the presence of NRG1, samples from schizophrenia patients display reduced tyrosine phosphorylation of the NR2A subunit (Hahn et al., 2006). Given that the influence of NRG1 on glutamatergic signaling is multi-faceted, subtle changes in NRG1 or erbB receptor expression or function could affect glutamatergic signaling and, in turn, other neurotransmitter systems.

Numerous studies point to complex alterations in GABAergic neurotransmission in schizophrenia (reviewed in Blum and Mann 2002). High-resolution autoradiographic analysis of ligand binding showed that $GABA_A$ receptors are upregulated in the cingulate cortex, prefrontal cortex and hippocampus of patients (Benes and Berretta 2001). The levels of $\alpha1$, $\alpha2$ and $\alpha5$ $GABA_A$ receptor subunit mRNA was increased in the prefrontal cortex in patients whereas the levels of the short isoform of the $GABA_A$ γ subunit were decreased (Benes and Berretta 2001; Huntsman et al. 1998; Impagnatiello et al. 1998). Two studies using microarray and RT PCR analysis of tissue from schizophrenia patients found an increase in GABA receptor genes and genes involved in the synthesis (*GAD65, 67*) and uptake (*GAT1*) of GABA (Hakak et al. 2001; Mirnics et al. 2000). In contrast, another research group, using *in situ* hybridization, reported decreased expression of *GAD67* and *GAT1* in layers I-V of the prefrontal cortex in patients with schizophrenia compared to control subjects (Volk et al. 2000, 2001). Alterations in this inhibitory neurotransmitter pathway could alter the ability of neurons to receive or respond appropriately to inputs particularly given that GABA is known to inhibit dopamine release from synaptic terminals.

The evidence for NRG1 influence on GABAergic neurotransmission is also strong. Rieff et al. (1999) showed that NRG1 treatment leads to increased expression of $GABA_A$ receptor $\beta2$ subunit mRNA and increased GABA-induced currents in cerebellar granule cells in culture (Rieff et al. 1999). More recently, treatment of CA1 hippocampal neurons with NRG1 decreased the mRNA level of $GABA_A$ receptor α subunits as well as the amplitude of mIPSCs in (Okada and Corfas, 2003). Thus, if NRG1-erbB signaling is abnormal GABAergic neurotransmission may be altered increased or decreased in a regional and receptor subunit dependent manner.

The possible link between nicotinic acetylcholine receptors (nAChRs) and schizophrenia is based on genetic and expression studies. The gene for the α7 subunit of the AChR is located in a region of chromosome 15 that has been mapped as a possible schizophrenia locus. Binding studies have shown reductions in the number of α7-containing AChR in the hippocampus, reticular formation and cortex of patients (Freedman et al. 1995).

NRG1 was initially isolated for its role in AChR synthesis in muscle cells but has since been shown to play a role in regulating AChRs in the central nervous system (Falls et al. 1993). Using in vitro cultures Fischbach and colleagues demonstrated that NRG1 regulates the expression and function of AChRs in hippocampal interneurons in diverse ways. On one hand, acute application of NRG1 reduces fast inward ACh currents and the number of surface α-bungarotoxin (BTX) binding sites (Chang and Fischbach 2006). On the other hand, longer-term NRG1 treatment increases the number of BTX sites and peak amplitude of ACh-induced currents (Law et al. 2006; Liu et al. 2001). These findings suggest that alterations in NRG1 may lead to changes in ACh neurotransmission. Nicotinic AChRs are generally localized to presynaptic structures and could therefore play an important role in modulation of neurotransmission and ultimately impact the disease.

Collectively, the studies on glutamatergic, GABAergic and cholinergic neurotransmission suggest that defects in NRG1 function may lead to altered excitatory and inhibitory neurotransmission. This could influence information processing and, thus contribute to schizophrenia.

4.7 Myelination

While myelination in the human brain starts during embryonic development and is mostly complete by 2 years of age, significant myelination continues to occur until adolescence or early adulthood (Bartzokis et al. 2003; Benes et al. 1994), a period corresponding to the peak time for onset of schizophrenia. This late myelination is most evident in the frontal and temporal lobes (Bartzokis et al. 2003). Recent histological and imaging studies support the notion that white matter changes exist in schizophrenia (reviewed in Davis et al. 2003). For example, light and electron microscopic analysis uncovered damaged myelin sheath lamellae, decreased nuclear area and a reduction in the density of oligodendrocyte mitochondria in the prefrontal cortex (Uranova et al. 2001, 2004). Signs of apoptosis and necrosis of oligodendrocytes in the prefrontal cortex and caudate nucleus of patients have been reported (Uranova et al. 2001) while reductions in the number of oligodendrocytes in cortical layers III and VI as well as gyral white matter of the premotor cortex have also been observed in patients (Hof et al. 2002, 2003). Advanced imaging techniques such as MRI and DTI (a technique that allows for the quantification of anisotropic diffusion in white matter) have revealed alterations in white matter integrity as well as a reduction of myelin content in the temporal and frontal lobes (Agartz et al. 2001, Ardekani et al. 2003; Bartzokis et al. 2003; Buchsbaum et al. 1998; Burns et al. 2003; Flynn et al. 2003; Foong et al. 2000, 2002; Kubicki et al. 2003, 2005a,b; Lim et al. 1999, Mimmack et al. 2002; Shenton et al. 2001; Sun et al. 2003a,b; Wang et al. 2003).

Molecular analysis has shown downregulation of oligodendrocyte-related mRNA species within the DLPFC of affected patients (Hakak et al. 2001). For example, one study found reduction in the levels of expression of genes involved in the compaction of myelin, axon-glia interactions and erbB3 (Hakak et al. 2001). Other studies showed that similar alterations can be found in Katsel et al (Katsel et al. 2005) several brain regions and that genes expressed by oligodendrocyte precursors may be also affected (Aberg et al. 2006; Dracheva et al. 2006; Katsel et al. 2005; Pongrac et al. 2002; Tkachev et al. 2003). While the alterations in myelin gene expression provide a correlation between the disease and white matter, the finding that some myelin genes are linked to the disease raises the possibility that alterations in myelin may be involved in the pathogenesis of schizophrenia. Significant association between schizophrenia and the genes expressed by oligodendrocytes such as myelin and oligodendrocyte gene (*MOG*), myelin-associated glycoprotein (*MAG*), *NOGO*, *CNP* (cyclic nucleotide phosphodiesterase), *OLIG2* and *PLP* (proteolipoprotein) have been found (Georgieva et al. 2006; Liu et al. 2005; Peirce et al. 2006; Qin et al. 2005; Sinibaldi et al. 2004; Tan et al. 2005; Wan et al. 2005; Yang et al. 2003).

Initial evidence that NRG1-erbB signaling plays a role in oligodendrocyte development came from *in vitro* experiments. Like many other glial cell types, oligodendrocytes express erbB receptors and respond to NRG1. However, the reported effects of NRG1 on cells of the oligodendrocyte lineage in culture are inconsistent. An early study showed that NRG1 induces forebrain oligodendrocyte precursors to differentiate *in vitro* (Vartanian et al. 1994) while others suggested that it has the opposite effect, i.e. inhibit their differentiation (Calaora et al. 2001; Shi et al. 1998). Other studies implicated NRG1 in the proliferation and survival of oligodendrocyte precursors (Canoll et al. 1996; Fernandez et al. 2000). Experiments using embryonic spinal cord explants from mice lacking NRG1, erbB2 or erbB4, confirmed that alterations in this signaling pathway can produce abnormalities in oligodendrocyte development and that the contributions of these molecules are complex. While explants from NRG1 -/- mice do not generate oligodendrocytes (Vartanian et al. 1999), explants from erbB2-deficient mice can produce these cells but fewer of them reach maturity (Vartanian et al. 1999). In contrast, loss of erbB4 increases the number of mature oligodendrocytes without affecting the number of less mature oligodendrocytes (Park et al. 2001; Sussman et al. 2005). Thus, it appears that NRG1 signaling through different erbB receptor combinations may control different aspects of oligodendrocyte development.

More recently, a mouse model in which erbB signaling is abolished in cells of the oligodendrocyte lineage provided a first glimpse into the roles of this signaling pathway in oligodendrocytes in vivo (Roy et al. 2007). Analysis of mice in which cells of the oligodendrocyte lineage express a dominant-negative erbB receptor showed that lack of erbB signaling results in oligodendrocytes that have simpler arborization and thinner myelin. Interestingly, these mice also have more oligodendrocytes, indicating that this signaling pathway is not necessary for oligodendrocyte survival in vivo. Most relevant to schizophrenia was the finding that dopamine neurotransmission and behaviors related to this disease are altered in these mice (Roy et al. 2007). Thus, these results indicate that alterations NRG1-erbB signaling can lead to alterations in white matter, and that these changes in oligodendrocytes can produce altered neurotransmission and behavior relevant to schizophrenia.

4.8 Hormonal Control of Puberty

The age at which the first episode of schizophrenia occurs is typically associated with the post-pubertal period, but mild symptoms including deficits in social, motor and cognitive function may be observed at earlier stages (Lewis and Lieberman 2000). These are periods of dramatic hormonal changes, suggesting that neuroendocrine maturational events may be involved in the pathogenesis of schizophrenia. While this line of inquiry has been explored to a lesser extent than the ones described above, some insightful observations have been recorded (Walker and Bollini 2002). For example, an association between early puberty and a later onset of the first schizophrenic episode and hospitalization has been reported for female patients (Cohen et al. 1999). It has also been hypothesized that late onset of puberty could be associated with over-pruning and insufficient synaptic density in schizophrenia (Walker and Bollini 2002).

A role for NRG1-erbB signaling in the neuroendocrine control of puberty and hormonal production and release has been demonstrated in transgenic mice expressing a dominant negative form of *erbB* in hypothalamic astrocytes (Prevot et al. 2003). These mice have a specific delay in the onset of puberty and in reproductive development due to the reduced release of LHRH by hypothalamic neurons, an event that initiates puberty. Thus, defects in NRG1 signaling could disrupt the control of hormonal levels during puberty, which would affect not only reproductive development but also neurodevelopmental processes that, when altered, may contribute to schizophrenia.

4.9 Cellular Basis of Schizophrenia

The hypotheses regarding the cellular and molecular bases of schizophrenia are numerous and, in many cases, contradictory. The evidence that defects in NRG1-erbB signaling contribute to this disease provides an opportunity to focus research efforts on specific aspects of schizophrenia. Most critical at this time will be to determine how changes to the levels or function of NRG1 or its receptors contribute to the disease and to further study variants of NRG1, as well erbB receptors, for association with schizophrenia. It would also be worthwhile to continue examining the genetic interaction between these and other loci for schizophrenia susceptibility.

Given that a large number of exquisitely timed steps are required for normal brain development and function, it is likely that schizophrenia originates from alterations in more than one developmental process. How could perturbations NRG1-erbB signaling lead to the disturbed brain function observed in schizophrenia? We have outlined several means by which alterations in NRG1/erbB signaling may contribute to the development of the disorder. Defects in all of these developmental processes may play a role; however, the evidence supporting the role of oligodendrocyte abnormalities is compelling. Postulating a cellular abnormality in oligodendrocytes in schizophrenia should lead to detailed analysis of these cells at various stages of the disorder and may ultimately implicate factors that promote oligodendrocyte survival function as protective in schizophrenia.

In summary, further examination of the NRG1-erbB pathway by genetic and molecular methods, and by the use of animal models may lead to a deeper understanding of the roles of these molecules in normal development and in the pathogenesis of this devastating disease.

References

Aberg K, Saetre P, Lindholm E, Ekholm B, Pettersson U, Adolfsson R, Jazin E (2006) Human QKI, a new candidate gene for schizophrenia involved in myelination. Am J Med Genet B Neuropsychiatr Genet 141:84-90.

Agartz I, Andersson JL, Skare S (2001) Abnormal brain white matter in schizophrenia: a diffusion tensor imaging study. Neuroreport 12:2251-2254.

Akbarian S, Kim JJ, Potkin S G, Hetrick WP, Bunney WE, Jr., & Jones EG (1996) Maldistribution of interstitial neurons in prefrontal white matter of the brains of schizophrenic patients. Arch Gen Psychiatry 53:425-436.

Akbarian S, Sucher N J, Bradley D, Tafazzoli A, Trinh D, Hetrick W P, Potkin S G, Sandman C A, Bunney W E, Jr., Jones E G (1996) Selective alterations in gene expression for NMDA receptor subunits in prefrontal cortex of schizophrenics. J Neurosci 16:19-30.

Anton ES, Ghashghaei HT, Weber JL, McCann C, Fischer TM, Cheung ID, Gassmann M, Messing A, Klein R, Schwab MH, Lloyd KC, Lai C (2004) Receptor tyrosine kinase ErbB4 modulates neuroblast migration and placement in the adult forebrain. Nat Neurosci 7:1319-1328.

Anton ES, Marchionni MA, Lee KF, Rakic P (1997) Role of GGF/neuregulin signaling in interactions between migrating neurons and radial glia in the developing cerebral cortex. Development 124:3501-3510.

Ardekani BA, Nierenberg J, Hoptman M J, Javitt DC, Lim KO (2003) MRI study of white matter diffusion anisotropy in schizophrenia. Neuroreport 14:2025-2029.

Arnold SE, Trojanowski JQ (1996) Recent advances in defining the neuropathology of schizophrenia. Acta Neuropathol (Berl) 92:217-231.

Bahn S (2002) Gene expression in bipolar disorder and schizophrenia: new approaches to old problems. Bipolar disorders 4 Suppl 1:70-72.

Bakker SC, Hoogendoorn ML, Selten JP, Verduijn W, Pearson PL, Sinke RJ, Kahn RS (2004) Neuregulin 1: genetic support for schizophrenia subtypes. Mol Psychiatry 9:1061-1063.

Bao J, Wolpowitz D, Role LW, Talmage DA (2003) Back signaling by the Nrg-1 intracellular domain. J Cell Biol 161:1133-1141.

Bartzokis G, Cummings J L, Sultzer D, Henderson VW, Nuechterlein KH, Mintz J (2003) White matter structural integrity in healthy aging adults and patients with Alzheimer disease: a magnetic resonance imaging study. Arch Neurol 60:393-398.

Bartzokis G, Nuechterlein KH, Lu PH, Gitlin M, Rogers S, Mintz J (2003) Dysregulated brain development in adult men with schizophrenia: a magnetic resonance imaging study. Biol Psychiatry 53:412-421.

Benes FM, Berretta S (2001) GABAergic interneurons: implications for understanding schizophrenia and bipolar disorder. Neuropsychopharmacology 25:1-27.

Benes FM, Turtle M, Khan Y, Farol P (1994) Myelination of a key relay zone in the hippocampal formation occurs in the human brain during childhood, adolescence, and adulthood. Arch Gen Psychiatry 51:477-484.

Bermingham-McDonogh O, McCabe KL, Reh TA (1996) Effects of GGF/neuregulins on neuronal survival and neurite outgrowth correlate with erbB2/neu expression in developing rat retina. Development 122:1427-1438.

Black JE, Kodish IM, Grossman AW, Klintsova AY, Orlovskaya D, Vostrikov V, Uranova N, Greenough WT (2004) Pathology of layer V pyramidal neurons in the prefrontal cortex of patients with schizophrenia. Am J Psychiatry 161:742-744.

Blum BP, Mann JJ (2002) The GABAergic system in schizophrenia. Int J Neuropsychopharmacol 5:159-179.

Bray NJ, Buckland PR, Owen MJ, O'Donovan MC (2003) Cis-acting variation in the expression of a high proportion of genes in human brain. Hum Genet 113:149-153.

Broadbelt K, Byne W, & Jones LB (2002) Evidence for a decrease in basilar dendrites of pyramidal cells in schizophrenic medial prefrontal cortex. Schizophr Res 58:75-81.

Buchsbaum MS, Tang CY, Peled S, Gudbjartsson H, Lu D, Hazlett EA, Downhill J, Haznedar M, Fallon JH, & Atlas SW (1998) MRI white matter diffusion anisotropy and PET metabolic rate in schizophrenia. Neuroreport 9:425-430.

Buonanno A, Fischbach GD (2001) Neuregulin and ErbB receptor signaling pathways in the nervous system. Curr Opin Neurobiol 11:287-296.

Burns J, Job D, Bastin ME, Whalley H, Macgillivray T, Johnstone EC, Lawrie SM (2003) Structural disconnectivity in schizophrenia: a diffusion tensor magnetic resonance imaging study. Br J Psychiatry 182:439-443.

Calaora V, Rogister B, Bismuth K, Murray K, Brandt H, Leprince P, Marchionni M, Dubois-Dalcq M (2001) Neuregulin signaling regulates neural precursor growth and the generation of oligodendrocytes in vitro. J Neurosci 21:4740-4751.

Cannella B, Pitt D, Marchionni M, Raine CS (1999) Neuregulin and erbB receptor expression in normal and diseased human white matter. J Neuroimmunol 100:233-242.

Canoll PD, Musacchio JM, Hardy R, Reynolds R, Marchionni MA, Salzer JL (1996) GGF/neuregulin is a neuronal signal that promotes the proliferation and survival and inhibits the differentiation of oligodendrocyte progenitors. Neuron 17:229-243.

Carraway KL, 3rd, Weber JL, Unger MJ, Ledesma J, Yu N, Gassmann M, Lai C (1997) Neuregulin-2, a new ligand of ErbB3/ErbB4-receptor tyrosine kinases. Nature 387:512-516.

Carteron C, Ferrer-Montiel A, Cabedo H (2006) Characterization of a neural-specific splicing form of the human neuregulin 3 gene involved in oligoden-drocyte survival. J Cell Sci 119:898-909.

Chang Q, Fischbach G D (2006) An acute effect of neuregulin 1 beta to suppress alpha 7-containing nicotinic acetylcholine receptors in hippocampal interneu-rons. J Neurosci 26:11295-11303.

Cohen RZ, Seeman MV, Gotowiec A, Kopala L (1999) Earlier puberty as a pre-dictor of later onset of schizophrenia in women. Am J Psychiatry 156:1059-1064.

Corfas G, Rosen KM, Aratake H, Krauss R, Fischbach GD (1995) Differential ex-pression of ARIA isoforms in the rat brain. Neuron 14:103-115.

Corvin AP, Morris DW, McGhee K, Schwaiger S, Scully P, Quinn J, Meagher D, Clair DS, Waddington JL, Gill M (2004) Confirmation and refinement of an 'at-risk' haplotype for schizophrenia suggests the EST cluster, Hs.97362, as a potential susceptibility gene at the Neuregulin-1 locus. Mol Psychiatry 9:208-213.

Cotter D, Wilson S, Roberts E, Kerwin R, Everall IP (2000) Increased dendritic MAP2 expression in the hippocampus in schizophrenia. Schizophr Res 41:313-323.

Davis KL, Stewart DG, Friedman JI, Buchsbaum M, Harvey PD, Hof PR, Bux-baum J, Haroutunian V (2003) White matter changes in schizophrenia: evi-dence for myelin-related dysfunction. Arch Gen Psychiatry 60:443-456.

Dracheva S, Davis KL, Chin B, Woo D A, Schmeidler J, Haroutunian V (2006) Myelin-associated mRNA and protein expression deficits in the anterior cin-gulate cortex and hippocampus in elderly schizophrenia patients. Neurobiol Dis 21:531-540.

Falls DL (2003) Neuregulins: functions, forms, and signaling strategies. Exp Cell Res 284: 14-30.

Falls DL, Rosen KM, Corfas G, Lane WS, Fischbach G D (1993) ARIA, a protein that stimulates acetylcholine receptor synthesis, is a member of the neu ligand family. Cell 72:801-815.

Fernandez PA, Tang DG, Cheng L, Prochiantz A, Mudge AW, Raff MC (2000) Evidence that axon-derived neuregulin promotes oligodendrocyte survival in the developing rat optic nerve. Neuron 28:81-90.

Flames N, Long JE, Garratt AN, Fischer TM, Gassmann M, Birchmeier C, Lai C, Rubenstein JL, Marin O (2004) Short- and long-range attraction of cortical GABAergic interneurons by neuregulin-1. Neuron 44:251-261.

Flynn SW, Lang DJ, Mackay AL, Goghari V, Vavasour IM, Whittall KP, Smith GN, Arango V, Mann JJ, Dwork AJ, Falkai P, Honer WG (2003) Abnormalities of myelination in schizophrenia detected in vivo with MRI, and postmortem with analysis of oligodendrocyte proteins. Mol Psychiatry 8:811-820.

Foong J, Maier M, Clark CA, Barker GJ, Miller DH, Ron MA (2000) Neuropathological abnormalities of the corpus callosum in schizophrenia: a diffusion tensor imaging study. J Neurol Neurosurg Psychiatry 68:242-244.

Foong J, Symms MR, Barker GJ, Maier M, Miller DH, Ron MA (2002) Investigating regional white matter in schizophrenia using diffusion tensor imaging. Neuroreport 13:333-336.

Freedman R, Hall M, Adler LE, Leonard S (1995) Evidence in postmortem brain tissue for decreased numbers of hippocampal nicotinic receptors in schizophrenia. Biol Psychiatry 38:22-33.

Fukui N, Muratake T, Kaneko N, Amagane H, Someya T (2006) Supportive evidence for neuregulin 1 as a susceptibility gene for schizophrenia in a Japanese population. Neurosci Lett 396:117-120.

Georgieva L, Moskvina V, Peirce T, Norton N, Bray N J, Jones L, Holmans P, Macgregor S, Zammit S, Wilkinson J, Williams H, Nikolov I, Williams N, Ivanov D, Davis KL, Haroutunian V, Buxbaum JD, Craddock N, Kirov G, Owen MJ, O'Donovan MC (2006) Convergent evidence that oligodendrocyte lineage transcription factor 2 (OLIG2) and interacting genes influence susceptibility to schizophrenia. Proc Natl Acad Sci USA 103:12469-12474.

Gerecke KM, Wyss JM, Carroll SL (2004) Neuregulin-1beta induces neurite extension and arborization in cultured hippocampal neurons. Mol Cell Neurosci 27:379-393.

Gerlai R, Pisacane P, Erickson S (2000) Heregulin, but not ErbB2 or ErbB3, heterozygous mutant mice exhibit hyperactivity in multiple behavioral tasks. Behav Brain Res 109:219-227.

Ghashghaei HT, Weber J, Pevny L, Schmid R, Schwab MH, Lloyd KC, Eisenstat DD, Lai C, Anton ES (2006) The role of neuregulin-ErbB4 interactions on the proliferation and organization of cells in the subventricular zone. Proc Natl Acad Sci USA 103:1930-1935.

Golub MS, Germann SL, Lloyd KC (2004) Behavioral characteristics of a nervous system-specific erbB4 knock-out mouse. Behav Brain Res 153:159-170.

Goodearl A, Viehover A, Vartanian T (2001) Neuregulin-induced association of Sos Ras exchange protein with HER2(erbB2)/HER3(erbB3) receptor complexes in Schwann cells through a specific Grb2-HER2(erbB2) interaction. Dev Neurosci 23: 25-30.

Gu Z, Jiang Q, Fu AK, Ip NY, Yan Z (2005) Regulation of NMDA receptors by neuregulin signaling in prefrontal cortex. J Neurosci 25:4974-4984.

Hahn CG, Wang HY, Cho DS, Talbot K, Gur RE, Berrettini WH, Bakshi K, Kamins J, Borgmann-Winter KE, Siegel SJ, Gallop R J, Arnold SE (2006) Altered neuregulin 1-erbB4 signaling contributes to NMDA receptor hypofunction in schizophrenia. Nat Med 12:824-828.

Hakak Y, Walker JR, Li C, Wong WH, Davis KL, Buxbaum JD, Haroutunian V, Fienberg AA (2001) Genome-wide expression analysis reveals dysregulation of myelination-related genes in chronic schizophrenia. Proc Natl Acad Sci U SA 98:4746-4751.

Harrison PJ (1999) The neuropathology of schizophrenia. A critical review of the data and their interpretation. Brain 122:593-624.

Hashimoto R, Straub RE, Weickert CS, Hyde TM, Kleinman JE, Weinberger DR (2004) Expression analysis of neuregulin-1 in the dorsolateral prefrontal cortex in schizophrenia. Mol Psychiatry 21:21.

Hof PR, Haroutunian V, Copland C, Davis KL, Buxbaum JD (2002) Molecular and cellular evidence for an oligodendrocyte abnormality in schizophrenia. Neurochem Res 27:1193-1200.

Hof PR, Haroutunian V, Friedrich V L, Byne W, Buitron C, Perl DP, & Davis KL (2003) Loss and Altered Spatial Distribution of Oligodendrocytes in the Superior Frontal Gyrus in Schizophrenia. Biol. Pych. 53:1075-1085.

Holmes WE, Sliwkowski MX, Akita RW, Henzel WJ, Lee J, Park JW, Yansura D, Abadi N, Raab H, Lewis GD, et al. (1992) Identification of heregulin, a specific activator of p185erbB2. Science 256:1205-1210.

Hong CJ, Huo SJ, Liao DL, Lee K, Wu JY, Tsai SJ (2004) Case-control and family-based association studies between the neuregulin 1 (Arg38Gln) polymorphism and schizophrenia. Neurosci Lett 366:158-161.

Huntsman MM, Tran BV, Potkin SG, Bunney WE, Jr., Jones EG (1998) Altered ratios of alternatively spliced long and short gamma2 subunit mRNAs of the gamma-amino butyrate type A receptor in prefrontal cortex of schizophrenics. Proc Natl Acad Sci USA 95:15066-15071.

Impagnatiello F, Guidotti AR, Pesold C, Dwivedi Y, Caruncho H, Pisu MG, Uzunov DP, Smalheiser NR, Davis JM, Pandey GN, Pappas GD, Tueting P, Sharma RP, Costa E (1998) A decrease of reelin expression as a putative vulnerability factor in schizophrenia. Proc Natl Acad Sci USA 95:15718-15723.

Ingason A, Soeby K, Timm S, Wang AG, Jakobsen KD, Fink-Jensen A, Hemmingsen R, Berg Rasmussen H, Werge T (2006) No significant association of the 5' end of neuregulin 1 and schizophrenia in a large Danish sample. Schizophr Res 83:1-5.

Iwata N, Suzuki T, Ikeda M, Kitajima T, Yamanouchi Y, Inada T, Ozaki N (2004) No association with the neuregulin 1 haplotype to Japanese schizophrenia. Mol Psychiatry 9:126-127.

Kalus P, Muller TJ, Zuschratter W, Senitz D (2000) The dendritic architecture of prefrontal pyramidal neurons in schizophrenic patients. Neuroreport 11:3621-3625.

Kampman O, Anttila S, Illi A, Saarela M, Rontu R, Mattila KM, Leinonen E, Lehtimaki T (2004) Neuregulin genotype and medication response in Finnish patients with schizophrenia. Neuroreport 15:2517-2520.

Katsel PL, Davis KL, Haroutunian V (2005) Large-scale microarray studies of gene expression in multiple regions of the brain in schizophrenia and Alzheimer's disease. Int Rev Neurobiol 63:41-82.

Kim JW, Lee YS, Cho E , Jang YL, Park DY, Choi KS, Jeun HO, Cho SH, Jang Y, & Hong KS (2006) Linkage and association of schizophrenia with genetic variations in the locus of neuregulin 1 in Korean population. Am J Med Genet B Neuropsychiatr Genet 141:281-286.

Konradi C, Heckers S (2003) Molecular aspects of glutamate dysregulation: implications for schizophrenia and its treatment. Pharmacol Ther 97:153-179.

Kubicki M, McCarley RW, Nestor PG, Huh T, Kikinis R, Shenton ME, Wible CG (2003) An fMRI study of semantic processing in men with schizophrenia. Neuroimage 20:1923-1933.

Kubicki M, McCarley RW, Shenton ME (2005) Evidence for white matter abnormalities in schizophrenia. Curr Opin Psychiatry 18:121-134.

Kubicki M, Park H, Westin CF, Nestor PG, Mulkern RV, Maier S, Niznikiewicz M, Connor E E, Levitt JJ, Frumin M, Kikinis R, Jolesz FA, McCarley RW, Shenton ME (2005) DTI and MTR abnormalities in schizophrenia: analysis of white matter integrity. Neuroimage 26: 1109-1118.

Kwon OB, Longart M, Vullhorst D, Hoffman DA, & Buonanno A (2005) Neuregulin-1 reverses long-term potentiation at CA1 hippocampal synapses. J Neurosci 25: 9378-9383.

Lauer M, Beckmann H, & Senitz D (2003) Increased frequency of dentate granule cells with basal dendrites in the hippocampal formation of schizophrenics. Psychiatry Res 122: 89-97.

Law AJ, Kleinman JE, Weinberger DR, Weickert CS (2007) Disease-associated intronic variants in the ErbB4 gene are related to altered ErbB4 splice-variant expression in the brain in schizophrenia. Hum Mol Genet 16:129-141.

Law AJ, Lipska BK, Weickert CS, Hyde TM, Straub RE, Hashimoto R, Harrison PJ, Kleinman J, Weinberger D R (2006) Neuregulin 1 transcripts are differentially expressed in schizophrenia and regulated by 5' SNPs associated with the disease. Proc Natl Acad Sci USA 103:6747-6752.

Law AJ, Shannon Weickert C, Hyde TM, Kleinman JE, & Harrison PJ (2004) Neuregulin-1 (NRG-1) mRNA and protein in the adult human brain. Neuroscience 127: 125-136.

Lewis DA, Lieberman JA (2000) Catching up on schizophrenia: natural history and neurobiology. Neuron 28:325-334.

Li T, Stefansson H, Gudfinnsson E, Cai G, Liu X, Murray RM, Steinthorsdottir V, Januel D, Gudnadottir VG, Petursson H, Ingason A, Gulcher JR, Stefansson K, Collier DA (2004) Identification of a novel neuregulin 1 at-risk haplotype in Han schizophrenia Chinese patients, but no association with the Icelandic/ Scottish risk haplotype. Mol Psychiatry 9:698-704.

Lim KO, Hedehus M, Moseley M, de Crespigny A, Sullivan EV, Pfefferbaum A (1999) Compromised white matter tract integrity in schizophrenia inferred from diffusion tensor imaging. Arch Gen Psychiatry 56:367-374.

Liu CM, Hwu HG, Fann CS, Lin CY, Liu YL, Ou-Yang WC, Lee SF (2005) Linkage evidence of schizophrenia to loci near neuregulin 1 gene on chromosome 8p21 in Taiwanese families. Am J Med Genet B Neuropsychiatr Genet 134:79-83.

Liu Y, Ford B, Mann MA, Fischbach GD (2001) Neuregulins increase alpha7 nicotinic acetylcholine receptors and enhance excitatory synaptic transmission in GABAergic interneurons of the hippocampus. J Neurosci 21:5660-5669.

Longart M, Liu Y, Karavanova I, Buonanno A (2004) Neuregulin-2 is developmentally regulated and targeted to dendrites of central neurons. J Comp Neurol 472:156-172.

Marchionni MA, Goodearl AD, Chen MS, Bermingham-McDonogh O, Kirk C, Hendricks M, Danehy F, Misumi D, Sudhalter J, Kobayashi K, et al. (1993) Glial growth factors are alternatively spliced erbB2 ligands expressed in the nervous system. Nature 362:312-318.

Markova E, Markov I, Revishchin A, Okhotin V, Sulimov G (2000) 3-D Golgi and image analysis of the olfactory tubercle in schizophrenia. Anal Quant Cytol Histol 22:178-182.

Mimmack ML, Ryan M, Baba H, Navarro-Ruiz J, Iritani S, Faull RL, McKenna PJ, Jones PB, Arai H, Starkey M, Emson PC, Bahn S (2002) Gene expression analysis in schizophrenia: reproducible up-regulation of several members of the apolipoprotein L family located in a high-susceptibility locus for schizophrenia on chromosome 22. Proc Natl Acad Sci USA 99:4680-4685.

Minami T, Nobuhara K, Okugawa G, Takase K, Yoshida T, Sawada S, Ha-Kawa S, Ikeda K, Kinoshita T (2003) Diffusion tensor magnetic resonance imaging of disruption of regional white matter in schizophrenia. Neuropsychobiology 47:141-145.

Mirnics K, Middleton FA, Marquez A, Lewis DA, Levitt P (2000) Molecular characterization of schizophrenia viewed by microarray analysis of gene expression in prefrontal cortex. Neuron 28:53-67.

Mohn AR, Gainetdinov RR, Caron MG, Koller BH (1999) Mice with reduced NMDA receptor expression display behaviors related to schizophrenia. Cell 98:427-436.

Nicodemus KK, Luna A, Vakkalanka R, Goldberg T, Egan M, Straub RE, Weinberger DR (2006) Further evidence for association between ErbB4 and schizophrenia and influence on cognitive intermediate phenotypes in healthy controls. Mol Psychiatry 11:1062-1065.

Norton N, Moskvina V, Morris DW, Bray NJ, Zammit S, Williams NM, Williams HJ, Preece AC, Dwyer S, Wilkinson JC, Spurlock G, Kirov G, Buckland P, Waddington JL, Gill M, Corvin AP, Owen MJ, O'Donovan MC (2006) Evidence that interaction between neuregulin 1 and its receptor erbB4 increases susceptibility to schizophrenia. Am J Med Genet B Neuropsychiatr Genet 141:96-101.

O'Tuathaigh CM, Babovic D, O'Meara G, Clifford JJ, Croke DT, Waddington JL (2006) Susceptibility genes for schizophrenia: Characterisation of mutant mouse models at the level of phenotypic behaviour. Neurosci Biobehav Rev .

Okada M, Corfas G (2003) Neuregulin1 Down-regulates Postsynaptic GABAA Receptors at the Hippocampal Inhibitory Synapse. Hippocampus 14:337-344.

Ozaki M, Sasner M, Yano R, Lu HS, Buonanno A (1997) Neuregulin-beta induces expression of an NMDA-receptor subunit. Nature 390:691-694.

Park SK, Miller R, Krane I, Vartanian T (2001) The erbB2 gene is required for the development of terminally differentiated spinal cord oligodendrocytes. J Cell Biol 154:1245-1258.

Patten BA, Peyrin JM, Weinmaster G, Corfas G (2003) Sequential signaling through Notch1 and erbB receptors mediates radial glia differentiation. J Neurosci 23:6132-6140.

Patten BA, Sardi SP, Koirala S, Nakafuku M, Corfas G (2006) Notch1 signaling regulates radial glia differentiation through multiple transcriptional mechanisms. J Neurosci 26:3102-3108.

Peirce TR, Bray NJ, Williams NM, Norton N, Moskvina V, Preece A, Haroutunian V, Buxbaum JD, Owen MJ, O'Donovan MC (2006) Convergent evidence for 2',3'-cyclic nucleotide 3'-phosphodiesterase as a possible susceptibility gene for schizophrenia. Arch Gen Psychiatry 63:18-24.

Peles E, Bacus SS, Koski RA, Lu HS, Wen D, Ogden SG, Levy RB, Yarden Y (1992) Isolation of the neu/HER-2 stimulatory ligand: a 44 kd glycoprotein that induces differentiation of mammary tumor cells. Cell 69:205-216.

Petryshen TL, Middleton FA, Kirby A, Aldinger KA, Purcell S, Tahl AR, Morley CP, McGann L, Gentile KL, Rockwell GN, Medeiros HM, Carvalho C, Macedo A, Dourado A, Valente J, Ferreira CP, Patterson NJ, Azevedo MH, Daly MJ, Pato CN, Pato MT, Sklar P (2005) Support for involvement of neuregulin 1 in schizophrenia pathophysiology. Mol Psychiatry 10:366-374, 328.

Pongrac J, Middleton FA, Lewis DA, Levitt P, Mirnics K (2002) Gene expression profiling with DNA microarrays: advancing our understanding of psychiatric disorders. Neurochem Res 27:1049-1063.

Prevot V, Rio C, Cho GJ, Lomniczi A, Heger S, Neville CM, Rosenthal NA, Ojeda S R, Corfas G (2003) Normal female sexual development requires neuregulin-erbB receptor signaling in hypothalamic astrocytes. J Neurosci 23:230-239.

Qin W, Gao J, Xing Q, Yang J, Qian X, Li X, Guo Z, Chen H, Wang L, Huang X, Gu N, Feng G, He L (2005) A family-based association study of PLP1 and schizophrenia. Neurosci Lett 375:207-210.

Rakic P (2003) Elusive radial glial cells: historical and evolutionary perspective. Glia 43: 19-32.

Rieff HI Corfas G (2006) ErbB receptor signalling regulates dendrite formation in mouse cerebellar granule cells in vivo. Eur J Neurosci 23:2225-2229.

Rieff HI, Raetzman LT, Sapp DW, Yeh HH, Siegel RE, Corfas G (1999) Neuregulin induces GABA(A) receptor subunit expression and neurite outgrowth in cerebellar granule cells. J Neurosci 19:10757-10766.

Rimer M, Barrett DW, Maldonado MA, Vock VM, Gonzalez-Lima F (2005) Neuregulin-1 immunoglobulin-like domain mutant mice: clozapine sensitivity and impaired latent inhibition. Neuroreport 16:271-275.

Rimer M, Prieto AL, Weber JL, Colasante C, Ponomareva O, Fromm L, Schwab MH, Lai C, Burden SJ (2004) Neuregulin-2 is synthesized by motor neurons and terminal Schwann cells and activates acetylcholine receptor transcription in muscle cells expressing ErbB4. Mol Cell Neurosci 26:271-281.

Rio C, Rieff HI, Qi P, Khurana TS, Corfas G (1997) Neuregulin and erbB receptors play a critical role in neuronal migration. Neuron 19:39-50.

Rioux L, Nissanov J, Lauber K, Bilker WB, Arnold SE (2003) Distribution of microtubule-associated protein MAP2-immunoreactive interstitial neurons in the parahippocampal white matter in subjects with schizophrenia. Am J Psychiatry 160:149-155.

Roy K, Murtie JC, El-Khodor BF, Edgar N, Sardi SP, Hooks BM, Benoit-Marand M, Chen C, Moore H, O'Donnell P, Brunner D, & Corfas G (2007) Loss of erbB signaling in oligodendrocytes alters myelin and dopaminergic function: a potential mechanism neuropsychiatric disorders. Proc Natl Acad Sci USA 104: 8131-8136.

Sardi SP, Murtie J, Koirala S, Patten BA, Corfas G (2006) Presenilin-dependent ErbB4 nuclear signaling regulates the timing of astrogenesis in the developing brain. Cell 127:185-197.

Schlaepfer TE, Harris G J, Tien A Y, Peng L W, Lee S, Federman EB, Chase GA, Barta P E, & Pearlson G D (1994) Decreased regional cortical gray matter volume in schizophrenia. Am J Psychiatry 151:842-848.

Schmid RS, McGrath B, Berechid BE, Boyles B, Marchionni M, Sestan N, Anton E (2003) Neuregulin 1-erbB2 signaling is required for the establishment of radial glia and their transformation into astrocytes in cerebral cortex. Proc Natl Acad Sci U S A 100: 4251-4256.

Schmucker J, Ader M, Brockschnieder D, Brodarac A, Bartsch U, Riethmacher D (2003) erbB3 is dispensable for oligodendrocyte development in vitro and in vivo. Glia 44: 67-75.

Senitz D, Beckmann H (2003) Granule cells of the dentate gyrus with basal and recurrent dendrites in schizophrenic patients and controls. A comparative Golgi study. J Neural Transm 110: 317-326.

Shenton ME, Dickey CC, Frumin M, McCarley RW (2001) A review of MRI findings in schizophrenia. Schizophr Res 49:1-52.

Shi J, Marinovich A, Barres BA (1998) Purification and characterization of adult oligodendrocyte precursor cells from the rat optic nerve. J Neurosci 18: 4627-4636.

Silberberg G, Darvasi A, Pinkas-Kramarski R, Navon R (2006) The involvement of ErbB4 with schizophrenia: association and expression studies. Am J Med Genet B Neuropsychiatr Genet 141:142-148.

Sinibaldi L, De Luca A, Bellacchio E, Conti E, Pasini A, Paloscia C, Spalletta G, Caltagirone C, Pizzuti A, Dallapiccola B (2004) Mutations of the Nogo-66 receptor (RTN4R) gene in schizophrenia. Hum Mutat 24:534-535.

Sklar P (2002) Linkage analysis in psychiatric disorders: the emerging picture. Annu Rev Genomics Hum Genet 3: 371-413.

Stefansson H, Sigurdsson E, Steinthorsdottir V, Bjornsdottir S, Sigmundsson T, Ghosh S, Brynjolfsson J, Gunnarsdottir S, Ivarsson O, Chou TT, Hjaltason O, Birgisdottir B, Jonsson H, Gudnadottir VG, Gudmundsdottir E, Bjornsson A, Ingvarsson B, Ingason A, Sigfusson S, Hardardottir H, Harvey R P, Lai D,

Zhou M, Brunner D, Mutel V, Gonzalo A, Lemke G, Sainz J, Johannesson G, Andresson T, Gudbjartsson D, Manolescu A, Frigge ML, Gurney ME, Kong A, Gulcher JR, Petursson H, Stefansson K (2002) Neuregulin 1 and susceptibility to schizophrenia. Am J Hum Genet 71:877-892.

Steinthorsdottir V, Stefansson H, Ghosh S, Birgisdottir B, Bjornsdottir S, Fasquel AC, Olafsson O, Stefansson K, Gulcher JR (2004) Multiple novel transcription initiation sites for NRG1. Gene 342:97-105.

Suddath R, Christison GW, Torrey EF, Casanova MF, & Weinberger DR (1990) Anatomical abnormalities in the brains of monozygotic twins discordant for schizophrenia. N Engl J Med 322:789-794.

Sun SW, Neil JJ, Song SK (2003) Relative indices of water diffusion anisotropy are equivalent in live and formalin-fixed mouse brains. Magn Reson Med 50:743-748.

Sun Z, Wang F, Cui L, Breeze J, Du X, Wang X, Cong Z, Zhang H, Li B, Hong N, Zhang D (2003) Abnormal anterior cingulum in patients with schizophrenia: a diffusion tensor imaging study. Neuroreport 14:1833-1836.

Sussman CR, Vartanian T, Miller RH (2005) The ErbB4 neuregulin receptor mediates suppression of oligodendrocyte maturation. J Neurosci 25: 5757-5762.

Tan E, Chong SA, Wang H, Chew-Ping Lim E, Teo YY (2005) Gender-specific association of insertion/deletion polymorphisms in the nogo gene and chronic schizophrenia. Brain Res Mol Brain Res 139: 212-216.

Tang JX, Chen WY, He G, Zhou J, Gu NF, Feng GY, He L (2004) Polymorphisms within 5' end of the Neuregulin 1 gene are genetically associated with schizophrenia in the Chinese population. Mol Psychiatry 9: 11-12.

Thiselton DL, Webb BT, Neale BM, Ribble RC, O'Neill FA, Walsh D, Riley B P, & Kendler K S (2004) No evidence for linkage or association of neuregulin-1 (NRG1) with disease in the Irish study of high-density schizophrenia families (ISHDSF). Mol Psychiatry 9:777-783; image 729.

Thuret S, Alavian KN, Gassmann M, Lloyd CK, Smits SM, Smidt MP, Klein R, Dyck RH, & Simon HH (2004) The neuregulin receptor, ErbB4, is not required for normal development and adult maintenance of the substantia nigra pars compacta. J Neurochem 91: 1302-1311.

Tkachev D, Mimmack ML, Ryan MM, Wayland M, Freeman T, Jones PB, Starkey M, Webster MJ, Yolken RH, & Bahn S (2003) Oligodendrocyte dysfunction in schizophrenia and bipolar disorder. Lancet 362: 798-805.

Tsuang MT, Stone WS, & Faraone SV (2001) Genes, environment and schizophrenia. Br J Psychiatry Suppl 40: s18-24.

Uranova N, Orlovskaya D, Vikhreva O, Zimina I, Kolomeets N, Vostrikov V, & Rachmanova V (2001) Electron microscopy of oligodendroglia in severe mental illness. Brain Res Bull 55: 597-610.

Uranova NA, Vostrikov VM, Orlovskaya DD, Rachmanova VI (2004) Oligodendroglial density in the prefrontal cortex in schizophrenia and mood disorders: a study from the Stanley Neuropathology Consortium. Schizophr Res 67: 269-275.

Vartanian T, Corfas G, Li Y, Fischbach GD, Stefansson K (1994) A role for the acetylcholine receptor-inducing protein ARIA in oligodendrocyte development. Proc Natl Acad Sci USA 91:11626-11630.

Vartanian T, Fischbach G, Miller R (1999) Failure of spinal cord oligodendrocyte development in mice lacking neuregulin. Proc Natl Acad Sci USA 96: 731-735.

Volk D, Austin M, Pierri J, Sampson A, Lewis D (2001) GABA transporter-1 mRNA in the prefrontal cortex in schizophrenia: decreased expression in a subset of neurons. Am J Psychiatry 158:256-265.

Volk D W, Austin M C, Pierri JN, Sampson AR, Lewis D A (2000) Decreased glutamic acid decarboxylase67 messenger RNA expression in a subset of prefrontal cortical gamma-aminobutyric acid neurons in subjects with schizophrenia. Arch Gen Psychiatry 57: 237-245.

Walker E & Bollini A M (2002) Pubertal neurodevelopment and the emergence of psychotic symptoms. Schizophr Res 54: 17-23.

Walss-Bass C, Liu W, Lew D F, Villegas R, Montero P, Dassori A, Leach R J, Almasy L, Escamilla M, & Raventos H (2006) A Novel Missense Mutation in the Transmembrane Domain of Neuregulin 1 is Associated with Schizophrenia. Biol Psychiatry

Wan C, Yang Y, Feng G, Gu N, Liu H, Zhu S, He L, & Wang L (2005) Polymorphisms of myelin-associated glycoprotein gene are associated with schizophrenia in the Chinese Han population. Neurosci Lett 388: 126-131.

Wang F, Sun Z, Du X, Wang X, Cong Z, Zhang H, Zhang D, Hong N (2003) A diffusion tensor imaging study of middle and superior cerebellar peduncle in male patients with schizophrenia. Neurosci Lett 348:135-138.

Williams NM, Preece A, Spurlock G, Norton N, Williams HJ, Zammit S, O'Donovan MC, Owen M J (2003) Support for genetic variation in neuregulin 1 and susceptibility to schizophrenia. Mol Psychiatry 8:485-487.

Yang JZ, Si TM, Ruan Y, Ling YS, Han YH, Wang XL, Zhou M, Zhang H Y, Kong Q M, Liu C, Zhang D R, Yu Y Q, Liu S Z, Ju G Z, Shu L, Ma D L, Zhang D (2003) Association study of neuregulin 1 gene with schizophrenia. Mol Psychiatry 8: 706-709.

Zavitsanou K, Huang XF (2002) Decreased [(3)H]spiperone binding in the anterior cingulate cortex of schizophrenia patients: an autoradiographic study. Neuroscience 109: 709-716.

Zavitsanou K, Ward P B, & Huang X F (2002) Selective alterations in ionotropic glutamate receptors in the anterior cingulate cortex in schizophrenia. Neuropsychopharmacology 27:826-833.

Zhang L, Fletcher-Turner A, Marchionni MA, Apparsundaram S, Lundgren KH, Yurek DM, & Seroogy KB (2004) Neurotrophic and neuroprotective effects of the neuregulin glial growth factor-2 on dopaminergic neurons in rat primary midbrain cultures. J Neurochem 91:1358-1368.

Zhao X, Shi Y, Tang J, Tang R, Yu L, Gu N, Feng G, Zhu S, Liu H, Xing Y, Zhao S, Sang H, Guan Y, St Clair D, & He L (2004) A case control and family based association study of the neuregulin1 gene and schizophrenia. J Med Genet 41:31-34.

Chapter 5

Nicotinic Cholinergic Cortical Dysfunction in Schizophrenia

Jason R. Tregellas and Robert Freedman

Cortical deficits in schizophrenia have been related to diverse psychophysiological, neurobiological, and genetic deficits. This chapter reviews specific psychophysiological deficits in sensory and motor gating, their neurobiological substrate in nicotinic acetylcholine-mediated neurotransmission, and their genetic basis. Detailed correlations between these several levels of observation are necessary to parse schizophrenia into discrete and hopefully more treatable elements.

5.1 Sensory Gating Dysfunction in Schizophrenia

The inability to filter out irrelevant information from the environment is a common feature of schizophrenia. (McGhie and Chapman 1961). Over the last two decades, investigators have developed experimental methods to reliably assess the brain's ability to filter out irrelevant sensory stimuli, often termed "sensory gating" (Adler et al. 1982). One such measure is a conditioning-testing paradigm, in which evoked potentials are measured from repeated pairs of clicks, separated by 500 ms. This measure is often called the P50 auditory evoked potential because the most commonly measured component is a middle latency waveform that occurs approximately 50 ms post-stimulus. If inhibitory pathways are functioning properly, there is a response to the first (conditioning) stimulus, but a greatly diminished response to the second (test) stimulus of the same intensity because the first stimulus activates an inhibitory pathway which is still active at the time the second stimulus is presented (Eccles 1969). Only if the two stimuli are far apart temporally (6-10 sec), do they act as independent stimuli and the responses elicited by the conditioning and test stimuli become equal. As shown in Fig. 5.1, evoked responses to the second click in the click pairs are not diminished in patients with schizophrenia, suggesting these patients have a defect in an inhibitory gating mechanism (Adler et al. 1998). Over the past two decades, this finding has been replicated by most (Adler et al. 1982; Freedman et al. 1983; Boutros et al. 1991; Judd et al. 1992; Adler et al. 1993; Freedman et al. 1996; Clementz et al. 1997; Erwin et al. 1998; Raux et al. 2002; Ghisolfi et al. 2002; Thoma et al. 2003) but not all studies (Kathmann and Engel 1990; Jin et al. 1998). Inhibitory failure during sensory gating has also been found in first degree relatives of schizophrenia patients (Waldo et al. 1988; Adler et al. 1992; Siegel et al. 1984; Clementz 1998; Myles-Worsley 2002; Wegrzyn and Wciorka 2004; Louchart-de la Chapelle et al. 2005). The fact that these gating abnormalities occur in approximately half of all first degree relatives in an apparent autosomal-dominant pattern suggests these deficits represent factors that increase vulnerability to schizophrenia (Freedman et al. 1997).

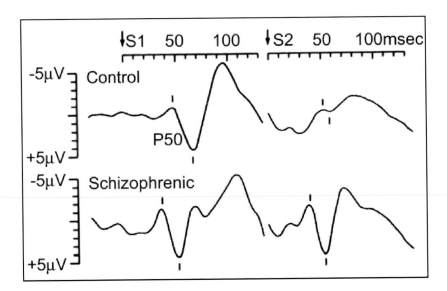

Fig. 5.1. Sensory gating deficits in schizophrenia. The P50 auditory evoked response to clicks 500 ms apart is indicated by the tick below the tracing. The ratio between the amplitude of the second or test response and the first or conditioning response is used as a measure for sensory gating. The control subject has a diminished response to the second stimulus. The subject with schizophrenia lacks inhibitory gating mechanisms. After Adler et al. 1998.

5.2 Effects of Neuroleptics on Sensory Gating

Typical neuroleptic treatment does not ameliorate sensory gating deficits in schizophrenia (Freedman et al. 1983;Light et al. 2000;Adler et al. 2004). Because these drugs primarily affect the dopamine system, it is unlikely that this system plays a key role in the deficit. Recent studies of patients treated with atypical neuroleptics, however, suggest effects on P50 gating. Studies have shown that clozapine treatment improves P50 gating in patients who respond to the medication (Freedman et al. 1996). Furthermore, P50 gating had been abnormal in these patients during previous treatment with typical neuroleptics. In a subsequent follow-up over a 5–27-month period of observation, clozapine's amelioration of the P50 auditory gating deficit was stable (Nagamoto et al. 1999). Interestingly, the improvement in P50 gating was paralleled by a corresponding decrease in symptoms as rated by the Brief Psychiatric Rating Scale.

Light and colleagues (Light et al. 2000) also found a significant enhancement of P50 auditory gating in medicated schizophrenia outpatients treated with atypical antipsychotics. Post hoc division of their subjects treated with atypical antipsychotics suggested that this improvement appeared to be mainly based on the effects of clozapine, and to a lesser degree olanzapine (Light et al. 2000). Subjects receiving risperidone continued to have sensory gating in the range seen with patients receiving typical neuroleptics. Yee and colleagues (1998) reported that risperidone did not have a significant effect on suppression of the P50 ratio in patients with recent-onset schizophrenia, although they did find some improvement in the suppression of the response to the second click as compared with subjects receiving typical neuroleptics (Yee et al. 1998). Arango and colleagues (2003) in a recent study that compared haloperidol and olanzapine found no significant differences on P50 auditory gating—neither was effective (Arango et al. 2003).

We have recently replicated prior results showing various effects of atypicals on sensory gating deficits, with clozapine treatment being associated with the only significant improvement in gating (Adler et al. 2004). The consistent finding that clozapine improves sensory gating suggests that an aspect of the drug's mechanism of action may be related to the pathophysiology of the deficit. Like other neuroleptics, clozapine is a dopamine D_2 receptor antagonist. It also has a high affinity for the serotonin 5-HT receptor family, where it has a potent inhibitory effect (Meltzer et al. 1989). We have recently explored further the involvement of the serotonergic system in gating by studying the effect of ondansetron, a highly selective $5\text{-}HT_3$ antagonist, on sensory gating in schizophrenia. We observed a highly significant improvement in P50 gating after ondansetron treatment, compared with placebo (Adler et al. 2005).

Antagonism of 5-HT receptors may improve auditory gating via any one of several inhibitory pathways. $5\text{-}HT_2$ receptors are found on GABAergic interneurons in the forebrain, including the hippocampus (Morilak et al. 1994), where a subpopulation of interneurons receives afferent innervation from brainstem serotonergic neurons (Freund et al. 1990). These interneurons may be involved in inhibitory sensory gating. $5\text{-}HT_3$ receptors are found on cholinergic neurons, where they have an inhibitory effect on the release of acetylcholine (Barnes et al. 1989). In rats, 5-HT$_3$ receptor blockade causes release of acetylcholine (Ramirez et al. 1996). The involvement of serotonin receptors in cholinergic systems is of particular interest, because direct stimulation of nicotinic cholinergic receptors also improves sensory gating in schizophrenia.

5.3 Nicotinic Mechanisms in Sensory Gating Dysfunction

Nicotine, administered via cigarettes or chewing gum, temporarily normalizes auditory gating deficits in schizophrenia patients and their first degree relatives (Fig. 5.2) (Adler et al. 1992, 1993). This effect is also observed in a rodent model, where invasive pharmacological techniques support nicotinic cholinergic involvement in sensory gating deficits. The rat P20-N40 wave, recorded from the cornu ammonis (CA)3 and 4 regions of the hippocampus, shows suppression of response to repeated stimuli similar to the human P50 wave. This response inhibition is lost after lesion of the fimbria-fornix, the pathway containing cholinergic afferents from the septal nuclei to the hippocampus (Vinogradova 1975; Harrison et al. 1988). Inhibition is also lost upon administration of α-bungarotoxin, an antagonist at α7 low affinity nicotinic receptors, but not with administration of mecamylamine, an antagonist of high-affinity nicotinic receptors (Luntz-Leybman et al. 1992). Methyllycaconitine, another antagonist at low affinity α7 nicotinic receptors, also abolishes auditory sensory gating in an awake, behaving rat model (Davies et al. 1999).

Additional evidence for the specific involvement of the low affinity nicotinic receptor in gating deficits comes from the DBA/21b inbred mouse strain, which shows low levels of α-bungarotoxin binding (Marks et al. 1989) and exhibits an inhibitory deficit in auditory gating similar to that seen in schizophrenia (Stevens et al. 1996). Furthermore, these deficits are normalized by nicotine and by GTS-21, a selective agonist of α7-nicotinic receptors (Stevens and Wear 1997; Stevens et al. 1998). α7 nicotinic receptors are located on GABAergic inhibitory interneurons (Freedman et al. 1993). Electrophysiological evidence suggests there are functional receptors on both the soma of almost all inhibitory interneurons, as well as on presynaptic terminals (Gray et al. 1996; Frazier et al. 1998). A possible mechanism for nicotinic-mediated improvement in sensory gating involves modulation of the inhibitory interneurons (Freedman et al. 2000).

Evidence for the involvement of nicotinic dysfunction in schizophrenia also comes from receptor expression and genetic studies. Human post-mortem binding studies using receptor autoradiography have demonstrated reduced binding to both high and low affinity nicotinic receptors in the hippocampus of schizophrenia patients (Freedman et al. 1995; Breese et al. 2000). Reduced binding to the low affinity α7 nicotinic receptor in schizophrenia patients has also been shown in the reticular nucleus of the thalamus (Court et al. 1999) and the cortex (Guan et al. 1999). Genetic linkage studies have shown that both the P50 auditory gating deficit and abnormalities in smooth pursuit eye movement are genetically linked in schizophrenia families to a chromosomal locus at 15q14, the locus containing the α7 nicotinic receptor gene (Freedman et al. 1997; Olincy et al. 1997). Linkage to schizophrenia was also found at this locus (Leonard et al. 1998; Freedman et al. 2001; Kaufmann et al. 1998; Riley et al. 2000; Stassen et al. 2000). In addition, multiple polymorphisms in the promoter region of the α7 gene have been identified that are associated with schizophrenia (Leonard et al. 2002).

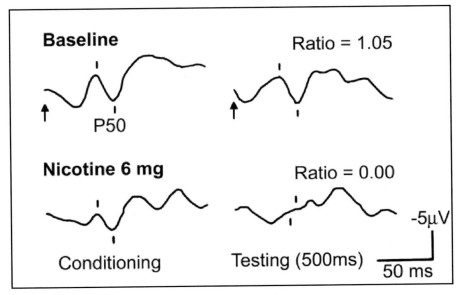

Fig. 5.2. Nicotine improves sensory gating deficits in schizophrenia. The P50 auditory evoked response to clicks 500 ms apart is indicated by the tick below the tracing. The ratio of the test amplitude to the conditioning amplitude is shown as a percentage at the right of each tracing. After Adler et al. 1993.

The above evidence of the involvement of nicotinic cholinergic pathology in schizophrenia, combined with the modulatory effects of nicotine on sensory deficits in the disease, suggest that a cholinergic mechanism may be involved in both disease pathology and sensory gating deficits. Because 5-HT receptors are located on cholinergic neurons and 5-HT antagonism causes acetylcholine release, we recently investigated the possibility that the effect of clozapine, a 5-HT antagonist, on sensory gating may involve a nicotinic cholinergic mechanism. We demonstrated that clozapine improves sensory gating in DBA/2 mice, which exhibit sensory gating deficits similar to schizophrenia. Furthermore, this dose-dependent effect was blocked by α -bungarotoxin, a low-affinity α7 nicotinic receptor antagonist, but not by dihydro-β-erythroidine, a high-affinity α4β2 nicotinic antagonist (Simosky et al. 2003). This suggests that clozapine's normalization of sensory gating in the rodent model may involve stimulation of α7 nicotinic receptors.

5.4 Smooth Pursuit Eye Movement Dysfunction in Schizophrenia

Nicotine also temporarily improves deficits in smooth pursuit eye movements, another commonly studied measure of inhibitory function in schizophrenia. In 1908, Diefendorf and Dodge first noticed abnormalities in smooth pursuit eye movements in patients with schizophrenia (Diefendorf and Dodge 1908). This phenomenon was rediscovered by Holzman in 1973 and has since become one of the most reproducible physiologic abnormalities associated with the disease (Levy et al. 1993). In addition, abnormalities during performance of a smooth pursuit task occur at high rates in relatives of patients with schizophrenia (Avila et al. 2003; Clementz et al. 1992; Iacono et al. 1992; Ross et al. 1998), suggesting that smooth pursuit deficits may be a marker for genetic risk for the disorder even in the absence of clinical symptoms (Levy et al. 1993; Rund and Landro 1990).

The smooth pursuit task involves the coordination of two neural systems: 1) the smooth pursuit system that matches the foveal area of the retina to the visual image of the target and 2) saccadic movements that rapidly bring the target image to the fovea. Many investigators have proposed that a primary feature of poor task performance in schizophrenia involves a failure to inhibit the saccade system. This failure results in the instruction of saccades during the pursuit task, an example of motor gating (Hommer et al. 1991; Rosenberg et al. 1997; Ross et al. 2002). This saccadic disinhibition may be part of generalized inhibitory dysfunction thought to be a core feature of schizophrenia (Ross et al. 2000).

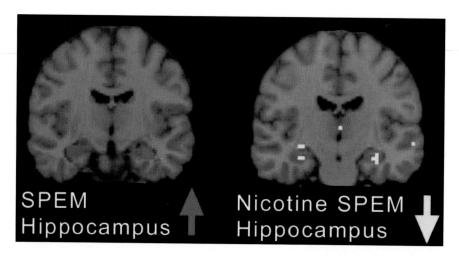

Fig. 5.3. Inhibitory dysfunction during a smooth pursuit eye movement task (SPEM) in schizophrenia. (Left) Increased activation in the right hippocampus during SPEM in patients with schizophrenia compared to controls. (Right) Decreased activation of the right hippocampus during SPEM following nicotine compared to placebo in schizophrenia.

Our group used fMRI to study brain responses associated with poor performance of a smooth pursuit eye movement task in schizophrenia (Tregellas et al.
2004). The main finding was greater activation of the hippocampus in schizophrenia subjects, compared to controls, during the task. Involvement of this region in
eye movement task has been demonstrated with single-unit recordings in primates
(Ringo et al. 1994; Sobotka et al. 1997) and with PET studies in humans (Hasebe
et al. 1999; Petit et al. 1993). Greater activation of this region may reflect inhibitory failure, consistent with much neuropathological and neurophysiological evidence for inhibitory dysfunction of these regions in the disease (Benes et al. 1998;
Freedman et al. 1994).

We also have studied the involvement of a nicotinic cholinergic mechanism in
eye movement deficits in schizophrenia. As in sensory gating, nicotine temporarily improves inhibitory function during the eye movement task in schizophrenia
(Olincy et al. 1998; Depatie et al. 2002; Sherr et al. 2002; Olincy et al. 2003; Avila
et al. 2003). We demonstrated that this improvement was associated with decreased activity in the hippocampus and thalamus, consistent with a normalization
of function in these brain regions (Fig. 5.3) (Tregellas et al. 2005). One mechanism for this could involve stimulation of nicotinic receptors on GABAergic inhibitory interneurons (Freedman et al. 2000).

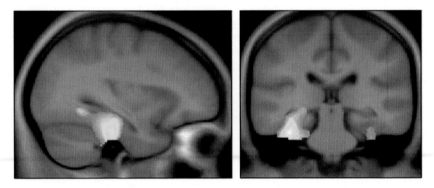

Fig. 5.4. Effect of α7 nicotinic receptor gene polymorphisms on gray matter volume.
Voxel-based morphometry reveals reduced gray matter volume in subjects with the less
common CHRNA7 promoter gene variants.

5.5 Effect of 7 Nicotinic Receptor Polymorphisms on Brain Structure

In light of converging physiological and genetic evidence for the involvement of
the α7 nicotinic receptor in the pathology of schizophrenia, we recently used voxel
based morphometry (VBM) to evaluate the effects of polymorphisms in the promoter region of the gene on differences in gray matter volume. Dividing a group

of schizophrenia patients and healthy comparison subjects into those with the less common variants in the promoter of CHRNA7 and those with the more common variant, we found that subjects with the less common variant (N = 5) showed reduced gray matter in the parahippocampus, hippocampus, and inferior temporal cortices, regardless of diagnosis, compared to those with the common genotype (N = 24; Fig. 5.4).

Reduced gray matter volume in these regions in subjects with α7 nicotinic receptor gene alterations is interesting in several regards. Firstly, the hippocampus is a region of high α7 nicotinic receptor density (Breese et al. 1997). It seems reasonable, therefore, that genetic differences that may effect receptor expression could lead to altered brain structure, especially given the likely role of α7 nicotinic receptors in regulating cell proliferation (Quik et al. 1994; Codignola et al. 1996), neurite outgrowth (Chan and Quik 1993; Pugh and Berg 1994; Coronas et al. 2000) and apoptosis (Hory-Lee and Frank 1995; Berger et al. 1998). Secondly, considering evidence for the involvement of the hippocampus in sensory gating deficits (Freedman et al. 1991; Miller and Freedman 1993; Grunwald et al. 2003; Boutros et al. 2005), temporary normalization of the deficit by nicotine (Adler et al. 1992; 1993), and the effect of α7 nicotinic receptor gene polymorphisms on sensory gating (Freedman et al. 2003), it is possible that one aspect of the link between genetics and brain function in this region is expressed as reduced gray matter volumes in this region in people with the less common genetic polymorphisms. Because VBM is only sensitive to macroscopic gray matter differences, future non-MRI studies are needed to determine if differences observed with VBM reflect differences in neuronal density or size, neuropil, dendritic or axonal arborization or cortical thickness (Mechelli et al. 2005).

5.6 Conclusion

Each of the many genes associated with schizophrenia likely has multiple physiological and developmental effects. Convergence of these effects with environmental influences produces subtle, but significant cortical deficits that are thought to underlie the symptoms of schizophrenia. Currently, only dopaminergic abnormalities are a major target of treatment with antipsychotic drugs. Expansion of efforts to other neurobiological targets, such as nicotinic receptors, may ultimately improve treatment outcome.

References

Adler LE, Cawthra EM, Donovan KA, Harris JG, Nagamoto HT, Olincy A, Waldo MC (2005) Improved p50 auditory gating with ondansetron in medicated schizophrenia patients. Am J Psychiatry 162:386-388.

Adler LE, Hoffer LD, Wiser A, Freedman R (1993) Normalization of auditory physiology by cigarette smoking in schizophrenic patients. Am J Psychiatry 150:1856-1861.

Adler LE, Hoffer LJ, Griffith J, Waldo MC, Freedman R (1992) Normalization by nicotine of deficient auditory sensory gating in the relatives of schizophrenics. Biol Psychiatry 32:607-616.

Adler LE, Olincy A, Cawthra EM, McRae KA, Harris JG, Nagamoto HT, Waldo MC, Hall MH, Bowles A, Woodward L, Ross RG, Freedman R (2004) Varied effects of atypical neuroleptics on P50 auditory gating in schizophrenia patients. Am J Psychiatry 161:1822-1828.

Adler LE, Olincy A, Waldo M, Harris JG, Griffith J, Stevens K, Flach K, Nagamoto H, Bickford P, Leonard S, Freedman R (1998) Schizophrenia, sensory gating, and nicotinic receptors. Schizophr Bull 24:189-202.

Adler LE, Pachtman E, Franks RD, Pecevich M, Waldo MC, Freedman R (1982) Neurophysiological evidence for a defect in neuronal mechanisms involved in sensory gating in schizophrenia. Biol Psychiatry 17:639-654.

Arango C, Summerfelt A, Buchanan RW (2003) Olanzapine effects on auditory sensory gating in schizophrenia. Am J Psychiatry 160:2066-2068.

Avila MT, Sherr JD, Hong E, Myers CS, Thaker GK (2003) Effects of nicotine on leading saccades during smooth pursuit eye movements in smokers and non-smokers with schizophrenia. Neuropsychopharmacology 28:2184-2191.

Barnes JM, Barnes NM, Costall B, Naylor RJ, Tyers MB (1989) 5-HT$_3$ receptors mediate inhibition of acetylcholine release in cortical tissue. Nature 338:762-763.

Benes FM, Kwok EW, Vincent SL, Todtenkopf MS (1998) A reduction of non-pyramidal cells in sector CA2 of schizophrenics and manic depressives. Biol Psychiatry 44:88-97.

Berger F, Gage FH, Vijayaraghavan S (1998) Nicotinic receptor-induced apoptotic cell death of hippocampal progenitor cells. J Neurosci 18:6871-6881.

Boutros NN, Trautner P, Rosburg T, Korzyukov O, Grunwald T, Schaller C, Elger CE, Kurthen M (2005) Sensory gating in the human hippocampal and rhinal regions. Clin Neurophysiol 116:1967-1974.

Boutros NN, Zouridakis G, Overall J (1991) Replication and extension of P50 findings in schizophrenia. Clin Electroencephalogr 22:40-45.

Breese CR, Adams C, Logel J, Drebing C, Rollins Y, Barnhart M, Sullivan B, Demasters BK, Freedman R, Leonard S (1997) Comparison of the regional expression of nicotinic acetylcholine receptor alpha7 mRNA and [^{125}I]-alpha-bungarotoxin binding in human postmortem brain. J Comp Neurol 387:385-398.

Breese CR, Lee MJ, Adams CE, Sullivan B, Logel J, Gillen KM, Marks MJ, Collins AC, Leonard S (2000) Abnormal regulation of high affinity nicotinic receptors in subjects with schizophrenia. Neuropsychopharmacology 23:351-364.

Chan J, Quik M (1993) A role for the nicotinic alpha-bungarotoxin receptor in neurite outgrowth in PC12 cells. Neuroscience 56:441-451.

Clementz BA, Geyer MA, Braff DL (1997) P50 suppression among schizophrenia and normal comparison subjects: a methodological analysis. Biol Psychiatry 41:1035-1044.

Clementz BA, Grove WM, Iacono WG, Sweeney JA (1992) Smooth-pursuit eye movement dysfunction and liability for schizophrenia: implications for genetic modeling. J Abnorm Psychol 101:117-129.

Clementz BA, Geyer MA Braff DL (1998) Poor P50 suppression among schizophrenia patients and their first-degree biological relatives. Am. J. Psychiatry 155:1691-1694.

Codignola A, McIntosh JM, Cattaneo MG, Vicentini LM, Clementi F, Sher E (1996) alpha-Conotoxin imperialis I inhibits nicotine-evoked hormone release and cell proliferation in human neuroendocrine carcinoma cells. Neurosci Lett 206:53-56.

Coronas V, Durand M, Chabot JG, Jourdan F, Quirion R (2000) Acetylcholine induces neuritic outgrowth in rat primary olfactory bulb cultures. Neuroscience 98:213-219.

Court J, Spurden D, Lloyd S, McKeith I, Ballard C, Cairns N, Kerwin R, Perry R, Perry E (1999) Neuronal nicotinic receptors in dementia with Lewy bodies and schizophrenia: alpha-bungarotoxin and nicotine binding in the thalamus. J Neurochem 73:1590-1597.

Dalack GW, Becks L, Hill E, Pomerleau OF, Meador-Woodruff JH (1999) Nicotine withdrawal and psychiatric symptoms in cigarette smokers with schizophrenia. Neuropsychopharmacology 21:195-202.

Davies AR, Hardick DJ, Blagbrough IS, Potter BV, Wolstenholme AJ, Wonnacott S (1999) Characterisation of the binding of [^{3}H]methyllycaconitine: a new radioligand for labelling alpha 7-type neuronal nicotinic acetylcholine receptors. Neuropharmacology 38:679-690.

de Leon J, Dadvand M, Canuso C, White AO, Stanilla JK, Simpson GM (1995) Schizophrenia and smoking: an epidemiological survey in a state hospital. Am J Psychiatry 152:453-455.

Depatie L, O'Driscoll GA, Holahan AL, Atkinson V, Thavundayil JX, Kin NN, Lal S (2002) Nicotine and behavioral markers of risk for schizophrenia: a double-blind, placebo-controlled, cross-over study. Neuropsychopharmacology 27:1056-1070.

Diefendorf AR, Dodge R (1908) An experimental study of the ocular reaction on the insane from photographic records. Brain 31:451-489.

Eccles JD (1969) The Inhibitory Pathways of the Central Nervous System. University Press, Liverpool.

Erwin RJ, Turetsky BI, Moberg P, Gur RC, Gur RE (1998) P50 abnormalities in schizophrenia: relationship to clinical and neuropsychological indices of attention. Schizophr Res 33:157-167.

Frazier CJ, Rollins YD, Breese CR, Leonard S, Freedman R, Dunwiddie TV (1998) Acetylcholine activates an alpha-bungarotoxin-sensitive nicotinic current in rat hippocampal interneurons, but not pyramidal cells. J Neurosci 18:1187-1195.

Freedman R, Adams CE, Leonard S (2000) The alpha7-nicotinic acetylcholine receptor and the pathology of hippocampal interneurons in schizophrenia. J Chem Neuroanat 20:299-306.

Freedman R, Adler LE, Bickford P, Byerley W, Coon H, Cullum CM, Griffith JM, Harris JG, Leonard S, Miller C (1994) Schizophrenia and nicotinic receptors. Harv Rev Psychiatry 2:179-192.

Freedman R, Adler LE, Myles-Worsley M, Nagamoto HT, Miller C, Kisley M, McRae K, Cawthra E, Waldo M (1996) Inhibitory gating of an evoked response to repeated auditory stimuli in schizophrenic and normal subjects. Human recordings, computer simulation, and an animal model. Arch Gen Psychiatry 53:1114-1121.

Freedman R, Adler LE, Waldo MC, Pachtman E, Franks RD (1983) Neurophysiological evidence for a defect in inhibitory pathways in schizophrenia: comparison of medicated and drug-free patients. Biol Psychiatry 18:537-551.

Freedman R, Coon H, Myles-Worsley M, Orr-Urtreger A, Olincy A, Davis A, Polymeropoulos M, Holik J, Hopkins J, Hoff M, Rosenthal J, Waldo MC, Reimherr F, Wender P, Yaw J, Young DA, Breese CR, Adams C, Patterson D, Adler LE, Kruglyak L, Leonard S, Byerley W (1997) Linkage of a neurophysiological deficit in schizophrenia to a chromosome 15 locus. Proc Natl Acad Sci USA 94:587-592.

Freedman R, Hall M, Adler LE, Leonard S (1995) Evidence in postmortem brain tissue for decreased numbers of hippocampal nicotinic receptors in schizophrenia. Biol Psychiatry 38:22-33.

Freedman R, Leonard S, Gault JM, Hopkins J, Cloninger CR, Kaufmann CA, Tsuang MT, Farone SV, Malaspina D, Svrakic DM, Sanders A, Gejman P (2001) Linkage disequilibrium for schizophrenia at the chromosome 15q13-14 locus of the alpha7-nicotinic acetylcholine receptor subunit gene (CHRNA7). Am J Med Genet 105:20-22.

Freedman R, Olincy A, Ross RG, Waldo MC, Stevens KE, Adler LE, Leonard S (2003) The genetics of sensory gating deficits in schizophrenia. Curr Psychiatry Rep 5:155-161.

Freedman R, Waldo M, Bickford-Wimer P, Nagamoto H (1991) Elementary neuronal dysfunctions in schizophrenia. Schizophr Res 4:233-243.

Freedman R, Wetmore C, Stromberg I, Leonard S, Olson L (1993) Alpha-bungarotoxin binding to hippocampal interneurons: immunocytochemical characterization and effects on growth factor expression. J Neurosci 13:1965-1975.

Freund TF, Gulyas AI, Acsady L, Gorcs T, Toth K (1990) Serotonergic control of the hippocampus via local inhibitory interneurons. Proc Natl Acad Sci USA 87:8501-8505.

Ghisolfi ES, Prokopiuk AS, Becker J, Ehlers JA, Belmonte-de-Abreu P, Souza DO, Lara DR (2002) The adenosine antagonist theophylline impairs p50 auditory sensory gating in normal subjects. Neuropsychopharmacology 27:629-637.

Glassman AH, Helzer JE, Covey LS, Cottler LB, Stetner F, Tipp JE, Johnson J (1990) Smoking, smoking cessation, and major depression. JAMA 264:1546-1549.

Gray R, Rajan AS, Radcliffe KA, Yakehiro M, Dani JA (1996) Hippocampal synaptic transmission enhanced by low concentrations of nicotine. Nature 383:713-716.

Grunwald T, Boutros NN, Pezer N, von Oertzen J, Fernandez G, Schaller C, Elger CE (2003) Neuronal substrates of sensory gating within the human brain. Biol Psychiatry 53:511-519.

Guan ZZ, Zhang X, Blennow K, Nordberg A (1999) Decreased protein level of nicotinic receptor alpha7 subunit in the frontal cortex from schizophrenic brain. Neuroreport. 10:779-1782.

Harrison JB, Buchwald JS, Kaga K, Woolf NJ, Butcher LL (1988) 'Cat P300' disappears after septal lesions. Electroencephalogr Clin Neurophysiol 69:55-64.

Hasebe H, Oyamada H, Kinomura S, Kawashima R, Ouchi Y, Nobezawa S, Tsukada H, Yoshikawa E, Ukai K, Takada R, Takagi M, Abe H, Fukuda H, Bando T (1999) Human cortical areas activated in relation to vergence eye movements-a PET study. Neuroimage 10:200-208.

Hommer DW, Clem T, Litman R, Pickar D (1991) Maladaptive anticipatory saccades in schizophrenia. Biol Psychiatry 30:779-794.

Hory-Lee F, Frank E (1995) The nicotinic blocking agents d-tubocurare and alpha-bungarotoxin save motoneurons from naturally occurring death in the absence of neuromuscular blockade. J Neurosci 15:6453-6460.

Hughes JR, Hatsukami DK, Mitchell JE, Dahlgren LA (1986) Prevalence of smoking among psychiatric outpatients. Am J Psychiatry 143:993-997.

Iacono WG, Moreau M, Beiser M, Fleming JA, Lin TY (1992) Smooth-pursuit eye tracking in first-episode psychotic patients and their relatives. J Abnorm Psychol 101:104-116.

Jin Y, Bunney WE, Sandman CA, Patterson JV, Fleming K, Moenter JR, Kalali AH, Hetrick WP, Potkin SG (1998) Is P50 suppression a measure of sensory gating in schizophrenia? Biol Psychiatry 43:873-878.

Judd LL, McAdams L, Budnick B, Braff DL (1992) Sensory gating deficits in schizophrenia: new results. Am J Psychiatry 149:488-493.

Kathmann N, Engel RR (1990) Sensory gating in normals and schizophrenics: a failure to find strong P50 suppression in normals. Biol Psychiatry 27:1216-1226.

Kaufmann CA, Suarez B, Malaspina D, Pepple J, Svrakic D, Markel PD, Meyer J, Zambuto CT, Schmitt K, Matise TC, Harkavy Friedman JM, Hampe C, Lee H, Shore D, Wynne D, Faraone SV, Tsuang MT, Cloninger CR (1998) NIMH Genetics Initiative Millenium Schizophrenia Consortium: linkage analysis of African-American pedigrees. Am J Med Genet 81:282-289.

Leonard S, Gault J, Moore T, Hopkins J, Robinson M, Olincy A, Adler LE, Cloninger CR, Kaufmann CA, Tsuang MT, Faraone SV, Malaspina D, Svrakic DM, Freedman R (1998) Further investigation of a chromosome 15 locus in schizophrenia: analysis of affected sibpairs from the NIMH Genetics Initiative. Am J Med Genet 81:308-312.

Levy DL, Holzman PS, Matthysse S, Mendell NR (1993) Eye tracking dysfunction and schizophrenia: a critical perspective. Schizophr Bull 19:461-536.

Light GA, Geyer MA, Clementz BA, Cadenhead KS, Braff DL (2000) Normal P50 suppression in schizophrenia patients treated with atypical antipsychotic medications. Am J Psychiatry 157:767-771.

Louchart-de la Chapelle S, Nkam I, Houy E, Belmont A, Menard JF, Roussignol AC, Siwek O, Mezerai M, Guillermou M, Fouldrin G, Levillain D, Dollfus S, Campion D, Thibaut F (2005) A concordance study of three electrophysiological measures in schizophrenia. Am J Psychiatry 162:466-474.

Luntz-Leybman V, Bickford PC, Freedman R (1992) Cholinergic gating of response to auditory stimuli in rat hippocampus. Brain Res 587:130-136.

Marks MJ, Romm E, Campbell SM, Collins AC (1989) Variation of nicotinic binding sites among inbred strains. Pharmacol Biochem Behav 33:679-689.

McGhie A, Chapman J (1961) Disorders of attention and perception in early schizophrenia. *Br* J Med Psychol 34:103-116.

Mechelli A, Price C, Friston K, Ashburner J (2005) Voxel-Based Morphometry of the Human Brain: Methods and Applications. Current Medical Imaging Reviews 1:1-9.

Meltzer HY, Bastani B, Ramirez L, Matsubara S (1989). Clozapine: new research on efficacy and mechanism of action. Eur Arch Psychiatry Neurol Sci 238:332-339.

Miller CL, Freedman R (1993) Medial septal neuron activity in relation to an auditory sensory gating paradigm. Neuroscience 55:373-380.

Morilak DA, Somogyi P, Lujan-Miras R, Ciaranello RD (1994) Neurons expressing 5-HT$_2$ receptors in the rat brain: neurochemical identification of cell types by immunocytochemistry. Neuropsychopharmacology 11:157-166.

Myles-Worsley M (2002) P50 sensory gating in multiplex schizophrenia families from a Pacific island isolate. Am J Psychiatry 159:2007-2012.

Nagamoto HT, Adler LE, McRae KA, Huettl P, Cawthra E, Gerhardt G, Hea R, Griffith J (1999) Auditory P50 in schizophrenics on clozapine: improved gating parallels clinical improvement and changes in plasma 3-methoxy-4-hydroxyphenylglycol. Neuropsychobiology 39:10-17.

Olincy A, Johnson LL, Ross RG (2003) Differential effects of cigarette smoking on performance of a smooth pursuit and a saccadic eye movement task in schizophrenia. Psychiatry Res 117:223-236.

Olincy A, Ross RG, Leonard S, Freedman R (1997) Preliminary linkage of eye movement abnormalities in schizophrenia. Biol.Psychiatry 14S.

Olincy A, Ross RG, Young DA, Roath M, Freedman R (1998) Improvement in smooth pursuit eye movements after cigarette smoking in schizophrenic patients. Neuropsychopharmacology 18:175-185.

Petit L, Orssaud C, Tzourio N, Salamon G, Mazoyer B, Berthoz A (1993) PET study of voluntary saccadic eye movements in humans: basal ganglia-thalamocortical system and cingulate cortex involvement. J Neurophysiol 69:1009-1017.

Pugh PC, Berg DK (1994) Neuronal acetylcholine receptors that bind alpha-bungarotoxin mediate neurite retraction in a calcium-dependent manner. J Neurosci 14:889-896.

Quik M, Chan J, Patrick J (1994) alpha-Bungarotoxin blocks the nicotinic receptor mediated increase in cell number in a neuroendocrine cell line. Brain Res 655:161-167.

Ramirez MJ, Cenarruzabeitia E, Las Heras B, Del Rio J (1996) Involvement of GABA systems in acetylcholine release induced by 5-HT3 receptor blockade in slices from rat entorhinal cortex. Brain Res 712:274-280.

Raux G, Bonnet-Brilhault F, Louchart S, Houy E, Gantier R, Levillain D, Allio G, Haouzir S, Petit M, Martinez M, Frebourg T, Thibaut F, Campion D (2002) The -2 bp deletion in exon 6 of the 'alpha 7-like' nicotinic receptor subunit gene is a risk factor for the P50 sensory gating deficit. Mol Psychiatry 7:1006-1011.

Riley BP, Makoff A, Mogudi-Carter M, Jenkins T, Williamson R, Collier D, Murray R (2000) Haplotype transmission disequilibrium and evidence for linkage of the CHRNA7 gene region to schizophrenia in Southern African Bantu families. Am J Med Genet 96:196-201.

Ringo JL, Sobotka S, Diltz MD, Bunce CM (1994) Eye movements modulate activity in hippocampal, parahippocampal, and inferotemporal neurons. J Neurophysiol 71:1285-1288.

Rosenberg DR, Sweeney JA, Squires-Wheeler E, Keshavan MS, Cornblatt BA, Erlenmeyer-Kimling L (1997) Eye-tracking dysfunction in offspring from the New York High-Risk Project: diagnostic specificity and the role of attention. Psychiatry Res 66:121-130.

Ross RG, Olincy A, Harris JG, Radant A, Adler LE, Freedman R (1998) Anticipatory saccades during smooth pursuit eye movements and familial transmission of schizophrenia. Biol Psychiatry 44:690-697.

Ross RG, Olincy A, Harris JG, Sullivan B, Radant A (2000) Smooth pursuit eye movements in schizophrenia and attentional dysfunction: adults with schizophrenia, ADHD, and a normal comparison group. Biol Psychiatry 48:197-203.

Ross RG, Olincy A, Mikulich SK, Radant AD, Harris JG, Waldo M, Compagnon N, Heinlein S, Leonard S, Zerbe GO, Adler L, Freedman R (2002) Admixture analysis of smooth pursuit eye movements in probands with schizophrenia and their relatives suggests gain and leading saccades are potential endophenotypes. Psychophysiology 39:809-819.

Rund BR, Landro NI (1990) Information processing: a new model for understanding cognitive disturbances in psychiatric patients. Acta Psychiatr Scand 81:305-316.

Sherr JD, Myers C, Avila MT, Elliott A, Blaxton TA, Thaker GK (2002) The effects of nicotine on specific eye tracking measures in schizophrenia. Biol Psychiatry 52:721-728.

Siegel C, Waldo M, Mizner G, Adler LE, Freedman R (1984) Deficits in sensory gating in schizophrenic patients and their relatives. Evidence obtained with auditory evoked responses. Arch Gen Psychiatry 41:607-612.

Sobotka S, Nowicka A, Ringo JL (1997) Activity linked to externally cued saccades in single units recorded from hippocampal, parahippocampal, and inferotemporal areas of macaques. J Neurophysiol 78:2156-2163.

Stassen HH, Bridler R, Hagele S, Hergersberg M, Mehmann B, Schinzel A, Weisbrod M, Scharfetter C (2000) Schizophrenia and smoking: evidence for a common neurobiological basis? Am J Med Genet 96:173-177.

Stevens KE, Freedman R, Collins AC, Hall M, Leonard S, Marks MJ, Rose GM (1996) Genetic correlation of inhibitory gating of hippocampal auditory evoked response and alpha-bungarotoxin-binding nicotinic cholinergic receptors in inbred mouse strains. Neuropsychopharmacology 15:152-162.

Stevens KE, Kem WR, Mahnir VM, Freedman R (1998) Selective alpha7-nicotinic agonists normalize inhibition of auditory response in DBA mice. Psychopharmacology (Berl) 136:320-327.

Stevens KE, Wear KD (1997) Normalizing effects of nicotine and a novel nicotinic agonist on hippocampal auditory gating in two animal models. Pharmacol Biochem Behav 57:869-874.

Strand JE, Nyback H (2005) Tobacco use in schizophrenia: a study of cotinine concentrations in the saliva of patients and controls. Eur Psychiatry 20:50-54.

Thoma RJ, Hanlon FM, Moses SN, Edgar JC, Huang M, Weisend MP, Irwin J, Sherwood A, Paulson K, Bustillo J, Adler LE, Miller GA, Canive JM (2003) Lateralization of auditory sensory gating and neuropsychological dysfunction in schizophrenia. Am J Psychiatry 160:1595-1605.

Tregellas JR, Tanabe JL, Martin LF, Freedman R (2005) FMRI of response to nicotine during a smooth pursuit eye movement task in schizophrenia. Am J Psychiatry 162:391-393.

Tregellas JR, Tanabe JL, Miller DE, Ross RG, Olincy A, Freedman R (2004) Neurobiology of smooth pursuit eye movement deficits in schizophrenia: an fMRI study. Am J Psychiatry 161:315-321.

Vinogradova O (1975) Functional Organization of the limbic system in the process of registration of information. R.L.Issacson and K.H.Pribram (Eds.), *The hippocampus: Neurophysiology and Behavior*. Plenum Press, New York, NY, pp. 3-69.

Waldo MC, Adler LE, Freedman R (1988) Defects in auditory sensory gating and their apparent compensation in relatives of schizophrenics. Schizophr Res 1:19-24.

Wegrzyn J, Wciorka J (2004) [P50 component of auditory evoked potentials in persons with schizophrenia and their first degree relatives]. Psychiatr Pol 38:395-408.

Yee CM, Nuechterlein KH, Morris SE, White PM (1998) P50 suppression in recent-onset schizophrenia: clinical correlates and risperidone effects. J Abnorm Psychol 107:691-698.

Chapter 6

A Role for Glutamate Receptors, Transporters, and Interacting Proteins in Cortical Dysfunction in Schizophrenia

Deborah Bauer, Robert E. McCullumsmith and James
H. Meador-Woodruff

Abstract. While there are myriad genetic, environmental, and epigenetic risk factors for schizophrenia, there is broad consensus that abnormal glutamate neurotransmission has a role in the ultimate clinical manifestations of this devastating illness. The glutamate hypothesis has evolved beyond the simplistic notion of NMDA receptor dysfunction, based in part on the dearth of consistently reproducible findings for altered NMDA receptor and binding site expression in postmortem studies. In this review, we review cortical glutamate receptor expression in this illness, and highlight postmortem data from novel families of glutamate receptor interacting proteins. We also discuss the vesicular and plasma membrane glutamate transporters, and review studies that have examined cortical glutamate transporter expression in this illness. Finally, we discuss the hypothesis that trafficking, assembly of signaling complexes, and receptor recycling might be dysregulated and represent important new targets for pharmacological manipulation in the treatment of schizophrenia.

6.1 Cortical Dysfunction in Schizophrenia

Converging lines of evidence implicate cortical dysfunction in schizophrenia (reviewed in Manoach 2003; Tamminga et al. 1992; Quintana et al. 2004; Liu et al 2006; Utsunomiya-Tate et al. 1996; Lehre et al. 1995; Hisano 2003; Varoqui et al. 2002). Cognitive deficits and other negative symptoms of schizophrenia are postulated to be the result of abnormalities in frontal and temporal cortices. Abnormalities of working memory, executive function, facial recognition and other cortical functions attributed to these regions have been found by a growing number of functional neuroimaging studies in this illness (Andreasen et al. 1998; Weinberger et al. 2001; Manoach 2003). Normal cortical function relies, in part, on carefully balanced glutamatergic (excitatory) and GABAergic (inhibitory) circuitry, strengthening or weakening of individual synapses, and modulatory input from myriad neurotransmitters and paracrine factors. Since these molecular mechanisms underlie normal cortical activity, they are high yield targets for investigation of cortical abnormalities in schizophrenia. In this review, we focus on one element of cortical pathophysiology in schizophrenia: postmortem studies that support the hypothesis of abnormal cortical glutamatergic neurotransmission in this illness.

6.2 Glutamate in Schizophrenia

For decades, schizophrenia research has focused on the dopamine hypothesis of schizophrenia, which postulates that dysregulated dopaminergic neurotransmission is a key feature of the pathophysiology of the illness. The dopamine hypothesis is based on the observation that antipsychotic efficacy for positive symptoms is associated with D2 dopamine receptor blockade. Although numerous studies point to dopaminergic abnormalities in schizophrenia, dopamine dysfunction does not completely account for all of the symptoms seen in schizophrenia, since antipsychotics typically are effective only for the positive symptoms of the illness, while negative symptoms and cognitive deficits are relatively refractory to treatment (Joyce and Meador-Woodruff 1997; Laruelle et al. 1999). Consequently, alternative neurotransmitter systems that may also be involved in the pathophysiology of schizophrenia have been sought, and a growing body of evidence now implicates glutamatergic dysfunction in this illness. The strongest evidence is that phencyclidine (PCP) and similar compounds, which are uncompetitive antagonists of the NMDA receptor, can induce both the positive and negative symptoms of schizophrenia, including cognitive deficits (Javitt and Zukin 1991; Tamminga 1999). Moreover, these compounds can exacerbate both positive and negative symptoms in schizophrenia (Lahti et al. 1995). Chronic administration of PCP-like compounds reduces frontal lobe blood flow and glucose utilization, which is similar to the "hypofrontality" described in schizophrenia (Hertzmann et al. 1990). More specifically, the most widely held hypotheses posit diminished NMDA receptor function in limbic brain structures, including the prefrontal cortex. There is compelling evidence for altered glutamatergic neurotransmission in the cortex in schizophrenia.

6.3 Glutamate Neurotransmission

The release, activity as a ligand, and reclamation of glutamate involves three distinct cell types: the astrocyte, presynaptic neuron and postsynaptic neuron (Salt and Eaton 1996). In the presynaptic neuron, glutamine can be converted to glutamate by the enzyme glutaminase, and packaged into vesicles by a family of vesicular glutamate transporters (VGLUT1-3) for release into the synapse (Bellocchio et al. 2000; Takamori et al. 2000). Glutamate may also be synthesized by the hydrolysis of N-acetyl-alpha-L-aspartyl-L-glutamate (NAAG) into NAA and glutamate by the enzyme carboxypeptidase II (GCP II)(Ghose et al. 2004). Once released into the synapse, glutamate may occupy and activate ionotropic (NMDA, AMPA, and kainate) or metabotropic (mGluR1-8) glutamate receptors located on both neurons and astrocytes (Hollmann and Heinemann 1994). Rapid removal of glutamate from the synapse is facilitated by a family of plasma membrane excitatory amino acid transporters (EAATs), generally localized to postsynaptic neurons and astrocytes (Masson et al. 1999). Recovered glutamate may enter the citric acid cycle via conversion to alpha-ketoglutarate by glutamate dehydrogenase, be converted to glutamine by glutamine synthetase and transported back into the synapse, or be released into the extracellular space by a cystine/ glutamate antiporter (Salt and Eaton 1996; Bassi et al. 2001; Kim et al. 2001). Finally, several families of novel glutamate receptor and transporter associated molecules mediate integration of intracellular signaling and glutamate reuptake (Jackson et al. 2001; Lin et al. 2001; Marie et al. 2002; Watanabe et al. 2003).

6.4 Glutamate Receptors

6.4.1 Ionotropic and Metabotropic Glutamate Receptors

There are four classes of functionally and pharmacologically distinct glutamate receptors. The ionotropic glutamate receptors, AMPA, kainate, and NMDA are comprised of four or five subunits that form ligand-gated ion channels, while the metabotropic glutamate receptors (mGluRs) are seven transmembrane domain, G-protein coupled receptors (Hollmann and Heinemann 1994; Wheal and Thomson 1995). The AMPA receptor subunits are derived from a family of four genes termed GluR1-GluR4 that confer heterogeneity in assembled AMPA receptors by alternative splicing and post-translational editing (Hollmann and Heinemann 1994; Wheal and Thomson 1995). Assembled AMPA receptors contain discrete binding sites for glutamate, competitive antagonists such as CNQX, and desensitization modulators such as aniracetam. Kainate receptors are also ligand gated ion channels composed of subunits derived from genes for the low affinity GluR5-GluR7 and high affinity KA1-KA2 subunits (Hollmann and Heinemann 1994; Wheal and Thomson 1995). These subunits also undergo alternative splicing and post-translational editing. Assembled kainate receptors may be composed of five identical subunits (homomers) or composed of low and high affinity subunits. The NMDA receptor subunits are encoded by six genes termed NR1, NR2A-NR2D, and NR3. NR1 is expressed as one of eight isoforms, due to the alternative splicing of exons 5, 21, and 22 (Nakanishi 1992; Durand et al. 1993; Hollmann and Heinemann 1994; Wheal and Thomson 1995). NMDA receptors exhibit subunit and splice variant specific properties, and pharmacological regulation of this receptor depends on the unique combination of glutamate, glycine/D-serine, polyamine, H^+, Zn^{2+}, and Mg^{2+} binding sites (Hollmann and Heinemann 1994; Wheal and Thomson 1995). In addition, there is an intrachannel binding site for uncompetitive antagonists of the NMDA receptor, such as PCP, ketamine, and MK801. The eight metabotropic glutamate receptors (mGluRs) are divided into three groups (I, II, and III) based on pharmacology, sequence homology, and which signal transduction pathways they activate in heterologous systems (Nakanishi 1992; Prezeau et al. 1994; Schoepp 1994; Pin and Duvoisin 1995; Corti et al. 1997; Saugstad et al. 1997; Corti et al. 1998). The mGluRs belong to a unique subset of G-protein coupled receptors with seven transmembrane domains and large extracellular amino terminal. Group I mGluRs have been shown to stimulate phospholipase C, phosphoinositide hydrolysis, and cAMP formation, while group II and III mGluRs inhibit forskolin-stimulated cAMP formation and adenylyl cyclase, possibly via a G protein (Aramori and Nakanishi 1992; Pin et al. 1992; Pickering et al. 1993; Schoepp 1994; Joly et al. 1995; McCool et al. 1996; Corti et al. 1997; Saugstad et al. 1997).

6.4.2 Glutamate Receptor Interacting Proteins

AMPA receptors may be recruited to the synapse by three distinct mechanisms: a constitutive pool, a regulated pool, and a Golgi-derived newly synthesized pool (Contractor and Heinemann 2002). Cycling and integration of AMPA receptors to the postsynaptic density by these mechanisms involve a number of recently characterized molecules that possess AMPA receptor subunit-specific protein binding domains and are regulated via phosphorylation or palmitoylation (Contractor and Heinemann 2002; McGee and Bredt 2003). GRIP, PICK, NSF, SAP97 stargazin, and ABP-L have all been implicated in AMPA receptor trafficking, and thus likely have critical roles in long term potentiation (LTP) (Contractor and Heinemann 2002; McGee and Bredt 2003). Receptor-associated molecules have also been identified for the NMDA receptor, including NF-L, SAP102, yotiao, PSD95, and PSD93 (Contractor and Heinemann 2002; McGee and Bredt 2003). Some of these molecules specifically bind C-terminal consensus sequences called PDZ domains, named for three proteins with this motif: PSD95, *Drosophila* disc-large tumor suppressor gene (Dlg-A) product, and ZO-1, a tight junction protein (Ehlers et al. 1998; Lin et al. 1998; Sheng and Pak 2000). PDZ and related binding domains link neurotransmitter receptors with kinases, phospholipases, and other signal transduction and receptor trafficking pathways.

Several of the AMPA interacting molecules, including GRIP1 and PICK1 (Toyooka et al. 2002; Dracheva et al. 2005a) as well as calcineurin (Kozlovsky et al. 2006), also interact with the kainate receptors. In addition, a family of molecules that interacts with the metabotropic receptors has been identified, including RGS4 (Mirnics et al. 2001; Erdely et al. 2006; Lipska et al. 2006) and Homer (Meador-Woodruff 2005).

It is notable that the glutamate receptor interacting proteins modulate myriad processes related to synthesis, trafficking, insertion, activation, recycling, and degradation of glutamate receptors. Thus, the expression and regulation of this family of molecules has a critical role in cortical glutamatergic synapses.

Table 6.1. NMDA receptor expression in schizophrenia.

Technique	Probe(s)*	Finding	Brain Region	Reference
Level of Gene Expression: Receptor binding studies				
homogenate binding	MK-801	unchanged	FC	Kornhuber et al 1989
homogenate binding	L-689,560	↑	TC	Grimwood et al 1999
	L-689,560	unchanged	MC	
	CGP 39653	unchanged	TC, MC	
	ifenprodil	↑	TC	
	ifenprodil	unchanged	MC	
homogenate binding	L-689,560	↑	TC	(Nudmamud and Reynolds, 2001)
	L-689,560	unchanged	PFC	
homogenate binding	glycine	↑	cerebral cortex	(Ishimaru et al., 1994)
autoradiography	MK-801	↑	ACC	(Zavitsanou et al., 2002)
autoradiography	MK-801	unchanged	PFC	(Noga et al., 2001)
autoradiography	MK-801	↑	PCC	(Newell et al., 2005)
autoradiography	MK-801	unchanged	DLPFC	(Scarr et al., 2005)
	CGP 39653	unchanged	DLPFC	
Level of gene expression: Subunit mRNA expression				
ISH	NR2D	↑	PFC	(Akbarian et al., 1996)
	NR1, NR2A-C	unchanged	PFC	
	NR1, NR2A-D	unchanged	PTC	
ISH	NR1 w/exon 5	↑	FC, OCC, TC	(Le Corre et al., 2000)
ISH	NR3A	↑	PFC	(Mueller and Meador-Woodruff, 2004)
		unchanged	ITC	
Northern blot	NR1	↓	TC	(Humphries et al., 1996)

qPCR	NR1	↓[#]	FC	(Sokolov, 1998)
	NR1	unchanged[@]	FC	
qPCR	NR1, NR2A	↑	PFC, OCC	(Dracheva et al., 2001)
	NR2B	unchanged	PFC, OCC	
Double ISH	NR2A/GAD67	unchanged	PFC	(Woo et al., 2004)
Level of gene expression: Sub-unit protein expression				
Western blot	NR1$^{C2'}$	↑	ACC	(Kristiansen et al., 2006)
	NR1$^{C2'}$	unchanged	DLPFC	
	NR1^{C2} NR2A-D	unchanged	ACC	
	NR1^{C2}, NR2A-D	unchanged	DLPFC	

*All binding studies utilized ^{3}[H]. #Neuroleptic free schizophrenics vs. controls. Neuroleptic treated schizophrenics vs. control. Abbreviations: in situ hybridization (ISH), quantitative polymerase chain reaction (qPCR) anterior cingulated cortex (ACC), dorsolateral prefrontal cortex (DLPFC), prefrontal cortex (PFC), occipital cortex (OCC), visual cortex (VC), motor cortex (MC), frontal cortex (FC), parietal cortex (PC), temporal cortex (TC), inferior temporal cortex (ITC), parietotemporal cortex (PTC), posterior cingulate cortex (PCC), (NR1C2') variably spliced NR1exon cassettes C2 and C2'.

6.5 Cortical Glutamate Receptor Abnormalities in Schizophrenia

Expression of NMDA receptor subunit transcripts and binding sites has been evaluated in the brain in schizophrenia (Table 6.1). Studies on the expression of the ionotropic glutamate receptors in the brain in psychiatric illnesses have often focused on schizophrenia, and have concentrated on cortical and medial temporal lobe structures. While there are discrepancies between some of these studies, alterations in NMDA receptor subunit and binding site expression are complex and region specific (Table 6.1). There is evidence for shifts in subunit stoichiometry and increased binding to at least some of the NMDA binding sites, primarily in cortical areas (Kornhuber et al. 1989; Akbarian et al. 1996; Humphries et al. 1996; Sokolov 1998; Grimwood et al. 1999; Dracheva et al. 2001).

The AMPA receptor has been extensively studied at multiple levels of gene expression in the brain in schizophrenia, as summarized in Table 6.2. Data from these studies result in one of the more robust and reproducible sets of findings in this field; many past studies have found that AMPA receptor expression is decreased in the hippocampus and related structures in schizophrenia, occurring at the levels of both transcript and subunit protein expression (Kerwin et al. 1990; Harrison et al. 1991; Breese et al. 1995; Eastwood et al. 1995; Eastwood et al.

1997a,b). With few exceptions, AMPA receptor expression tends to be unchanged in other cortical areas (Freed et al. 1993; Breese et al. 1995; Eastwood et al. 1995, 1997b; Sokolov 1998; Beneyto and Meador-Woodruff 2006).

Similar to the AMPA receptor, kainate receptor expression and binding has also been evaluated in cortical regions in schizophrenia. Unlike the AMPA receptor, findings for the kainate receptor are contradictory, with decreased binding reported in two studies, increased binding reported in three studies, and no changes reported in another study . In addition, one study found increased GluR7 mRNA in the prefrontal cortex (PFC) while another found decreased GluR7 in the frontal cortex (Meador-Woodruff et al. 2001). These disparate results might be due to differences in which PFC regions were examined and/or differences between brain collections.

Table 6.2. AMPA receptor expression in schizophrenia.

Technique	Probe(s)*	Finding	Brain Region	Reference
Level of Gene Expression: Receptor binding studies				
autoradiography	CNQX	↑	PFC	(Noga et al., 2001)
homogenate binding	AMPA	unchanged	FC	(Freed et al., 1993)
autoradiography	AMPA	↑	ACC	(Zavitsanou et al., 2002)
homogenate binding	AMPA	unchanged	FC, PC, OCC, TC, LC	(Kurumaji et al., 1992)
autoradiography	AMPA	unchanged	PCC	(Newell et al., 2005)
autoradiography	AMPA	unchanged	DLPFC	(Scarr et al., 2005; Beneyto and Meador-Woodruff, 2006)
Level of Gene Expression: Subunit mRNA expression				
qPCR	GluR1	↓#	FC	(Sokolov, 1998)
	GluR1	unchanged@	FC	
qPCR	GluR1, GluR4	↑	DLPFC	(Dracheva et al., 2005b)
	GluR2, GluR3	unchanged	DLPFC	
	GluR4	↑	OCC	
	GluR1-3	unchanged	OCC	
Microarray	GluR2	↓	PFC	(Vawter et al., 2002)
ISH	GluR1, GluR3	unchanged	DLPFC	(Beneyto and Meador-Woodruff, 2006)

	GluR2, GluR4	✦		

Level of Gene Expression:
Subunit protein expression

Western blot	GluR2, GluR3	unchanged	cingulate cortex	(Breese et al., 1995)

*All binding studies utilized ³[H]. #Neuroleptic free schizophrenics vs. controls. @Neuroleptic treated schizophrenics vs. control. Abbreviations: in situ hybridization (ISH), quantitative-polymerase chain reaction (qPCR) anterior cingulated cortex (ACC), dorsolateral prefrontal cortex (DLPFC), prefrontal cortex (PFC), frontal cortex (FC), parietal cortex (PC), occipital cortex (OCC), temporal cortex (TC), limbic cortex (LC), posterior cingulate cortex (PCC)

Table 6.3. Summary of kainate receptor binding and expression data in schizophrenia.

Technique	Probe(s)*	Finding	Location	Reference
Level of Gene Expression: Receptor Binding Sites				
autoradiography	KA	✦	PHG	
homogenate binding	KA	✦	PFC	
homogenate binding	KA	✦	FC	
	KA	unchanged	TC	
homogenate binding	KA	✦	PFC	
autoradiography	KA	✦	PFC	(Meador-Woodruff et al., 2001)
autoradiography	KA	unchanged	PCC	(Newell et al., 2005)
autoradiography	KA	✦	DLPFC	(Scarr et al., 2005)
autoradiography	KA	unchanged	ACC	(Zavitsanou et al., 2002)
Level of Gene Expression: Subunit mRNA Expression				
ISH	GluR7	✦	PFC	(Meador-Woodruff et al., 2001)
	KA2	✦	PFC	
	GluR5,6, KA1	unchanged	PFC	
RT-PCR	GluR7	✦#	FC	
	GluR7	unchanged@	FC	
	KA1	✦#	FC	
	KA1	unchanged@	FC	

[*]All binding studies utilized ^{3}[H]. [#]Neuroleptic free schizophrenics vs. controls. [@]Neuroleptic treated schizophrenics vs. control. Abbreviations: kainate (KA), in situ hybridization (ISH), reversed transcribed-polymerase chain reaction (RT-PCR), immunoreactive dendrites (IR) anterior cingulate cortex (ACC), dorsolateral prefrontal cortex (DLPFC), prefrontal cortex (PFC), frontal cortex (FC), occipital cortex (OCC), temporal cortex (TC), parahippocampal gyrus (PHG), posterior cingulate cortex (PCC)

Table 6.4. Summary of metabotropic glutamate receptor binding and expression data in schizophrenia.

Technique	Probe(s)	Finding	Location	Reference
Level of Gene Expression: Subunit mRNA expression				
Western	mGluR1a, mGluR2/3	▲	PFC	(Gupta et al., 2005)
	mGluR4a, mGluR5	unchanged	PFC	
ISH	mGluR3	unchanged	PFC	
	mGluR5	▲	PFC	

Abbreviations: in situ hybridization (ISH), prefrontal cortex (PFC)

Only two studies have examined mGluR expression in the cortex in schizophrenia. Increased mGluR1a and mGluR2/3 immunoreactivity was found in Brodmann areas 9, 11, 32, and 46 in schizophrenia (Gupta et al. 2005), while another group found increased mGluR5 mRNA in the PFC in this illness .

Most abnormalities of glutamate receptor expression in schizophrenia have been reported in limbic regions including the dorsolateral prefrontal cortex (DLPFC) and anterior cingulate cortex (ACC). With a few exceptions, findings of altered glutamate receptor expression in schizophrenia are typically region and brain collection specific, with many studies yielding conflicting results. Given this, we suggest that considering only the direction and magnitude of glutamate receptor expression will not lead to a full understanding of the complexities of glutamate receptor dysfunction in this devastating illness; other aspects need to be studied to fully appreciate how this neurotransmitter system is abnormal in schizophrenia.

6.6 Cortical Glutamate Receptor interacting Protein Abnormalities in Schizophrenia

A number of studies have moved beyond measuring receptors in the plasma membrane to examine the molecules which functionally link glutamate receptors to cellular processes. Published data on receptor interacting proteins revealed that cortical regions are differentially affected in schizophrenia (Table 6.5). For example, ABP mRNA was increased in the occipital cortex (OCC) but not DLPFC, changes in PSD93 mRNA were detected in the ACC but not DLPFC, and changes in NF-L mRNA were detected in DLPFC but not ACC. RGS4 protein was decreased in the frontal cortex but not insular cortex, and transcript was decreased in the cingulate gyrus, superior frontal gyrus, insular cortex, visual cortex, and motor cortex, but not inferior frontal sulcus or inferior frontal gyrus. Such regional specificity is not surprising, given that some functional deficits in schizophrenia may be attributed to specific cortical regions (Tamminga et al. 1992; Silbersweig et al. 1995; Andreasen et al. 1996; Humphries et al. 1996; Andreasen et al. 1997; Callicott et al. 2000; Perlstein et al. 2001; Manoach 2003; Quintana et al. 2004). These data support the hypothesis of regional specificity for cortical glutamatergic abnormalities in schizophrenia, and may suggest alterations in AMPA receptor trafficking (due to changes in ABP), NMDA receptor signaling (PSD93), NMDA receptor signaling complex assembly (NF-L), and mGluR activation (RGS4).

Another observation is that changes in transcript expression do not necessarily correspond to changes in protein expression. Our group found increased transcripts but decreased protein for PSD95, PSD93, and NF-L expression in the PFC (Kristiansen et al. 2006), while two other studies reported decreased NSF mRNA and increased NSF protein expression in the DLPFC (Mirnics et al. 2000; Prabakaran et al. 2004). In addition, no changes in SAP97 transcript expression were detected in the DLPFC or OCC (Dracheva et al. 2005a), but decreased SAP97 protein was detected in a different sample in the DLPFC (Toyooka et al. 2002) and similarly divergent results were found for cortical GRIP1 expression. There are several possibilities that might explain this observation. If the opposing changes are occurring in the same cells, these data may suggest complex regulation of gene expression. Changes in mRNA expression could be compensatory to altered protein levels. For example, if a gene is abnormally translated or is rapidly degraded, lower levels of protein might lead to feedback mechanisms that signal for increased transcription.

Another possibility is that these changes could be in different cell populations. Proteins that are presynaptic might be expressed in different regions than their transcripts, because the transcripts are typically located in the cell soma, while the projections (where some of the translated protein is located) may be in an entirely different cortical or subcortical region. Therefore, it is possible that an increase in transcript in one region could be accompanied by a decrease in protein (from efferent projections) in that same region. This is a less likely explanation for generally postsynaptic molecules such as PSD-95. It is also possible that different cell types within the same region have differing profiles with respect to the expression of the same molecule. For example, for the same molecule, transcript expression could be increased in pyramidal neurons and protein expression could be decreased in GABAergic interneurons. While quantitative cell level protein studies are currently not feasible in postmortem tissue, cell level transcripts studies have been performed for some glutamate receptor subunits and interacting proteins (Beneyto and Meador-Woodruff 2006). Our group found decreased PICK1 and increased stargazin mRNAs in DLPFC in schizophrenia in large cortical layer III cells with pyramidal neuron-like morphology, suggesting that some changes in gene expression may be cell type specific (Beneyto and Meador-Woodruff 2006).

Another emerging finding for studies of expression of glutamate receptor interacting proteins is the variability of results in samples from different brain banks. Conflicting results for NSF, RGS4, and PSD95 expression in the same region and at the same level of gene expression have been reported (Mirnics et al. 2001; Prabakaran et al. 2004; Quintana et al. 2004; Zhang et al. 2005; Kristiansen et al. 2006; Lipska et al. 2006). There are several possible explanations for these differing results, including differences in methodology (quantitative polymerase chain reaction (QPCR) vs. *in situ* hybridization (ISH)) and differences in subject characteristics of these samples.

Table 6.5. Cortical glutamate receptor interacting protein anomalies in schizophrenia.

Receptor-Interacting Protein	Associated Receptor(s)	Level of gene expression	Technique	Finding	Cortical Region	Reference
PSD95	NMDA, AMPA, kainate	mRNA	ISH	↓	DLPFC	(Ohnuma et al., 2000b)
		mRNA	ISH	↑	ACC	(Kristiansen et al., 2006)
				unchanged	DLPFC	
		protein	Western	unchanged	DLPFC	(Toyooka et al., 2002)
			Western	↓	ACC	(Kristiansen et al., 2006)
				unchanged	DLPFC	
PSD93	NMDA	mRNA	ISH	↑	ACC	(Kristiansen et al., 2006)
				unchanged	DLPFC	
		protein	Western	↓	ACC	
				unchanged	DLPFC	
			Western	unchanged	DLPFC	(Toyooka et al., 2002)
NF-L	NMDA	mRNA	ISH	unchanged	ACC	(Kristiansen et al., 2006)
				↑	DLPFC	
		protein	Western	unchanged	ACC	
				↓	DLPFC	
SAP-102	NMDA, AMPA	mRNA	ISH	unchanged	ACC	(Kristiansen et al., 2006)
				unchanged	DLPFC	

Receptor-Interacting Protein	Associated Receptor(s)	Level of gene expression	Technique	Finding	Cortical Region	Reference
		protein	Western	unchanged	ACC	
				unchanged	DLPFC	
		protein	Western	unchanged	DLPFC	(Toyooka et al., 2002)
NSF	AMPA	mRNA	ISH microarray	↓	DLPFC	(Mirnics et al., 2000)
		mRNA	ISH	unchanged	DLPFC	(Beneyto and Meador-Woodruff, 2006)
		mRNA	QPCR	unchanged	DLPFC	(Imai et al., 2001)
		protein	Western	unchanged	DLPFC, FP, PC	(Gray et al., 2006)
		protein	2D-DIGE	↑	DLPFC	(Prabakaran et al., 2004)
		protein	Western	unchanged	DLPFC	(Imai et al., 2001)
SAP97	AMPA	mRNA	QPCR	unchanged	DLPFC	(Dracheva et al., 2005a)
				unchanged	OCC	
		protein	Western	↓	DLPFC	(Toyooka et al., 2002)
GRIP1	AMPA, kainate, mGluR	mRNA	QPCR	↑	DLPFC	(Dracheva et al., 2005a)
				↑	OCC	
		protein	Western	unchanged	DLPFC	(Toyooka et al., 2002)
PICK1	AMPA, kainate	mRNA	QPCR	unchanged	DLPFC	(Dracheva et al., 2005a)
				unchanged	OCC	
		mRNA	ISH	↓	DLPFC	(Beneyto and Meador-Woodruff, 2006)
ABP	AMPA	mRNA	QPCR	unchanged	DLPFC	(Dracheva et al., 2005a)
				↑	OCC	

Receptor-Interacting Protein	Associated Receptor(s)	Level of gene expression	Technique	Finding	Cortical Region	Reference
stargazin	AMPA	mRNA	ISH	↓	DLPFC	(Beneyto and Meador-Woodruff, 2006)
syntenin	AMPA	mRNA	ISH	unchanged	DLPFC	(Beneyto and Meador-Woodruff, 2006)
calcineurin	kainate	protein	Western	unchanged	DLPFC	(Kozlovsky et al., 2006)
RGS4	mGluR	mRNA	QPCR	unchanged	DLPFC	(Lipska et al., 2006)
		mRNA	ISH	↓ / unchanged	CG, SFG, IC / IFS, IFG	(Erdely et al., 2006)
		mRNA	Microarray, ISH	↓	DLPFC, VC, MC	(Mirnics et al., 2001)
		protein	Western	↓ / unchanged	FC / IC	(Erdely et al., 2006)
Homer1b	mGluR1/5	mRNA	ISH	unchanged	ACC, DLPFC	(Meador-Woodruff, 2005)
Homer2	mGluR1/5	mRNA	ISH	↑	ACC, DLPFC	(Meador-Woodruff, 2005)

Abbreviations: in situ hybridization (ISH), quantitative real time polymerase chain reaction (QPCR), two-dimensional fluorescence difference gel electrophoresis (2D-DIGE), anterior cingulated cortex (ACC), dorsolateral prefrontal cortex (DLPFC), prefrontal cortex (PFC), occipital cortex (OCC), visual cortex (VC), motor cortex (MC), cingulated gyrus (CG), superior frontal gyrus (SFG), insular cortex (IC), frontal cortex (FC), inferior frontal sulcus (IFS), inferior frontal gyrus (IFG), frontal pole (FP), parietal cortex (PC)

Glutamate receptor-interacting proteins have not been extensively examined at the genomic level in schizophrenia. While RGS4 has become a molecule of particular interest in schizophrenia because several early linkage studies implicated it as a schizophrenia susceptibly gene (Chowdari et al. 2002; Chen et al. 2004b; Morris et al. 2004; Williams et al. 2004; Zhang et al. 2005), later studies incorporating larger, more homogenous samples, as well as meta-analyses have failed to replicate these findings (Cordeiro et al. 2005; Sobell et al. 2005; Guo et al. 2006; Levitt et al. 2006; Li and He 2006; Liu et al. 2006; Rizig et al. 2006; Talkowski et al. 2006). Transcript and protein studies in the cortex have shown decreases or no changes in RGS4 in schizophrenia (Mirnics et al. 2001; Erdely et al. 2006; Lipska et al. 2006). Similar to RGS4, Homer has been studied at the genomic level, but Homer was not implicated in schizophrenia by genetic studies (Norton et al., 2003). At the mRNA level, however, increased transcript expression of Homer2, but not Homer1b, was detected in the DLPFC and ACC in schizophrenia (Meador-Woodruff 2005). Taken together with the functional roles attributed to RGS4 and Homer, these data suggest there may be abnormalities in mGluR receptor signaling in the cortex in schizophrenia.

Studies of glutamate receptor interacting proteins suggest that glutamatergic dysfunction in schizophrenia is not simply a problem of too many or too few receptors, but is likely a problem of how these receptors are assembled, transported, and functionally linked to receptor signaling complexes. Glutamate receptor dysfunction may represent a primary pathophysiological process in schizophrenia, with compensatory changes in other functionally linked processes in glutamate synapses. Next, we discuss the possible role of glutamate transporters in schizophrenia, and refine the hypothesis of glutamatergic dysfunction in schizophrenia to include these important elements of glutamate synapses.

6.7 Glutamate Transporters

6.7.1 Excitatory Amino Acid Transporters

Extracellular glutamate levels are maintained at very low levels by functionally distinct low and high affinity systems that facilitate glutamate reuptake. High affinity Na^+-dependent synaptic reuptake of glutamate and aspartate is mediated by a family of excitatory amino acid transporters (EAAT1-EAAT5), which are localized in the plasma membrane and transport glutamate into neurons and glia (Arriza et al. 1993; Utsunomiya-Tate et al. 1996). EAATs mediate glutamate transport by an electrogenic exchange of 3 Na^+, 1 H^+, and 1 glutamate molecule into the cell and 1 K^+ ion out of the cell, with the net inward movement of one positive charge (Zerangue and Kavanaugh 1996; Levy et al. 1998). EAATs are expressed natively as homomeric trimers comprised of non-covalently linked subunits that have 6-10 transmembrane domains (Haugeto et al. 1996; Danbolt 2001; Gendreau et al. 2004). The transporters have specific patterns of cellular localization: EAAT1 and EAAT2 have primarily been localized to astroglia, whereas EAAT3 and EAAT4 are primarily localized to neurons (Wheal and Thomson 1995; Nakanishi 1992; Durand et al. 1993). EAAT5 expression is limited to the retina. EAAT expression may be regulated at multiple levels, including transcription, translation, post-translational modification, trafficking, substrate affinity and transport kinetics (Danbolt 2001). Expression, intracellular localization, regulation and function of EAAT1-4 are discussed below.

EAAT1

Called GLAST in the rodent, EAAT1 is expressed on astrocytes throughout the CNS, with the highest levels of expression in the Bergmann glia in the cerebellum and the retina (Lehre et al. 1995; Bar-Peled et al. 1997; Schmitt et al. 1997; Banner et al. 2002). Immunocytochemical studies suggest that GLAST/EAAT1 expression is limited to the plasma membranes of astrocyte cell bodies and processes, although one study suggests that EAAT1 is expressed in a subset of cortical neurons in subjects with Alzheimer's-type pathology (Lehre et al. 1995; Scott et al. 2002). GLAST/EAAT1 expression is enriched in regions of the plasma membrane proximate to nerve terminals (Chaudhry et al. 1995b). Immunogold labeling demonstrated relatively little intracellular GLAST/EAAT1 protein expression, suggesting that localization of GLAST/EAAT1 to the plasma membrane is not regulated via mobilization of an intracellular pool of transporter protein (Chaudhry et al. 1995b). Cell surface expression of GLAST/EAAT1 was increased by insulin-like growth factor-1 via a protein kinase dependent mechanism, and region-specific GLAST/EAAT1 expression is increased by a number of peptide growth factors likely mediated by serine and tyrosine kinases (Conradt and Stoffel 1997; Figiel and Engele 2000; Gamboa and Ortega 2002; Schluter et al. 2002; Figiel et al. 2003). Recent studies have identified consensus tyrosine kinase phosphorylation sites targeted by specific glucocorticoid-sensitive kinase isoforms that facilitate increased EAAT1 activity.

Table 6.6. Summary of EAAT characteristics

Species Names		Cellular Localization	Subcellular Localization	Identified interacting proteins
Human	Rodent		Plasma Membrane /Cytoplasm	Rodent Genes
EAAT1	GLAST	Astrocytes	Yes/No	Sept2
EAAT2	GLT-1	Astrocytes	Yes/No	Ajuba, GPS-1, Caspase3
EAAT2B	GLT-1B?	Neurons	Yes/Unknown	None identified
EAAT3	EAAC1	Neurons	Yes/Yes	GTRAP3-18, GPS-1, JM4, Syntaxin 1a, DOR, HCCS
EAAT4	EAAT4	Neurons	Yes/No	GTRAP-41, GTRAP-48, GPS-1

Table 6.7. Cortical glutamate transporter abnormalities in schizophrenia

Transporter	Level of gene expression	Technique	Finding	Cortical Region	Reference
EAAT1	mRNA	QPCR	un-changed	DLPFC, PVC	(Lauriat et al., 2006)
EAAT2	mRNA	QPCR	un-changed	DLPFC, PVC	(Lauriat et al., 2006)
	mRNA	ISH	♦	PFC	(Ohnuma et al., 1998)
	mRNA protein	QPCR ICC	♠ ♠	PFC PFC	(Matute et al., 2005)
	gluta-mate transport	expression in oocytes	♠	PFC	
EAAT3	mRNA	QPCR	un-changed	DLPFC, PVC	(Lauriat et al., 2006)
VGLUT1	mRNA	ISH	un-changed	STC	(Eastwood and Harrison, 2005)
			♦	DLPFC	

Abbreviations: quantitative real time polymerase chain reaction (QPCR), *in situ* hybridization (ISH), immunocytochemistry (ICC), dorsolateral prefrontal cortex (DLPFC), primary visual cortex (PVC), prefrontal cortex (PFC), superior temporal cortex (STC)

EAAT1 interacting proteins

Disruption of putative c-terminal EAAT1 protein-protein interaction domains with a c-terminal peptide increased the affinity of EAAT1 for glutamate in salamander retinal glial cells, suggesting that protein-protein interactions modulate EAAT1 function (Marie and Attwell 1999). Consistent with this finding, another study has demonstrated that sept2, a member of the septin family of GTPases, negatively modulates GLAST/EAAT1 by direct binding of the carboxy-terminal region (Kinoshita et al. 2004).

EAAT2

Called GLT-1 in the rodent, EAAT2 is the most comprehensively studied of the glutamate transporters. GLT-1/EAAT2 accounts for approximately 90% of rodent forebrain glutamate reuptake (Rothstein et al. 1996; Tanaka et al. 1997). Expression of EAAT2 protein and mRNA has been observed throughout the human brain, but is highest in the forebrain (Bar-Peled et al. 1997; Milton et al. 1997). Astrocytes express EAAT2 transcripts and protein, while EAAT2 mRNA is detectable in neurons found in the neocortex and thalamus (Rothstein et al. 1994; Lehre et al. 1995; Furuta et al. 1997a; Berger and Hediger 1998). EAAT2 protein is expressed in neurons of the retina and embryonic neuronal cultures; in one recent report, protein expression of an EAAT2 splice variant (GLT-1a) was detected in rodent hippocampal neurons, while a PDZ-domain containing variant (called EAAT2b) was expressed in human neurons (Rauen and Kanner 1994; Danbolt 2001; Schmitt et al. 2002; Chen et al. 2004a). GLT-1/EAAT2 protein is enriched in the portion of the astrocytic plasma membrane facing neuropil (Chaudhry et al. 1995b). Less than 5% of GLT-1/EAAT2 immunoreactivity is found in structures other than the plasma membrane, including mitochondria and the endoplasmic reticulum (Chaudhry et al. 1995b). GLT-1/EAAT2 mediated glutamate uptake is modulated by a number of paracrine factors including arachidonic acid, cytokines, oxygen radicals, and peptide growth factors (Volterra et al. 1994; Dunlop et al. 1999; Hu et al. 2000; Figiel et al. 2003).

EAAT2 interacting proteins

A number of studies have found effects of the modulation of signaling mechanisms on GLT-1/EAAT2 expression and function, suggesting that there may be intermediary regulatory molecules that facilitate transporter localization and activation. PKC alpha, however, has been shown to directly interact with GLT-1/EAAT2, mediating transporter internalization (Gonzalez et al. 2005). Using the yeast-two hybrid technique, at least two additional EAAT2 interacting proteins have been identified. Ajuba is a cytosolic Lin11, Isl-1, and Mec-3 (LIM) protein that translocates to the plasma membrane and colocalizes with GLT-1/EAAT2 (Marie et al. 2002). Ajuba interacts directly with F-actin, and may contribute to cell-cell junctions by bridging cadherin adhesive complexes (Marie et al. 2003). Transcript expression of ajuba, however, was undetectable in the cortex by in situ hybridization (our unpublished observations). Another GLT-1/EAAT2 interacting protein, G-protein pathway suppressor 1 (GPS-1), is a subunit of the COP9 signalsome (Watanabe et al. 2003). Transfection with GPS-1 of HEK cells stably expressing GLT-1/EAAT2 downregulated glutamate transport activity (Watanabe et al. 2003). GPS-1 also interacts with EAAT3 and EAAT4 (Watanabe et al. 2003). Caspase 3 also interacts with EAAT2. Caspase 3 can bind to and cleave EAAT2 at aspartate 505, located in the cytoplasmic C-terminal domain, strongly and selectively impairing transport activity (Boston-Howes et al. 2006).

EAAT3

Called EAAC1 in the rodent, EAAT3 is expressed on neurons in the neocortex, striatum, thalamus, cerebellum and other structures (Conti et al. 1998; Kugler and Schmitt 1999). EAAT3 protein expression has been demonstrated on glutamatergic, GABAergic and aminergic neurons, as well as astrocytes of the cortex and in white matter (Conti et al. 1998; Kugler and Schmitt 1999). In contrast to EAAT1 and EAAT2, subcellular distribution of EAAT3 includes the cytoplasm, where it is available for rapid mobilization to perisynaptic regions of the plasma membrane (Conti et al. 1998; Kugler and Schmitt 1999; He et al. 2000, 2001). Dendritic trafficking of EAAT3 is mediated by a novel c-terminal sorting motif, but mutation of this motif did not impair clustering of EAAT3 on dendritic spines (Cheng et al. 2002). Surface expression of EAAT3 (and glutamate uptake) is rapidly increased by activation of PKC in C6 glioma cells constitutively expressing this transporter, and PKC activation increased glutamate uptake in cortical neuron cultures (Dowd et al. 1996; Dowd and Robinson 1996; Davis et al. 1998; Lortet et al. 1999). Platelet-derived growth factor (PDGF) increased surface expression of EAAT3 and glutamate uptake in C6 glioma cells, an effect mediated by activation of PI3K, and blocked by the PI3K inhibitor wortmanin (Davis et al. 1998; Sims et al. 2000). Activation of the neurotensin receptor NTS1 also increased EAAC surface expression, an effect that did not require PKC or PI3K activity (Najimi et al. 2002). The transcription factor RFX1 has also been shown to increase EAAT3 protein expression (Ma et al. 2006). Similar to EAAT2, EAAT3 appears to be regulated by converging signaling pathways that utilize independent signaling substrates (Danbolt 2001).

EAAT3 interacting proteins

Like EAAT2, EAAT3 interacts with PKC alpha (Gonzalez et al. 2003), but unlike EAAT2, EAAT3 membrane expression is increased by PKC alpha activation. GTRAP3-18 has been identified as an EAAT3 interacting protein. GTRAP3-18 has 95% nucleic acid homology with the human gene JWA, and specifically interacts with the intracellular c-terminus of EAAT3 (Lin et al. 2001). Increases in GTRAP3-18 expression lowered EAAT3 substrate affinity, an effect mediated by modulation of transporter N-linked glycosylation (Lin et al. 2001; Ruggiero et al. 2003). GTRAP3-18 transcripts and protein are ubiquitously expressed in the rodent, in both neuronal and non-neuronal cells (Butchbach et al. 2002). Recently, Jena-Muenchen 4 (JM4), a JWA homolog sharing 62% homology with JWA and GTRAP3-18, has been shown to heterodimerize with JWA, and may interact with EAAT3 (Schweneker et al. 2005). Interaction with the delta opiate receptor has been shown to decrease EAAT3 function (Xia et al. 2006). The SNARE protein Syntaxin 1A has been shown to interact with EAAT3 when coexpressed in oocytes, increasing membrane expression through trafficking, but decreasing transport activity by disrupting the structure of the conductance pathway (Zhu et al. 2005; Yu et al. 2006). However, another group showed that syntaxin 1A is not necessary for EAAT3 membrane trafficking in C6 glioma cells, and that another SNARE protein SNAP-23 was involved in the constitutive recycling of EAAT3 (Fournier and Robinson 2006). EAAT3 has also been shown to interact with the mitochondrial protein holocytochrome c synthetase (HCCS), serving as an anti-apoptotic mechanism, but no effect on transport activity has been shown (Kiryu-Seo et al. 2006).

EAAT4

In the adult CNS, human and rodent EAAT4 transcripts and protein are only robustly expressed in the Purkinje cells of the cerebellar molecular layer (Yamada et al. 1996; Bar-Peled et al. 1997; Furuta et al. 1997b; Nagao et al. 1997). We have detected EAAT4 transcript expression in the human striatum by in situ hybridization, and low levels of EAAT4 mRNA and protein have been detected in rodent cortical neurons and astrocyte cultures (Massie et al. 2001; McCullumsmith and Meador-Woodruff 2002; Hu et al. 2003). Unlike the other transporters, EAAT4 protein has an uneven subcellular distribution and is typically localized in extrasynaptic regions (Nagao et al. 1997; Dehnes et al. 1998). Rat EAAT4 interacts with GTRAP41 and GTRAP48, both of which increase EAAT4 mediated glutamate uptake by increasing and stabilizing cell surface transporter expression (Jackson et al. 2001). GTRAP-41 links EAAT4 with the cytoskeleton, while GTRAP-48 is associated with Rho-GTPase signaling (Jackson et al. 2001). Transcripts for the human isoforms of GTRAP-41 (KIAA0302) and GTRAP-48 (ARHGEF11) are robustly expressed throughout the CNS, including the thalamus and neocortex (our unpublished observations). The mismatch in the expression of EAAT4 and these interacting proteins suggests they may facilitate intracellular trafficking of other elements of the glutamate synapse.

EAATs, glutamate receptors, and synaptic plasticity

Glutamate neurotransmission is mediated via activation of specific pre- and post-synaptic receptors. To a large extent, the level of synaptic glutamate determines glutamate receptor activation (Danbolt 2001). High synaptic glutamate levels and excessive activation of glutamate receptors may lead to seizures, while low levels of synaptic glutamate prior to vesicular glutamate release provide a high signal to noise ratio, optimizing receptor function (Claudio et al. 2000; Danbolt 2001). Modulation of glutamate reuptake alters glutamate mediated synaptic currents, directly implicating EAATs in synaptic plasticity (Kinney et al. 1997; Brasnjo and Otis 2001; Grassi et al. 2002; Zakharenko et al. 2002; Sokolov et al. 2003). Since glutamate receptors have a central role in the molecular correlates of synaptic plasticity, such as long-term potentiation and long-term depression, alterations in EAAT-mediated glutamate reuptake in disease states may directly affect synaptic plasticity. Supporting this, induction of long-term potentiation by high frequency stimulation increased both surface expression of EAAC and glutamate reuptake in the CA1 hippocampal subfield, an effect that was dependent on NMDA receptor activation (Levenson et al. 2002). These findings highlight the complex interactions between neurons and glia that modulate glutamate reuptake, interactions that if altered, may significantly impact cortical glutamate transmission in schizophrenia.

6.7.2 Vesicular Glutamate Transporter

While the EAATs are responsible for the vast majority of glutamate reuptake from the synapse, another family of glutamate transporters, the vesicular glutamate transporters (VGLUTs), are involved in packaging of glutamate into synaptic vesicles. Vesicular uptake of glutamate is an ATP–dependent process independent of sodium and potassium. It has a biphasic dependence on chloride concentration, and is driven by an internal positive membrane potential. The VGLUTs are highly selective for L-glutamate, but have a lower affinity for glutamate in comparison to the EAATs (Danbolt et al. 1998; Danbolt 2001). Alterations in synaptic activity can be induced by the modulation of the amount of glutamate released from synaptic vesicles. Therefore, VGLUT mediated transport of glutamate into vesicles for release into the synaptic cleft is a pivotal control point for synaptic activity (Wilson et al. 2005a). Three vesicular transporters (VGLUT1, VGLUT2, and VGLUT3) have been cloned and characterized, and are discussed below (Hisano 2003).

VGLUT1

Originally identified as a brain-specific sodium-dependent inorganic phosphate co-transporter (BNPi), VGLUT1 is a membrane-bound protein consisting of 560 amino acids (Aihara et al. 2000) with six putative transmembrane portions and eight hydrophobic domains (Hisano 2003). The intracellular loops include phosphorylation sites for tyrosine kinase, protein kinase C (PKC) and calmodulin-dependent kinase II (Shigeri et al. 2004). VGLUT1-mediated glutamate transport is inhibited by proton ionophores, indicating an electrochemical proton gradient is the driving force behind transport (Takamori et al. 2000). VGLUT1 transcripts are expressed in the cortex, hippocampus, and cerebellum, and are associated with many synapses that exhibit activity-dependent synaptic plasticity such as long-term potentiation (Takamori et al. 2000).

VGLUT2

Originally called differentiation-associated Na^+-dependent inorganic phosphate co-transporter (DNPi), VGLUT2 has 583 amino acids and six putative transmembrane portions with eight hydrophobic domains, similar to VGLUT1 (Varoqui et al. 2002; Hisano 2003). Although VGLUT2 shares an 82% amino acid identity and 92% nucleic acid similarity to VGLUT1, VGLUT2 contains a different c-terminal amino acid sequence, suggesting there are differences in regulation and intracellular trafficking (Kaneko and Fujiyama 2002). The expression pattern of VGLUT2 in the brain strongly complements that of VGLUT1 (Kaneko and Fujiyama 2002). It is particularly expressed by neurons in the thalamus, hypothalamus, and brainstem, and is expressed primarily in sensory and autonomic pathways that display high-fidelity neurotransmission (Takamori et al. 2000; Varoqui et al. 2002).

VGLUT3

VGLUT3 has 589 amino acids and shares a 70% amino acid identity with VGLUT1 and VGLUT2 (Gras et al. 2002). VGLUT3 is the least abundant of the VGLUTs, and is expressed at relatively low levels in scattered glutamatergic neurons throughout the brain, at higher levels in cholinergic neurons in the striatum, and in serotoninergic neurons in the raphe nuclei (Gras et al. 2002).

6.8 Abnormalities in Glutamate Transporters and Transporter Interacting Proteins in Schizophrenia

Studies of glutamate transporter expression in the cortex have yielded conflicting results. Of the transporters, EAAT2 has been the most extensively studied. Three groups have investigated changes in mRNA expression of EAAT2 in the PFC in schizophrenia, all with different results. Ohnuma et al. found decreased EAAT2 mRNA in the PFC using *in situ* hybridization (Ohnuma et al. 1998), while Matute et al. found increased and Lauriat et al. (Lauriat et al. 2006) found no changes in EAAT2 mRNA in the PFC using QPCR (Matute et al. 2005). These discrepant findings might be explained by differences in subjects studied, as the samples varied widely in age, postmortem interval (PMI), and neuroleptic treatment. The differences could also be explained by the use of different methodologies, as QPCR is more sensitive than ISH for detecting mRNA expression.

Another possible source of variation in these studies is the type of probe used to detect EAAT2. EAAT2 is differentially spliced, and the majority of EAAT splice variants are either EAAT2a or EAAT2b, which differ in their carboxyl termini. EAAT2a is more abundant and is generally found in astrocytes (Chaudhry et al. 1995a; Lehre et al. 1995), whereas EAAT2b is less abundant and generally found in neurons (Schmitt et al. 2002; Kugler and Schmitt 2003). These variants appear to be abnormally expressed in disease states such as amyotrophic lateral sclerosis (Maragakis et al., 2004). Since Ohnuma et al. (Ohnuma et al., 2000a) detected EAAT2 by ISH using a probe near the 3' end of EAAT2, they likely detected only the EAAT2a isoforms, while Matute et al. (Matute et al., 2005) did not report the location of the EAAT2 probe used in their QPCR study. It is possible that the groups were measuring different mRNA isoforms altogether, accounting for their discrepant results.

In addition to these conflicting EAAT2 data, the only other abnormality in glutamate transporter expression detected in the cortex in schizophrenia thus far is in the presynaptic vesicular glutamate transporter VGLUT1, which is involved in packaging of presynaptic glutamate into vesicles. Decreased VGLUT1 mRNA expression was found in superficial layers of the DLPFC in schizophrenia (Eastwood and Harrison 2005). This finding suggests a decrease in glutamatergic neurotransmission in the DLPFC, given that VGLUT1 expression has been shown to directly influence quantal glutamate release (Wilson et al. 2005b). Such a change in presynaptic innervation might be consistent with deficits in cortical function in schizophrenia.

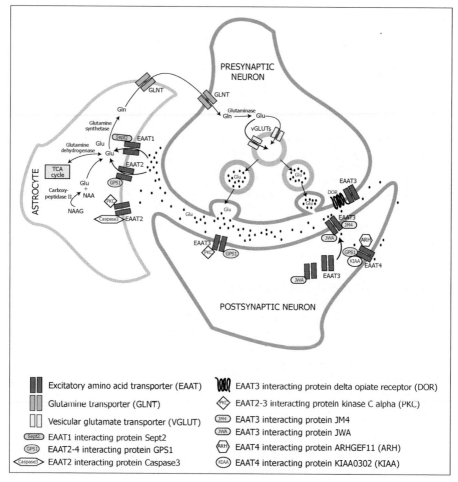

Fig. 6.1. The glutamate synapse. Glutamate is packaged into presynaptic glutamate transporters (VGLUTs), and is released into the synaptic cleft where it can activate glutamate receptors. Glutamate is rapidly removed from the synapse by excitatory amino acid transporters
(EAATs) localized on astrocytes (EAAT1-2) and neurons (EAAT3-4).

Although there is ongoing work on the glutamate transporters in the cortex in
schizophrenia, to our knowledge, there are no published studies of mRNA expression for EAAT4, VGLUT2, or VGLUT3. In addition, the transporters have not
been tested in other cortical regions known to be affected by schizophrenia, only
one study has published any findings on cortical changes in transporter protein
levels, and there are no published studies of transporter interacting proteins in
the cortex. Given the critical role of glutamate transmission in the cortex and the
body of evidence implicating cortical dysfunction in schizophrenia, study of the

glutamate transporters and their interacting proteins presents a promising area for future research in this illness.

6.9 Summary and Implications

We have summarized postmortem data in the cortex in schizophrenia for glutamate receptors, transporters and interacting molecules. Glutamate receptor expression varies by cortical region, brain bank from which samples are studied, and methodology, making mechanistic interpretations difficult when considering the data in total. Examination of receptor interacting protein data in the cortex reveals a consistent pattern of disparate mRNA and protein expression for the same genes. The relevance of this pattern could be answered with cell level studies, to determine if these findings are limited to a subset of cells, or reflect a change in glutamate transmission for an entire region. Cell level transcript studies suggest that there are neuron-specific alterations in mRNA expression for glutamatergic genes, while quantitative cell level protein studies are currently not feasible in postmortem tissue.

Examination of glutamate receptor interacting molecules represents a promising new approach to understanding glutamatergic abnormalities in schizophrenia. Studies of these genes suggest abnormalities of specific cell biological processes, such as receptor trafficking and assembly of receptor signaling complexes. Taken together, studies of the receptors and their interacting proteins suggest that there are abnormalities of trafficking, assembly and function of glutamate receptor signaling complexes in the cortex in schizophrenia.

While the receptors, and to a lesser extent, the receptor interacting proteins, have been comprehensively studied in the cortex in schizophrenia, the glutamate transporters have not. Only three studies have examined the plasma membrane transporters in the cortex, and these only measured EAAT2 expression. The EAAT interacting proteins have not been studied in the cortex in schizophrenia.

Postmortem data for glutamatergic molecules in cortex suggest that glutamate dysfunction in schizophrenia involves more than NMDA receptor disturbances, but also involves other critical elements of glutamate synapses as well as several cell biological processes associated with regulation of the glutamate receptor and other synaptic molecules. We propose that receptor trafficking, assembly of signaling complexes, and receptor recycling might be dysregulated in schizophrenia, and represent important new targets for pharmacological manipulation in this illness. In addition, we suggest that secondary to alterations in glutamate receptor signaling complexes, there are changes in the cellular systems that manage glutamate levels, including the expression and regulation of glutamate transporters. Finally, since glutamate transmission has a central role in normal cortical function, these postmortem studies support the hypothesis that glutamatergic abnormalities contribute to cortical dysfunction in schizophrenia.

Acknowledgments

Supported by MH53327 and MH074016.

References

Aihara Y, Mashima H, Onda H, Hisano S, Kasuya H, Hori T, Yamada S, Tomura H, Yamada Y, Inoue I, Kojima I, Takeda J (2000) Molecular cloning of a novel brain-type Na(+)-dependent inorganic phosphate cotransporter. J Neurochem 74:2622-2625.

Akbarian S, Sucher NJ, Bradley D, Tafazzoli A, Trinh D, Hetrick WP, Potkin SG, Sandman CA, Bunney WE, Jr., Jones EG (1996) Selective alterations in gene expression for NMDA receptor subunits in prefrontal cortex of schizophrenics. J Neurosci 16:19-30.

Andreasen NC, Paradiso S, O'Leary DS (1998) "Cognitive dysmetria" as an integrative theory of schizophrenia: a dysfunction in cortical-subcortical-cerebellar circuitry? Schizophr Bull 24:203-218.

Andreasen NC, O'Leary DS, Flaum M, Nopoulos P, Watkins GL, Boles Ponto LL, Hichwa RD (1997) Hypofrontality in schizophrenia: distributed dysfunctional circuits in neuroleptic-naive patients. Lancet 349:1730-1734.

Andreasen NC, O'Leary DS, Cizadlo T, Arndt S, Rezai K, Ponto LL, Watkins GL, Hichwa RD (1996) Schizophrenia and cognitive dysmetria: a positron-emission tomography study of dysfunctional prefrontal-thalamic-cerebellar circuitry. Proc Natl Acad Sci USA 93:9985-9990.

Aramori I, Nakanishi S (1992) Signal transduction and pharmacological characteristics of a metabotropic glutamate receptor, mGluR1, in transfected CHO cells. Neuron 8:757-765.

Arriza JL, Kavanaugh MP, Fairman WA, Wu YN, Murdoch GH, North RA, Amara SG (1993) Cloning and expression of a human neutral amino acid transporter with structural similarity to the glutamate transporter gene family. J Biol Chem 268:15329-15332.

Banner SJ, Fray AE, Ince PG, Steward M, Cookson MR, Shaw PJ (2002) The expression of the glutamate re-uptake transporter excitatory amino acid transporter 1 (EAAT1) in the normal human CNS and in motor neurone disease: an immunohistochemical study. Neuroscience 109:27-44.

Bar-Peled O, Ben-Hur H, Biegon A, Groner Y, Dewhurst S, Furuta A, Rothstein JD (1997) Distribution of glutamate transporter subtypes during human brain development. J Neurochem 69:2571-2580.

Bassi MT, Gasol E, Manzoni M, Pineda M, Riboni M, Martin R, Zorzano A, Borsani G, Palacin M (2001) Identification and characterisation of human xCT that co-expresses, with 4F2 heavy chain, the amino acid transport activity system xc. Pflugers Arch 442:286-296.

Bellocchio EE, Reimer RJ, Fremeau RT, Jr., Edwards RH (2000) Uptake of glutamate into synaptic vesicles by an inorganic phosphate transporter. Science 289:957-960.

Beneyto M, Meador-Woodruff JH (2006) Lamina-Specific abnormalities of AMPA receptor trafficking and signaling molecule transcripts in the prefrontal cortex in schizophrenia. Synapse:In press.

Berger UV, Hediger MA (1998) Comparative analysis of glutamate transporter expression in rat brain using differential double in situ hybridization. Anat Embryol (Berl) 198:13-30.

Boston-Howes W, Gibb SL, Williams EO, Pasinelli P, Brown RH, Jr., Trotti D (2006) Caspase-3 cleaves and inactivates the glutamate transporter EAAT2. J Biol Chem 281:14076-14084.

Brasnjo G, Otis TS (2001) Neuronal glutamate transporters control activation of postsynaptic metabotropic glutamate receptors and influence cerebellar long-term depression. Neuron 31:607-616.

Breese CR, Freedman R, Leonard SS (1995) Glutamate receptor subtype expression in human postmortem brain tissue from schizophrenics and alcohol abusers. Brain Res 674:82-90.

Butchbach ME, Lai L, Lin CL (2002) Molecular cloning, gene structure, expression profile and functional characterization of the mouse glutamate transporter (EAAT3) interacting protein GTRAP3-18. Gene 292:81-90.

Callicott JH, Bertolino A, Egan MF, Mattay VS, Langheim FJ, Weinberger DR (2000) Selective relationship between prefrontal N-acetylaspartate measures and negative symptoms in schizophrenia. Am J Psychiatry 157:1646-1651.

Chaudhry FA, Lehre KP, van Lookeren Campagne M, Ottersen OP, Danbolt NC, Storm-Mathisen J (1995a) Glutamate transporters in glial plasma membranes: highly differentiated localizations revealed by quantitative ultrastructural immunocytochemistry. Neuron 15:711-720.

Chaudhry FA, Lehre KP, van Lookeren Campagne M, Ottersen OP, Danbolt NC, Storm-Mathisen J (1995b) Glutamate transporters in glial plasma membranes: highly differentiated localizations revealed by quantitative ultrastructural immunocytochemistry. Neuron 15:711-720.

Chen W, Mahadomrongkul V, Berger UV, Bassan M, DeSilva T, Tanaka K, Irwin N, Aoki C, Rosenberg PA (2004a) The glutamate transporter GLT1a is expressed in excitatory axon terminals of mature hippocampal neurons. J Neurosci 24:1136-1148.

Chen X, Dunham C, Kendler S, Wang X, O'Neill FA, Walsh D, Kendler KS (2004b) Regulator of G-protein signaling 4 (RGS4) gene is associated with schizophrenia in Irish high density families. Am J Med Genet B Neuropsychiatr Genet 129:23-26.

Cheng C, Glover G, Banker G, Amara SG (2002) A novel sorting motif in the glutamate transporter excitatory amino acid transporter 3 directs its targeting in Madin-Darby canine kidney cells and hippocampal neurons. J Neurosci 22:10643-10652.

Chowdari KV, Mirnics K, Semwal P, Wood J, Lawrence E, Bhatia T, Deshpande SN, B KT, Ferrell RE, Middleton FA, Devlin B, Levitt P, Lewis DA, Nimgaonkar VL (2002) Association and linkage analyses of RGS4 polymorphisms in schizophrenia. Hum Mol Genet 11:1373-1380.

Claudio OI, Ferchmin P, Velisek L, Sperber EF, Moshe SL, Ortiz JG (2000) Plasticity of excitatory amino acid transporters in experimental epilepsy. Epilepsia 41 Suppl 6:S104-110.

Conradt M, Stoffel W (1997) Inhibition of the high-affinity brain glutamate transporter GLAST-1 via direct phosphorylation. J Neurochem 68:1244-1251.

Conti F, DeBiasi S, Minelli A, Rothstein JD, Melone M (1998) EAAC1, a high-affinity glutamate tranporter, is localized to astrocytes and gabaergic neurons besides pyramidal cells in the rat cerebral cortex. Cereb Cortex 8:108-116.

Contractor A, Heinemann SF (2002) Glutamate receptor trafficking in synaptic plasticity. Sci STKE 2002:RE14.

Cordeiro Q, Talkowski ME, Chowdari KV, Wood J, Nimgaonkar V, Vallada H (2005) Association and linkage analysis of RGS4 polymorphisms with schizophrenia and bipolar disorder in Brazil. Genes Brain Behav 4:45-50.

Corti C, Cavanni P, Caveggion E, Ferraguti F, Corsi M, Trist DG (1997) Different levels of receptor expression as a new procedure to estimate agonist affinity constant. Application to the metabotropic receptors. Ann N Y Acad Sci 812:231-233.

Corti C, Restituito S, Rimland JM, Brabet I, Corsi M, Pin JP, Ferraguti F (1998) Cloning and characterization of alternative mRNA forms for the rat metabotropic glutamate receptors mGluR7 and mGluR8. Eur J Neurosci 10:3629-3641.

Danbolt NC (2001) Glutamate uptake. Prog Neurobiol 65:1-105.

Danbolt NC, Chaudhry FA, Dehnes Y, Lehre KP, Levy LM, Ullensvang K, Storm-Mathisen J (1998) Properties and localization of glutamate transporters. Prog Brain Res 116:23-43.

Davis KE, Straff DJ, Weinstein EA, Bannerman PG, Correale DM, Rothstein JD, Robinson MB (1998) Multiple signaling pathways regulate cell surface expression and activity of the excitatory amino acid carrier 1 subtype of Glu transporter in C6 glioma. J Neurosci 18:2475-2485.

Dehnes Y, Chaudhry FA, Ullensvang K, Lehre KP, Storm-Mathisen J, Danbolt NC (1998) The glutamate transporter EAAT4 in rat cerebellar Purkinje cells: a glutamate-gated chloride channel concentrated near the synapse in parts of the dendritic membrane facing astroglia. J Neurosci 18:3606-3619.

Dowd LA, Robinson MB (1996) Rapid stimulation of EAAC1-mediated Na+-dependent L-glutamate transport activity in C6 glioma cells by phorbol ester. J Neurochem 67:508-516.

Dowd LA, Coyle AJ, Rothstein JD, Pritchett DB, Robinson MB (1996) Comparison of Na+-dependent glutamate transport activity in synaptosomes, C6 glioma, and Xenopus oocytes expressing excitatory amino acid carrier 1 (EAAC1). Mol Pharmacol 49:465-473.

Dracheva S, McGurk SR, Haroutunian V (2005a) mRNA expression of AMPA receptors and AMPA receptor binding proteins in the cerebral cortex of elderly schizophrenics. J Neurosci Res 79:868-878.

Dracheva S, McGurk SR, Haroutunian V (2005b) mRNA expression of AMPA receptors and AMPA receptor binding proteins in the cerebral cortex of elderly schizophrenics. J Neurosci Res 79:868-878.

Dracheva S, Marras SA, Elhakem SL, Kramer FR, Davis KL, Haroutunian V (2001) N-methyl-D-aspartic acid receptor expression in the dorsolateral prefrontal cortex of elderly patients with schizophrenia. Am J Psychiatry 158:1400-1410.

Dunlop J, Lou Z, Zhang Y, McIlvain HB (1999) Inducible expression and pharmacology of the human excitatory amino acid transporter 2 subtype of L-glutamate transporter. Br J Pharmacol 128:1485-1490.

Durand GM, Bennett MV, Zukin RS (1993) Splice variants of the N-methyl-D-aspartate receptor NR1 identify domains involved in regulation by polyamines and protein kinase C. Proc Natl Acad Sci U S A 90:6731-6735.

Eastwood SL, Harrison PJ (2005) Decreased expression of vesicular glutamate transporter 1 and complexin II mRNAs in schizophrenia: further evidence for a synaptic pathology affecting glutamate neurons. Schizophr Res 73:159-172.

Eastwood SL, Burnet PW, Harrison PJ (1997a) GluR2 glutamate receptor subunit flip and flop isoforms are decreased in the hippocampal formation in schizophrenia: a reverse transcriptase- polymerase chain reaction (RT-PCR) study. Brain Res Mol Brain Res 44:92-98.

Eastwood SL, Kerwin RW, Harrison PJ (1997b) Immunoautoradiographic evidence for a loss of alpha-amino-3-hydroxy-5- methyl-4-isoxazole propionate-preferring non-N-methyl-D-aspartate glutamate receptors within the medial temporal lobe in schizophrenia. Biol Psychiatry 41:636-643.

Eastwood SL, McDonald B, Burnet PW, Beckwith JP, Kerwin RW, Harrison PJ (1995) Decreased expression of mRNAs encoding non-NMDA glutamate receptors GluR1 and GluR2 in medial temporal lobe neurons in schizophrenia. Brain Res Mol Brain Res 29:211-223.

Ehlers MD, Fung ET, O'Brien RJ, Huganir RL (1998) Splice variant-specific interaction of the NMDA receptor subunit NR1 with neuronal intermediate filaments. J Neurosci 18:720-730.

Erdely HA, Tamminga CA, Roberts RC, Vogel MW (2006) Regional alterations in RGS4 protein in schizophrenia. Synapse 59:472-479.

Figiel M, Engele J (2000) Pituitary adenylate cyclase-activating polypeptide (PACAP), a neuron-derived peptide regulating glial glutamate transport and metabolism. J Neurosci 20:3596-3605.

Figiel M, Maucher T, Rozyczka J, Bayatti N, Engele J (2003) Regulation of glial glutamate transporter expression by growth factors. Exp Neurol 183:124-135.

Fournier KM, Robinson MB (2006) A dominant-negative variant of SNAP-23 decreases the cell surface expression of the neuronal glutamate transporter EAAC1 by slowing constitutive delivery. Neurochem Int 48:596-603.

Freed WJ, Dillon-Carter O, Kleinman JE (1993) Properties of [3H]AMPA binding in post-mortem human brain from psychotic subjects and controls: increases in caudate nucleus associated with suicide. Exp Neurol 121:48-56.

Furuta A, Rothstein JD, Martin LJ (1997a) Glutamate transporter protein subtypes are expressed differentially during rat CNS development. J Neurosci 17:8363-8375.

Furuta A, Martin LJ, Lin CL, Dykes-Hoberg M, Rothstein JD (1997b) Cellular and synaptic localization of the neuronal glutamate transporters excitatory amino acid transporter 3 and 4. Neuroscience 81:1031-1042.

Gamboa C, Ortega A (2002) Insulin-like growth factor-1 increases activity and surface levels of the GLAST subtype of glutamate transporter. Neurochem Int 40:397-403.

Gendreau S, Voswinkel S, Torres-Salazar D, Lang N, Heidtmann H, Detro-Dassen S, Schmalzing G, Hidalgo P, Fahlke C (2004) A trimeric quaternary structure is conserved in bacterial and human glutamate transporters. J Biol Chem 279:39505-39512.

Ghose S, Weickert CS, Colvin SM, Coyle JT, Herman MM, Hyde TM, Kleinman JE (2004) Glutamate carboxypeptidase II gene expression in the human frontal and temporal lobe in schizophrenia. Neuropsychopharmacology 29:117-125.

Gonzalez MI, Bannerman PG, Robinson MB (2003) Phorbol myristate acetate-dependent interaction of protein kinase Calpha and the neuronal glutamate transporter EAAC1. J Neurosci 23:5589-5593.

Gonzalez MI, Susarla BT, Robinson MB (2005) Evidence that protein kinase Calpha interacts with and regulates the glial glutamate transporter GLT-1. J Neurochem 94:1180-1188.

Gras C, Herzog E, Bellenchi GC, Bernard V, Ravassard P, Pohl M, Gasnier B, Giros B, El Mestikawy S (2002) A third vesicular glutamate transporter expressed by cholinergic and serotoninergic neurons. J Neurosci 22:5442-5451.

Grassi S, Frondaroli A, Pettorossi VE (2002) Different metabotropic glutamate receptors play opposite roles in synaptic plasticity of the rat medial vestibular nuclei. J Physiol 543:795-806.

Gray L, Scarr E, Dean B (2006) N-Ethylmaleimide sensitive factor in the cortex of subjects with schizophrenia and bipolar I disorder. Neurosci Lett 391:112-115.

Grimwood S, Slater P, Deakin JF, Hutson PH (1999) NR2B-containing NMDA receptors are up-regulated in temporal cortex in schizophrenia. Neuroreport 10:461-465.

Guo S, Tang W, Shi Y, Huang K, Xi Z, Xu Y, Feng G, He L (2006) RGS4 polymorphisms and risk of schizophrenia: An association study in Han Chinese plus meta-analysis. Neurosci Lett.

Gupta DS, McCullumsmith RE, Beneyto M, Haroutunian V, Davis KL, Meador-Woodruff JH (2005) Metabotropic glutamate receptor protein expression in the prefrontal cortex and striatum in schizophrenia. Synapse 57:123-131.

Harrison PJ, McLaughlin D, Kerwin RW (1991) Decreased hippocampal expression of a glutamate receptor gene in schizophrenia. Lancet 337:450-452.

Haugeto O, Ullensvang K, Levy LM, Chaudhry FA, Honore T, Nielsen M, Lehre KP, Danbolt NC (1996) Brain glutamate transporter proteins form homomultimers. J Biol Chem 271:27715-27722.

He Y, Janssen WG, Rothstein JD, Morrison JH (2000) Differential synaptic localization of the glutamate transporter EAAC1 and glutamate receptor subunit GluR2 in the rat hippocampus. J Comp Neurol 418:255-269.

He Y, Hof PR, Janssen WG, Rothstein JD, Morrison JH (2001) Differential synaptic localization of GluR2 and EAAC1 in the macaque monkey entorhinal cortex: a postembedding immunogold study. Neurosci Lett 311:161-164.

Hertzmann M, Reba RC, Kotlyarov EV (1990) Single photon emission computed tomography in phencyclidine and related drug abuse. Am J Psychiatry 147:255-256.

Hisano S (2003) Vesicular glutamate transporters in the brain. Anat Sci Int 78:191-204.

Hollmann M, Heinemann S (1994) Cloned glutamate receptors. Annu Rev Neurosci 17:31-108.

Hu S, Sheng WS, Ehrlich LC, Peterson PK, Chao CC (2000) Cytokine effects on glutamate uptake by human astrocytes. Neuroimmunomodulation 7:153-159.

Hu WH, Walters WM, Xia XM, Karmally SA, Bethea JR (2003) Neuronal glutamate transporter EAAT4 is expressed in astrocytes. Glia 44:13-25.

Humphries C, Mortimer A, Hirsch S, de Belleroche J (1996) NMDA receptor mRNA correlation with antemortem cognitive impairment in schizophrenia. Neuroreport 7:2051-2055.

Imai C, Sugai T, Iritani S, Niizato K, Nakamura R, Makifuchi T, Kakita A, Takahashi H, Nawa H (2001) A quantitative study on the expression of synapsin II and N-ethylmaleimide-sensitive fusion protein in schizophrenic patients. Neurosci Lett 305:185-188.

Ishimaru M, Kurumaji A, Toru M (1994) Increases in strychnine-insensitive glycine binding sites in cerebral cortex of chronic schizophrenics: evidence for glutamate hypothesis. Biol Psychiatry 35:84-95.

Jackson M, Song W, Liu MY, Jin L, Dykes-Hoberg M, Lin CI, Bowers WJ, Federoff HJ, Sternweis PC, Rothstein JD (2001) Modulation of the neuronal glutamate transporter EAAT4 by two interacting proteins. Nature 410:89-93.

Javitt DC, Zukin SR (1991) Recent advances in the phencyclidine model of schizophrenia. Am J Psychiatry 148:1301-1308.

Joly C, Gomeza J, Brabet I, Curry K, Bockaert J, Pin JP (1995) Molecular, functional, and pharmacological characterization of the metabotropic glutamate receptor type 5 splice variants: comparison with mGluR1. J Neurosci 15:3970-3981.

Joyce JN, Meador-Woodruff JH (1997) Linking the family of D2 receptors to neuronal circuits in human brain: insights into schizophrenia. Neuropsychopharmacology 16:375-384.

Kaneko T, Fujiyama F (2002) Complementary distribution of vesicular glutamate transporters in the central nervous system. Neurosci Res 42:243-250.

Kerwin R, Patel S, Meldrum B (1990) Quantitative autoradiographic analysis of glutamate binding sites in the hippocampal formation in normal and schizophrenic brain post mortem. Neuroscience 39:25-32.

Kim JY, Kanai Y, Chairoungdua A, Cha SH, Matsuo H, Kim DK, Inatomi J, Sawa H, Ida Y, Endou H (2001) Human cystine/glutamate transporter: cDNA cloning and upregulation by oxidative stress in glioma cells. Biochim Biophys Acta 1512:335-344.

Kinney GA, Overstreet LS, Slater NT (1997) Prolonged physiological entrapment of glutamate in the synaptic cleft of cerebellar unipolar brush cells. J Neurophysiol 78:1320-1333.

Kinoshita N, Kimura K, Matsumoto N, Watanabe M, Fukaya M, Ide C (2004) Mammalian septin Sept2 modulates the activity of GLAST, a glutamate transporter in astrocytes. Genes Cells 9:1-14.

Kiryu-Seo S, Gamo K, Tachibana T, Tanaka K, Kiyama H (2006) Unique anti-apoptotic activity of EAAC1 in injured motor neurons. Embo J 25:3411-3421.

Kornhuber J, Mack-Burkhardt F, Riederer P, Hebenstreit GF, Reynolds GP, Andrews HB, Beckmann H (1989) [3H]MK-801 binding sites in postmortem brain regions of schizophrenic patients. J Neural Transm 77:231-236.

Kozlovsky N, Scarr E, Dean B, Agam G (2006) Postmortem brain calcineurin protein levels in schizophrenia patients are not different from controls. Schizophr Res 83:173-177.

Kristiansen LV, Beneyto M, Haroutunian V, Meador-Woodruff JH (2006) Changes in NMDA receptor subunits and interacting PSD proteins in dorsolateral prefrontal and anterior cingulate cortex indicate abnormal regional expression in schizophrenia. Mol Psychiatry 11:737-747, 705.

Kugler P, Schmitt A (1999) Glutamate transporter EAAC1 is expressed in neurons and glial cells in the rat nervous system. Glia 27:129-142.

Kugler P, Schmitt A (2003) Complementary neuronal and glial expression of two high-affinity glutamate transporter GLT1/EAAT2 forms in rat cerebral cortex. Histochem Cell Biol 119:425-435.

Kurumaji A, Ishimaru M, Toru M (1992) Alpha-[3H]amino-3-hydroxy-5-methylisoxazole-4-propionic acid binding to human cerebral cortical membranes: minimal changes in postmortem brains of chronic schizophrenics. J Neurochem 59:829-837.

Lahti AC, Holcomb HH, Medoff DR, Tamminga CA (1995) Ketamine activates psychosis and alters limbic blood flow in schizophrenia. Neuroreport 6:869-872.

Laruelle M, Abi-Dargham A, Gil R, Kegeles L, Innis R (1999) Increased dopamine transmission in schizophrenia: relationship to illness phases. Biol Psychiatry 46:56-72.

Lauriat TL, Dracheva S, Chin B, Schmeidler J, McInnes LA, Haroutunian V (2006) Quantitative analysis of glutamate transporter mRNA expression in prefrontal and primary visual cortex in normal and schizophrenic brain. Neuroscience 137:843-851.

Le Corre S, Harper CG, Lopez P, Ward P, Catts S (2000) Increased levels of expression of an NMDAR1 splice variant in the superior temporal gyrus in schizophrenia. Neuroreport 11:983-986.

Lehre KP, Levy LM, Ottersen OP, Storm-Mathisen J, Danbolt NC (1995) Differential expression of two glial glutamate transporters in the rat brain: quantitative and immunocytochemical observations. J Neurosci 15:1835-1853.

Levenson J, Weeber E, Selcher JC, Kategaya LS, Sweatt JD, Eskin A (2002) Long-term potentiation and contextual fear conditioning increase neuronal glutamate uptake. Nat Neurosci 5:155-161.

Levitt P, Ebert P, Mirnics K, Nimgaonkar VL, Lewis DA (2006) Making the Case for a Candidate Vulnerability Gene in Schizophrenia: Convergent Evidence for Regulator of G-Protein Signaling 4 (RGS4). Biol Psychiatry.

Levy LM, Warr O, Attwell D (1998) Stoichiometry of the glial glutamate transporter GLT-1 expressed inducibly in a Chinese hamster ovary cell line selected for low endogenous Na+-dependent glutamate uptake. J Neurosci 18:9620-9628.

Li D, He L (2006) Association study of the G-protein signaling 4 (RGS4) and proline dehydrogenase (PRODH) genes with schizophrenia: a meta-analysis. Eur J Hum Genet.

Lin CI, Orlov I, Ruggiero AM, Dykes-Hoberg M, Lee A, Jackson M, Rothstein JD (2001) Modulation of the neuronal glutamate transporter EAAC1 by the interacting protein GTRAP3-18. Nature 410:84-88.

Lin JW, Wyszynski M, Madhavan R, Sealock R, Kim JU, Sheng M (1998) Yotiao, a novel protein of neuromuscular junction and brain that interacts with specific splice variants of NMDA receptor subunit NR1. J Neurosci 18:2017-2027.

Lipska BK, Mitkus S, Caruso M, Hyde TM, Chen J, Vakkalanka R, Straub RE, Weinberger DR, Kleinman JE (2006) RGS4 mRNA Expression in Postmortem Human Cortex Is Associated with COMT Val158Met Genotype and COMT Enzyme Activity. Hum Mol Genet.

Liu YL, Shen-Jang Fann C, Liu CM, Wu JY, Hung SI, Chan HY, Chen JJ, Lin CY, Liu SK, Hsieh MH, Hwang TJ, Ouyang WC, Chen CY, Lin JJ, Chou FH, Chueh CM, Liu WM, Tsuang MM, Faraone SV, Tsuang MT, Chen WJ, Hwu HG (2006) Evaluation of RGS4 as a candidate gene for schizophrenia. Am J Med Genet B Neuropsychiatr Genet 141:418-420.

Lortet S, Samuel D, Had-Aissouni L, Masmejean F, Kerkerian-Le Goff L, Pisano P (1999) Effects of PKA and PKC modulators on high affinity glutamate uptake in primary neuronal cell cultures from rat cerebral cortex. Neuropharmacology 38:395-402.

Ma K, Zheng S, Zuo Z (2006) The transcription factor regulatory factor X1 increases the expression of neuronal glutamate transporter type 3. J Biol Chem 281:21250-21255.

Manoach DS (2003) Prefrontal cortex dysfunction during working memory performance in schizophrenia: reconciling discrepant findings. Schizophr Res 60:285-298.

Maragakis NJ, Dykes-Hoberg M, Rothstein JD (2004) Altered expression of the glutamate transporter EAAT2b in neurological disease. Ann Neurol 55:469-477.

Marie H, Attwell D (1999) C-terminal interactions modulate the affinity of GLAST glutamate transporters in salamander retinal glial cells. J Physiol 520 Pt 2:393-397.

Marie H, Billups D, Bedford FK, Dumoulin A, Goyal RK, Longmore GD, Moss SJ, Attwell D (2002) The amino terminus of the glial glutamate transporter GLT-1 interacts with the LIM protein Ajuba. Mol Cell Neurosci 19:152-164.

Marie H, Pratt SJ, Betson M, Epple H, Kittler JT, Meek L, Moss SJ, Troyanovsky S, Attwell D, Longmore GD, Braga VM (2003) The LIM protein Ajuba is recruited to cadherin-dependent cell junctions through an association with alpha-catenin. J Biol Chem 278:1220-1228.

Massie A, Vandesande F, Arckens L (2001) Expression of the high-affinity glutamate transporter EAAT4 in mammalian cerebral cortex. Neuroreport 12:393-397.

Masson J, Sagne C, Hamon M, El Mestikawy S (1999) Neurotransmitter transporters in the central nervous system. Pharmacol Rev 51:439-464.

Matute C, Melone M, Vallejo-Illarramendi A, Conti F (2005) Increased expression of the astrocytic glutamate transporter GLT-1 in the prefrontal cortex of schizophrenics. Glia 49:451-455.

McCool BA, Pin JP, Brust PF, Harpold MM, Lovinger DM (1996) Functional coupling of rat group II metabotropic glutamate receptors to an omega-conotoxin GVIA-sensitive calcium channel in human embryonic kidney 293 cells. Mol Pharmacol 50:912-922.

McCullumsmith RE, Meador-Woodruff JH (2002) Striatal excitatory amino acid transporter transcript expression in schizophrenia, bipolar disorder, and major depressive disorder. Neuropsychopharmacology 26:368-375.

McGee AW, Bredt DS (2003) Assembly and plasticity of the glutamatergic postsynaptic specialization. Curr Opin Neurobiol 13:111-118.

Meador-Woodruff JH, Davis KL, Haroutunian V (2001) Abnormal kainate receptor expression in prefrontal cortex in schizophrenia. Neuropsychopharmacology 24:545-552.

Meador-Woodruff JH, Reyes, E., Haroutunian, V, Kristiansen, LV (2005) Abnormal Expression of Transcripts for GKAP and Shank Suggest Alterations of Interactions Between Metabotropic and Ionotropic Glutamate Receptors in Schizophrenic Brain. In: ACNP 44th Annual Meeting (Nemeroff CB, ed), p S247. Waikoloa, Hawaii: Neuropsychopharmacology.

Milton ID, Banner SJ, Ince PG, Piggott NH, Fray AE, Thatcher N, Horne CH, Shaw PJ (1997) Expression of the glial glutamate transporter EAAT2 in the human CNS: an immunohistochemical study. Brain Res Mol Brain Res 52:17-31.

Mirnics K, Middleton FA, Marquez A, Lewis DA, Levitt P (2000) Molecular characterization of schizophrenia viewed by microarray analysis of gene expression in prefrontal cortex. Neuron 28:53-67.

Mirnics K, Middleton FA, Stanwood GD, Lewis DA, Levitt P (2001) Disease-specific changes in regulator of G-protein signaling 4 (RGS4) expression in schizophrenia. Mol Psychiatry 6:293-301.

Morris DW, Rodgers A, McGhee KA, Schwaiger S, Scully P, Quinn J, Meagher D, Waddington JL, Gill M, Corvin AP (2004) Confirming RGS4 as a susceptibility gene for schizophrenia. Am J Med Genet B Neuropsychiatr Genet 125:50-53.

Mueller HT, Meador-Woodruff JH (2004) NR3A NMDA receptor subunit mRNA expression in schizophrenia, depression and bipolar disorder. Schizophr Res 71:361-370.

Nagao S, Kwak S, Kanazawa I (1997) EAAT4, a glutamate transporter with properties of a chloride channel, is predominantly localized in Purkinje cell dendrites, and forms parasagittal compartments in rat cerebellum. Neuroscience 78:929-933.

Najimi M, Maloteaux JM, Hermans E (2002) Cytoskeleton-related trafficking of the EAAC1 glutamate transporter after activation of the G(q/11)-coupled neurotensin receptor NTS1. FEBS Lett 523:224-228.

Nakanishi S (1992) Molecular diversity of glutamate receptors and implications for brain function. Science 258:597-603.

Newell KA, Zavitsanou K, Huang XF (2005) Ionotropic glutamate receptor binding in the posterior cingulate cortex in schizophrenia patients. Neuroreport 16:1363-1367.

Noga JT, Hyde TM, Bachus SE, Herman MM, Kleinman JE (2001) AMPA receptor binding in the dorsolateral prefrontal cortex of schizophrenics and controls. Schizophr Res 48:361-363.

Norton N, Williams HJ, Williams NM, Spurlock G, Zammit S, Jones G, Jones S, Owen R, O'Donovan MC, Owen MJ (2003) Mutation screening of the Homer gene family and association analysis in schizophrenia. Am J Med Genet B Neuropsychiatr Genet 120:18-21.

Nudmamud S, Reynolds GP (2001) Increased density of glutamate/N-methyl-D-aspartate receptors in superior temporal cortex in schizophrenia. Neurosci Lett 304:9-12.

Ohnuma T, Augood SJ, Arai H, McKenna PJ, Emson PC (1998) Expression of the human excitatory amino acid transporter 2 and metabotropic glutamate receptors 3 and 5 in the prefrontal cortex from normal individuals and patients with schizophrenia. Brain Res Mol Brain Res 56:207-217.

Ohnuma T, Tessler S, Arai H, Faull RL, McKenna PJ, Emson PC (2000a) Gene expression of metabotropic glutamate receptor 5 and excitatory amino acid transporter 2 in the schizophrenic hippocampus. Brain Res Mol Brain Res 85:24-31.

Ohnuma T, Kato H, Arai H, Faull RL, McKenna PJ, Emson PC (2000b) Gene expression of PSD95 in prefrontal cortex and hippocampus in schizophrenia. Neuroreport 11:3133-3137.

Perlstein WM, Carter CS, Noll DC, Cohen JD (2001) Relation of prefrontal cortex dysfunction to working memory and symptoms in schizophrenia. Am J Psychiatry 158:1105-1113.

Pickering DS, Thomsen C, Suzdak PD, Fletcher EJ, Robitaille R, Salter MW, MacDonald JF, Huang XP, Hampson DR (1993) A comparison of two alternatively spliced forms of a metabotropic glutamate receptor coupled to phosphoinositide turnover. J Neurochem 61:85-92.

Pin JP, Duvoisin R (1995) The metabotropic glutamate receptors: structure and functions. Neuropharmacology 34:1-26.

Pin JP, Waeber C, Prezeau L, Bockaert J, Heinemann SF (1992) Alternative splicing generates metabotropic glutamate receptors inducing different patterns of calcium release in Xenopus oocytes. Proc Natl Acad Sci USA 89:10331-10335.

Prabakaran S, Swatton JE, Ryan MM, Huffaker SJ, Huang JT, Griffin JL, Wayland M, Freeman T, Dudbridge F, Lilley KS, Karp NA, Hester S, Tkachev D, Mimmack ML, Yolken RH, Webster MJ, Torrey EF, Bahn S (2004) Mitochondrial dysfunction in schizophrenia: evidence for compromised brain metabolism and oxidative stress. Mol Psychiatry 9:684-697, 643.

Prezeau L, Carrette J, Helpap B, Curry K, Pin JP, Bockaert J (1994) Pharmacological characterization of metabotropic glutamate receptors in several types of brain cells in primary cultures. Mol Pharmacol 45:570-577.

Quintana J, Wong T, Ortiz-Portillo E, Marder SR, Mazziotta JC (2004) Anterior cingulate dysfunction during choice anticipation in schizophrenia. Psychiatry Res 132:117-130.

Rauen T, Kanner BI (1994) Localization of the glutamate transporter GLT-1 in rat and macaque monkey retinae. Neurosci Lett 169:137-140.

Rizig MA, McQuillin A, Puri V, Choudhury K, Datta S, Thirumalai S, Lawrence J, Quested D, Pimm J, Bass N, Lamb G, Moorey H, Badacsonyi A, Kelly K, Morgan J, Punukollu B, Kandasami G, Curtis D, Gurling H (2006) Failure to confirm genetic association between schizophrenia and markers on chromosome 1q23.3 in the region of the gene encoding the regulator of G-protein signaling 4 protein (RGS4). Am J Med Genet B Neuropsychiatr Genet 141:296-300.

Rothstein JD, Martin L, Levey AI, Dykes-Hoberg M, Jin L, Wu D, Nash N, Kuncl RW (1994) Localization of neuronal and glial glutamate transporters. Neuron 13:713-725.

Rothstein JD, Dykes-Hoberg M, Pardo CA, Bristol LA, Jin L, Kuncl RW, Kanai Y, Hediger MA, Wang Y, Schielke JP, Welty DF (1996) Knockout of glutamate transporters reveals a major role for astroglial transport in excitotoxicity and clearance of glutamate. Neuron 16:675-686.

Ruggiero A, Vidensky S, Rothstein JD (2003) GTRAP3-18 protein is able to regulate the activity of excitatory amino acid transporters through alterations in ASN linked glycosyl processing. Abstract viewer, Society for Neuroscience:Program No. 372.315.

Rutter AR, Freeman FM, Stephenson FA (2002) Further characterization of the molecular interaction between PSD-95 and NMDA receptors: the effect of the NR1 splice variant and evidence for modulation of channel gating. J Neurochem 81:1298-1307.

Salt TE, Eaton SA (1996) Functions of ionotropic and metabotropic glutamate receptors in sensory transmission in the mammalian thalamus. Prog Neurobiol 48:55-72.

Saugstad JA, Kinzie JM, Shinohara MM, Segerson TP, Westbrook GL (1997) Cloning and expression of rat metabotropic glutamate receptor 8 reveals a distinct pharmacological profile. Mol Pharmacol 51:119-125.

Scarr E, Beneyto M, Meador-Woodruff JH, Deans B (2005) Cortical glutamatergic markers in schizophrenia. Neuropsychopharmacology 30:1521-1531.

Schluter K, Figiel M, Rozyczka J, Engele J (2002) CNS region-specific regulation of glial glutamate transporter expression. Eur J Neurosci 16:836-842.

Schmitt A, Asan E, Puschel B, Kugler P (1997) Cellular and regional distribution of the glutamate transporter GLAST in the CNS of rats: nonradioactive in situ hybridization and comparative immunocytochemistry. J Neurosci 17:1-10.

Schmitt A, Asan E, Lesch KP, Kugler P (2002) A splice variant of glutamate transporter GLT1/EAAT2 expressed in neurons: cloning and localization in rat nervous system. Neuroscience 109:45-61.

Schoepp DD (1994) Novel functions for subtypes of metabotropic glutamate receptors. Neurochem Int 24:439-449.

Schweneker M, Bachmann AS, Moelling K (2005) JM4 is a four-transmembrane protein binding to the CCR5 receptor. FEBS Lett 579:1751-1758.

Scott HL, Pow DV, Tannenberg AE, Dodd PR (2002) Aberrant expression of the glutamate transporter excitatory amino acid transporter 1 (EAAT1) in Alzheimer's disease. J Neurosci 22:RC206.

Sheng M, Pak DT (2000) Ligand-gated ion channel interactions with cytoskeletal and signaling proteins. Annu Rev Physiol 62:755-778.

Shigeri Y, Seal RP, Shimamoto K (2004) Molecular pharmacology of glutamate transporters, EAATs and VGLUTs. Brain Research - Brain Research Reviews 45:250-265, 2004 Jul.

Silbersweig DA, Stern E, Frith C, Cahill C, Holmes A, Grootoonk S, Seaward J, McKenna P, Chua SE, Schnorr L, et al. (1995) A functional neuroanatomy of hallucinations in schizophrenia. Nature 378:176-179.

Sims KD, Straff DJ, Robinson MB (2000) Platelet-derived growth factor rapidly increases activity and cell surface expression of the EAAC1 subtype of glutamate transporter through activation of phosphatidylinositol 3-kinase. J Biol Chem 275:5228-5237.

Smith RE, Haroutunian V, Davis KL, Meador-Woodruff JH (2002) Expression of glutaminase transcripts
in the thalamus in schizophrenia. Biological Psychiatry 51:25.

Sobell JL, Richard C, Wirshing DA, Heston LL (2005) Failure to confirm association between RGS4 haplotypes and schizophrenia in Caucasians. Am J Med Genet B Neuropsychiatr Genet 139:23-27.

Sokolov BP (1998) Expression of NMDAR1, GluR1, GluR7, and KA1 glutamate receptor mRNAs is decreased in frontal cortex of "neuroleptic-free" schizophrenics: evidence on reversible up-regulation by typical neuroleptics. J Neurochem 71:2454-2464.

Sokolov MV, Rossokhin AV, A MK, Gasparini S, Berretta N, Cherubini E, Voronin LL (2003) Associative mossy fibre LTP induced by pairing presynaptic stimulation with postsynaptic hyperpolarization of CA3 neurons in rat hippocampal slice. Eur J Neurosci 17:1425-1437.

Takamori S, Rhee JS, Rosenmund C, Jahn R (2000) Identification of a vesicular glutamate transporter that defines a glutamatergic phenotype in neurons. Nature 407:189-194.

Takamori S, Malherbe P, Broger C, Jahn R (2002) Molecular cloning and functional characterization of human vesicular glutamate transporter 3. EMBO Rep 3:798-803.

Talkowski ME, Seltman H, Bassett AS, Brzustowicz LM, Chen X, Chowdari KV, Collier DA, Cordeiro Q, Corvin AP, Deshpande SN, Egan MF, Gill M, Kendler KS, Kirov G, Heston LL, Levitt P, Lewis DA, Li T, Mirnics K, Morris DW, Norton N, O'Donovan MC, Owen MJ, Richard C, Semwal P, Sobell JL, St Clair D, Straub RE, Thelma BK, Vallada H, Weinberger DR, Williams NM, Wood J, Zhang F, Devlin B, Nimgaonkar VL (2006) Evaluation of a susceptibility gene for schizophrenia: genotype based meta-analysis of RGS4 polymorphisms from thirteen independent samples. Biol Psychiatry 60:152-162.

Tamminga C (1999) Glutamatergic aspects of schizophrenia. Br J Psychiatry Suppl:12-15.

Tamminga CA, Thaker GK, Buchanan R, Kirkpatrick B, Alphs LD, Chase TN, Carpenter WT (1992) Limbic system abnormalities identified in schizophrenia using positron emission tomography with fluorodeoxyglucose and neocortical alterations with deficit syndrome. Arch Gen Psychiatry 49:522-530.

Tanaka K, Watase K, Manabe T, Yamada K, Watanabe M, Takahashi K, Iwama H, Nishikawa T, Ichihara N, Kikuchi T, Okuyama S, Kawashima N, Hori S, Takimoto M, Wada K (1997) Epilepsy and exacerbation of brain injury in mice lacking the glutamate transporter GLT-1. Science 276:1699-1702.

Tarsy D, Baldessarini RJ, Tarazi FI (2002) Effects of newer antipsychotics on extrapyramidal function. CNS Drugs 16:23-45.

Theberge J, Bartha R, Drost DJ, Menon RS, Malla A, Takhar J, Neufeld RW, Rogers J, Pavlosky W, Schaefer B, Densmore M, Al-Semaan Y, Williamson PC (2002) Glutamate and glutamine measured with 4.0 T proton MRS in never-treated patients with schizophrenia and healthy volunteers. Am J Psychiatry 159:1944-1946.

Toyooka K, Iritani S, Makifuchi T, Shirakawa O, Kitamura N, Maeda K, Nakamura R, Niizato K, Watanabe M, Kakita A, Takahashi H, Someya T, Nawa H (2002) Selective reduction of a PDZ protein, SAP-97, in the prefrontal cortex of patients with chronic schizophrenia. J Neurochem 83:797-806.

Utsunomiya-Tate N, Endou H, Kanai Y (1996) Cloning and functional characterization of a system ASC-like Na+- dependent neutral amino acid transporter. J Biol Chem 271:14883-14890.

Varoqui H, Schafer MK, Zhu H, Weihe E, Erickson JD (2002) Identification of the differentiation-associated Na+/PI transporter as a novel vesicular glutamate transporter expressed in a distinct set of glutamatergic synapses. J Neurosci 22:142-155.

Vawter MP, Crook JM, Hyde TM, Kleinman JE, Weinberger DR, Becker KG, Freed WJ (2002) Microarray analysis of gene expression in the prefrontal cortex in schizophrenia: a preliminary study. Schizophr Res 58:11-20.

Volterra A, Trotti D, Racagni G (1994) Glutamate uptake is inhibited by arachidonic acid and oxygen radicals via two distinct and additive mechanisms. Mol Pharmacol 46:986-992.

Watanabe M, Robinson MB, Kalandadze A, Rothstein JD (2003) GPS1, interacting protein with GLT-1. Program No 37216 2003 SFN Abstract.

Weinberger DR, Egan MF, Bertolino A, Callicott JH, Mattay VS, Lipska BK, Berman KF, Goldberg TE (2001) Prefrontal neurons and the genetics of schizophrenia. Biol Psychiatry 50:825-844.

Wheal HV, Thomson AM (1995) Excitatory Amino Acids and Synaptic Transmission (2nd ed.). Academis Press, New York.

Williams NM, Preece A, Spurlock G, Norton N, Williams HJ, McCreadie RG, Buckland P, Sharkey V, Chowdari KV, Zammit S, Nimgaonkar V, Kirov G, Owen MJ, O'Donovan MC (2004) Support for RGS4 as a susceptibility gene for schizophrenia. Biol Psychiatry 55:192-195.

Wilson NR, Kang J, Hueske EV, Leung T, Varoqui H, Murnick JG, Erickson JD, Liu G
(2005a) Presynaptic regulation of quantal size by the vesicular glutamate transporter
VGLUT1. Journal of Neuroscience 25:6221-6234, 2005 Jun 6229.

Wilson NR, Kang J, Hueske EV, Leung T, Varoqui H, Murnick JG, Erickson JD, Liu G
(2005b) Presynaptic regulation of quantal size by the vesicular glutamate transporter
VGLUT1. J Neurosci 25:6221-6234.

Woo TU, Walsh JP, Benes FM (2004) Density of glutamic acid decarboxylase 67 messen-
ger RNA-containing neurons that express the N-methyl-D-aspartate receptor subunit
NR2A in the anterior cingulate cortex in schizophrenia and bipolar disorder. Arch Gen
Psychiatry 61:649-657.

Xia P, Pei G, Schwarz W (2006) Regulation of the glutamate transporter EAAC1 by expression and activation of delta-opioid receptor. Eur J Neurosci 24:87-93.

Yamada K, Watanabe M, Shibata T, Tanaka K, Wada K, Inoue Y (1996) EAAT4 is a postsynaptic glutamate transporter at Purkinje cell synapses. Neuroreport 7:2013-2017.

Yu YX, Shen L, Xia P, Tang YW, Bao L, Pei G (2006) Syntaxin 1A promotes the endo-
cytic sorting of EAAC1 leading to inhibition of glutamate transport. J Cell Sci
119:3776-3787.

Zakharenko SS, Zablow L, Siegelbaum SA (2002) Altered presynaptic vesicle release and
cycling during mGluR-dependent LTD. Neuron 35:1099-1110.

Zavitsanou K, Ward PB, Huang XF (2002) Selective alterations in ionotropic glutamate receptors in the anterior cingulate cortex in schizophrenia. Neuropsychopharmacology
27:826-833.

Zerangue N, Kavanaugh MP (1996) ASCT-1 is a neutral amino acid exchanger with chloride channel activity. J Biol Chem 271:27991-27994.

Zhang F, St Clair D, Liu X, Sun X, Sham PC, Crombie C, Ma X, Wang Q, Meng H, Deng
W, Yates P, Hu X, Walker N, Murray RM, Collier DA, Li T (2005) Association analy-
sis of the RGS4 gene in Han Chinese and Scottish populations with schizophrenia.
Genes Brain Behav 4:444-448.

Zhu Y, Fei J, Schwarz W (2005) Expression and transport function of the glutamate transporter EAAC1 in Xenopus oocytes is regulated by syntaxin 1A. J Neurosci Res 79:503-
508.

Chapter 7

Cortical Circuit Dysfunction in the NMDA Receptor Hypofunction Model for the Psychosis of Schizophrenia

Robert Greene

7.1 What Local Circuits are Relevant?

The psychosis of schizophrenia can often be striking in its presentation, in large part due to the bizarreness of psychotic behavior and thinking. It is nonetheless an extremely subtle dysfunction that has proven remarkably difficult to define. The most widely recognized cognitive deficit associated with schizophrenia is a working memory abnormality. Cognitive deficits assessed by the more traditional neuropsychological tests, including the Wisconsin Card Sort and Stroop tests indicate abnormalities consistent with dysfunction of working memory systems although these tests may be affected by the functionality of other systems. Tests, selective for working memory dysfunction, revealed deficits in spatial but not verbal working memory, in schizophrenic patients (Park et al. 1995). A dysfunctional spatial working memory may contribute to the core psychotic symptoms of thought disorder and delusions but does not seem sufficient to account for them.

Hippocampal dependent processing is essential for episodic memory in animal models and is a fundamentally necessary component of declarative memory in humans (Eichenbaum 2004). If declarative memory were abnormal (rather than absent as with amnesia) then abnormalities of thought might be forthcoming as with thought disorder of psychosis. In fact, a number of recent studies using fMRI and hippocampal related learning and memory tasks have demonstrated an abnormal activation of the hippocampus in schizophrenic patients (Heckers et al. 1998; Heckers 2001; Titone et al. 2004; Weiss et al. 2004).

The encoding of an unambiguous and unique configural representation of an episode requires an intact hippocampus (Leutgeb et al. 2005; Rajji et al. 2006).

Encoding of this kind of representation may be essential for acquisition of salient episodic configurations (i.e., contextual cues and their associations) and may be considered a form of contextual information processing. Patients suffering from schizophrenia are especially deficient in performance of tasks requiring contextual information processing (Bazin et al. 2000; Martins et al. 2001; Servan-Schreiber et al. 1996; Stratta et al. 2000; Titone et al. 2000), consistent with dysfunctional hippocampal processing.

7.2 What are the Potentially Relevant Characteristics of Local Circuits?

It is now well known that a psychotic state clinically indistinguishable from the psychosis of schizophrenia may be acutely induced by antagonists of the NMDA receptor (NMDAR) (Javitt and Zukin 1991). Further, formal neuropsychological testing shows similar deficits for the two kinds of psychotic states (Krystal et al. 1994; Newcomer and Krystal 2001). The possibility is then raised that the reduction of NMDAR-dependent function induces an acute psychosis by a similar or related mechanism to the pathophysiological mechanism of schizophrenia.

Because of the acute and reversible effects of NMDAR antagonists on humans, this mechanism cannot be attributed to loss of neurons or a pathological neurodevelopmental effect. Nonetheless, other NMDAR antagonists models have posited these chronic effects of NMDAR antagonism as important contributors to schizophrenic pathology (Farber et al. 1995; Jentsch and Roth 1999). The repercussions of long term exposure to NMDAR antagonists compared to acute exposures, on local circuit function have not been well characterized. For the purposes of this short review, the focus will be on acute effects. This seems reasonable since acute exposure to NMDAR antagonists is sufficient to induce a reversible state of schizophrenic-like psychosis and thus the mechanism(s) of this induction is/are clearly of interest.

Since NMDARs have a functional role in most circuits in the CNS, it seems a surprise that a subtle functional deficit, expressed as psychosis, is for the most part, all that is expressed by blockade of these receptors. It should be kept in mind that the psychotomimetic properties are dose dependent and that at higher dose; the not so subtle effect is that of a dissociative anesthesia with loss of consciousness. This is obviously not the same effect that is induced by a lower dose leading to a state of psychosis.

Taken together, these behavioral clues of response to NMDAR antagonism indicate that the circuits relevant to psychosis related behavior and state could have a particular vulnerability to NMDAR antagonists and further that this vulnerability may provide insight into psychosis related mechanisms. As noted above, a good place to look for sensitive NMDAR antagonists effects that could lead to psychotic thought disorder, due to dysfunctional declarative memory is the hippocampus and its circuits.

In the CA1 field of the hippocampus, NMDAR activation on pyramidal neurons leads to a depolarization-dependent influx of calcium, ultimately resulting in the best characterized form of Long Term Potentiation (LTP) of excitatory synapses (Bliss and Collingridge 1993). The required timely coupling of post-synaptic NMDAR activation with depolarization is thought to underlie a Hebbian form of learning. Regionally constrained knock out models suggest that intact NMDAR function in the hippocampus is necessary for episodic learning and memory (McHugh et al. 1996; Nakazawa et al. 2004; Rajji et al. 2006).

NMDAR activation also occurs on interneurons in CA1(Hajos et al. 2002; Sah et al. 1990) and contributes to activation of interneurons in the recurrent inhibitory circuit (Grunze et al. 1996). The NMDAR role may account for 20-30% of the inhibitory postsynaptic potential (IPSP) amplitude recorded in CA1 pyramidal cells in response to a single stimulation activation of their recurrent collaterals onto the interneurons (Grunze et al. 1996). This suggests less of a voltage dependence of interneuron NMDARs compared to pyramidal cells, since the contribution to the IPSP did not require coincident post-synaptic activation and is consistent with a different sub-unit composition of NMDARs on interneurons compared to pyramidal cells (Hajos et al. 2002; Monyer et al. 1994; Nyiri et al. 2003).

The consequence of a reduction of NMDAR activity at interneuron glutamatergic synapses is likely to disinhibit pyramidal cells (I-P connection) as well as other interneurons (through I-I connections) in a dynamic manner. Because of the slower decay of NMDAR generated synaptic potentials, not only may the amplitude of a single glutamatergic EPSP be reduced by NMDAR antagonism, but the duration of the depolarization may be also be reduced so that integration of multiple EPSPs is less likely (Maccaferri and Dingledine 2002). The integration mediated by NMDAR currents may be important for maintenance of a GABA tone on post-synaptic targets that is superimposed on top of precisely timed AMPAR driven GABA IPSPs.

The most sensitive circuit effect that we have observed in response to NMDAR antagonism is a loss of activity dependent long term enhancement (LTP) of recurrent inhibition in CA1 neurons. This was measured as LTP of an IPSP recorded in pyramidal cells in response to a tetanic stimulation of the alveus, that activates, recurrent CA1 pyramidal cell to interneuron, axons (Grunze et al. 1996). This LTP of the recurrent inhibitory circuit was blocked by NMDA antagonists, including the competitive blocker, 2-amino-5-phosphonovalerate (APV). Field recordings from the same CA1 pyramidal neurons indicated that blockade of LTP of the recurrent inhibitory circuit was 10 times more sensitive to APV than LTP of the direct feed forward CA3 to CA1, Pyramidal cell to Pyramidal cell circuit. Thus, it is possible to selectively reduce the recurrent inhibitory circuit plasticity function while at the same time preserving that of the feedforward circuits.

The mechanism for the APV mediated blockade of LTP of the recurrent inhibitory circuit may not be a result of a direct action on NMDAR-dependent LTP. Feedforward inhibition undergoes a Hebbian form of LTP at the glutamate synapse driving the interneuron (P-I synapse) but this does not appear to be NMDAR dependent (Perez et al., 2001; Topolnik et al. 2006). Recently, a new form of LTP of interneurons in CA1 that participate in recurrent inhibitory circuits has been described (Lamsa et al. 2007). It is remarkable for requiring post-synaptic inactivation (or hyperpolarization) to be paired with incoming bursts of glutamatergic synaptic activity, and has been accordingly labeled an anti-Hebbian form of LTP. However, this form of LTP is also NMDAR independent.

How then, might an NMDAR-dependent LTP of the recurrent inhibitory circuit occur? The answer may lie with the role of interneuron NMDAR EPSPs to integrate bursts of excitatory input into a more tonic GABAergic output as follows. NMDAR activation of interneuronal activity differs from that mediated by AMPAR activation in part because of the longer time course of decay of the NMDAR-dependent component of the glutamatergic EPSP. The rapid decay time of AMPARs may be important for fast timing of synchronizing oscillatory activity of interneurons (Bartos et al. 2002; Geiger et al. 1997), whereas the slower decay of the NMDAR currents may be required for the integration of multiple P-I EPSPs that can contribute to a GABAergic tone on downstream targets of interneurons. Other interneurons are primary targets, thus following a burst of excitatory P-I input, as with a theta burst (that can induce LTP), downstream target neurons may undergo a tonic GABAergic inhibition, induced by NMDAR-dependent integration of excitatory P-I input. In P cells this will prevent or reduce a Hebbian form of LTP, but in I cells, including those involved in recurrent inhibition, anti-Hebbian LTP would be enhanced (Lamsa et al. 2007). Thus, a reduced NMDAR-dependent activation of I cells could reduce this anti-Hebbian form of LTP, and account for the experimentally observed reduction of recurrent inhibitory LTP in the presence of the NMDAR antagonist, APV. In addition, a tonic GABAergic might narrow the time window of AMPAR excitation to effectively induce an action potential in either P or I cells thus further sharpening the synchronized activity of hippocampal cells.

The changes in circuit function brought about by acutely reduced NMDAR activity will predictably alter the way that information is processed in the hippocampus. Based on electrophysiological realistic modeling of hippocampal networks, the outcome is likely to involve an aberrant lateral spread of excitation that is normally constrained by NMDAR dependent inhibitory interneuron dependent activation. Thus, a less constrained ensemble pattern of neuronal activation might be less distinguishable form another ensemble pattern. This kind of deficit in neuronal processing may be associated with psychotic thought disorder, to the extent that the disorder depends on disambiguation of distinct ensemble patterns of neuronal firing. In any case, an understanding of the effects of reduced NMDAR function can draw attention to neuronal mechanisms, including NMDAR function, interneuron function and other modulatory effects on interneurons as outlined above that are likely to be relevant to the expression of psychosis.

References

Bartos M, Vida I, Frotscher M, Meyer A, Monyer H, Geiger JR, Jonas P (2002) Fast synaptic inhibition promotes synchronized gamma oscillations in hippocampal interneuron networks. Proc Natl Acad Sci USA 99:13222-13227.

Bazin N, Perruchet P, Hardy-Bayle MC, Feline A (2000) Context-dependent information processing in patients with schizophrenia. Schizophr Res 45:93-101.

Bliss TV, Collingridge GL (1993) A synaptic model of memory: long-term potentiation in the hippocampus. Nature 361:31-39.

Eichenbaum H (2004) Hippocampus: cognitive processes and neural representations that underlie declarative memory. Neuron 44:109-120.

Farber NB, Wozniak DF, Price MT, Labruyere J, Huss J, St.Peter H, Olney JW (1995) Age-specific neurotoxicity in the rat associated with NMDA receptor blockade: Potential relevance to schizophrenia. Biol Psychiatry 38:788-796.

Geiger JR, Lubke J, Roth A, Frotscher M, Jonas P (1997) Submillisecond AMPA receptor-mediated signaling at a principal neuron-interneuron synapse. Neuron 18:1009-1023.

Grunze HC, Rainnie DG, Hasselmo ME, Barkai E, Hearn EF, McCarley RW, Greene RW (1996) NMDA-dependent modulation of CA1 local circuit inhibition. J Neurosci 16:2034-2043.

Hajos N, Freund TF, Mody I (2002) Comparison of single NMDA receptor channels recorded on hippocampal principal cells and oriens/alveus interneurons projecting to stratum lacunosum-moleculare (O-LM cells). Acta Biol Hung 53:465-472.

Heckers S (2001) Neuroimaging studies of the hippocampus in schizophrenia. Hippocampus 11:520-528.

Heckers S, Rauch SL, Goff D, Savage CR, Schacter DL, Fischman AJ, Alpert NM (1998) Impaired recruitment of the hippocampus during conscious recollection in schizophrenia. Nat Neurosci 1:318-323.

Javitt DC, Zukin SR (1991) Recent advances in the phencyclidine model of schizophrenia. Am J Psychiatry 148:1301-1308.

Jentsch JD, Roth RH (1999) The neuropsychopharmacology of phencyclidine: from NMDA receptor hypofunction to the dopamine hypothesis of schizophrenia. Neuropsychopharmacology 20:201-225.

Krystal JH, Karper LP, Seibyl JP, Freeman GK, Delaney R, Bremner JD, Heninger GR, Bowers MB, Charney DS (1994) Subanesthetic effects of the noncompetitive NMDA antagonist, ketamine, in humans. Psychotomimetic, perceptual, cognitive, and neuroendocrine responses. Arch Gen Psychiatry 51:199-214.

Lamsa KP, Heeroma JH, Somogyi P, Rusakov DA, Kullmann DM (2007) Anti-Hebbian long-term potentiation in the hippocampal feedback inhibitory circuit. Science 315:1262-1266.

Leutgeb S, Leutgeb JK, Moser MB, Moser EI (2005) Place cells, spatial maps and the population code for memory. Curr Opin Neurobiol 15:738-746.

Maccaferri G, Dingledine R (2002) Control of feedforward dendritic inhibition by NMDA receptor-dependent spike timing in hippocampal interneurons. J Neurosci 22:5462-5472.

Martins SA, Jones SH, Toone B, Gray JA (2001) Impaired associative learning in chronic schizophrenics and their first-degree relatives: a study of latent inhibition and the Kamin blocking effect. Schizophr Res 48:273-289.

McHugh TJ, Blum KI, Tsien JZ, Tonegawa S, Wilson MA (1996) Impaired hippocampal representation of space in CA1-specific NMDAR1 knockout mice. Cell 87:1339-1349.

Monyer H, Burnashev N, Laurie DJ, Sakmann B, Seeburg PH (1994) Developmental and regional expression in the rat brain and functional properties of four NMDA receptors. Neuron 12:529-540.

Nakazawa K, McHugh TJ, Wilson MA, Tonegawa S (2004) NMDA receptors, place cells and hippocampal spatial memory. Nat Rev Neurosci 5:361-372.

Newcomer JW, Krystal JH (2001) NMDA receptor regulation of memory and behavior in humans. Hippocampus 11:529-542.

Nyiri G, Stephenson FA, Freund TF, Somogyi P (2003) Large variability in synaptic N-methyl-D-aspartate receptor density on interneurons and a comparison with pyramidal-cell spines in the rat hippocampus. Neuroscience 119:347-363.

Park S, Holzman PS, Goldman-Rakic PS (1995) Spatial working memory deficits in the relatives of schizophrenic patients. Arch Gen Psychiatry 52:821-828.

Perez Y, Morin F, Lacaille JC (2001) A hebbian form of long-term potentiation dependent on mGluR1a in hippocampal inhibitory interneurons. Proc Natl Acad Sci USA 98:9401-9406.

Rajji T, Chapman D, Eichenbaum H, Greene R (2006) The role of CA3 hippocampal NMDA receptors in paired associate learning. J Neurosci 26:908-915.

Sah P, Hestrin S, Nicoll RA (1990) Properties of excitatory postsynaptic currents recorded in vitro from rat hippocampal interneurones. J Physiol 430:605-616.

Servan-Schreiber D, Cohen JD, Steingard S (1996) Schizophrenic deficits in the processing of context. A test of a theoretical model. Arch Gen Psychiatry 53:1105-1112.

Stratta P, Daneluzzo E, Bustini M, Prosperini P, Rossi A (2000) Processing of context information in schizophrenia: relation to clinical symptoms and WCST performance. Schizophr Res 44:57-67.

Titone D, Ditman T, Holzman PS, Eichenbaum H, Levy DL (2004) Transitive inference in schizophrenia: impairments in relational memory organization. Schizophr Res 68:235-247.

Titone D, Prentice KJ, Wingfield A (2000) Resource allocation during spoken discourse processing: effects of age and passage difficulty as revealed by self-paced listening. Mem Cognit 28:1029-1040.

Topolnik L, Azzi M, Morin F, Kougioumoutzakis A, Lacaille JC (2006) mGluR1/5 subtype-specific calcium signalling and induction of long-term potentiation in rat hippocampal oriens/alveus interneurones. J Physiol 575:115-131.

Weiss AP, Zalesak M, DeWitt I, Goff D, Kunkel L, Heckers S (2004) Impaired hippocampal function during the detection of novel words in schizophrenia. Biol Psychiatry 55:668-675.

Chapter 8

Alterations in Hippocampal Function in Schizophrenia: its Genetic Associations and Systems Implications

Subroto Ghose and Carol A. Tamminga

8.1 Introduction: Domains of Dysfunction in Schizophrenia

Schizophrenia is characterized by a constellation of symptoms that include psychosis, cognitive deficits and negative symptoms. This has been verified in several investigations using factor analysis of large schizophrenic patient populations showing these distinct domains of dysfunction. When clinical symptoms are related to imaging findings, specific brain areas are found to be differentially involved in symptom domains in the illness (Liddle 1987; Carpenter and Buchanan 1989; Kay and Sevy 1990; Arndt et al. 1991; Lenzenweger et al. 1991; Barnes and Liddle 1990; Andreasen et al 1995), with prefrontal (PFC) most involved in executive function and medial temporal lobe (MTL) most associated with declarative memory. Studies defining the anatomy of schizophrenia show delimited regions of brain associated with the symptoms; these include the prefrontal cortex (both dorsolateral prefrontal cortex, DLPFC, and anterior cingulate cortex, ACC) and the MTL as well as the superior temporal gyrus (STG). In this chapter, we review a large and significant body of convergent evidence that demonstrates the importance of the medial temporal lobe as a brain region central to the pathophysiology of schizophrenia.

The extent to which symptom domains in schizophrenia are independent of each other is not clear; but heuristically, they are often taken to be such (Neuchterlein et al 2004). Here, we will use a discussion of the cognition domain as illustrative. Considerable evidence suggests that cognition is, at least in part, an independent symptom domain: cognitive disabilities in schizophrenia have a unique disease course, often beginning before psychosis onset and lasting without improvement across disease years (Cornblatt and Erlenmeyer-Kimling 1985). Cognitive dysfunction is similar in new and chronic schizophrenic persons (Bilder et al 1995; Gur et al 1999, 2001; Hicks and Marsh 1999; Saykin et al 1994), between adult-onset and adolescent-onset schizophrenia (Brickman et al 2004; Kravariti et al 2003); and, to some degree, similar in family members (Niendam et al 2003) and in high-risk populations (Brewer et al 1998). The profile of cognitive disabilities has been shown to be largely distinct in schizophrenia compared with non-schizophrenia psychosis (Barch et al 2004). Moreover, cognitive deficits show only small to moderate correlations with clinical state (Censits et al 1997), incomplete improvement with antipsychotic drug treatment (APD) (Goldberg and Weinberger 1996), and either no or small gender effects (Kurtz et al 2001; Tamminga and Holcomb 2004).

The nature of cognitive dysfunction has been extensively examined in schizophrenia (Neuchterlein et al 2004; Braff 1993; Dickinson et al 2004; Flashman and Green 2004; Gur et al 2001; Heinrichs and Zakzanis 1998). Persons with the illness present a wide range of cognitive disturbances thought to be basic to the illness (Green et al 2000), in addition to psychotic symptoms. It is particularly in the areas of memory and attention that deficits appear most pronounced (Flashman and Green 2004). Yet, cognitive deficits show diversity among affected individuals (Bruder et al 2004; Hill et al 2002). In schizophrenia, while a generalized compromise in cognition is widely acknowledged, certain domains of function stand out as particularly affected, including visual and verbal declarative memory, working memory and processing speed (Saykin et al 1994; Dickinson et al 2004; Heinrichs and Zakzanis 1998; Eichenbaum 2000). The degree to which these distinct cognitive profiles result from a primary or from distinct etiologies has been argued and is an issue important for therapeutics (Neuchterlein et al 2004).

Declarative memory is one of the most reliably impaired neuropsychological (NP) functions in schizophrenia (Neuchterlein et al 2004; Heinrichs and Zakzanis 1998; Saykin et al 1994; Keefe et al 1997; Stone et al 1998). Poor declarative memory performance has been associated with reductions in hippocampal volume and reductions in hippocampal activation during novelty encoding (McCarley et al 1993; Weiss et al 2004). In persons with ultra-high-risk for developing psychosis, it is the verbal memory index (due to lower logical memory scores) that identifies those who advance to psychosis (Brewer et al 1998). It is also the case that unaffected relatives show poorer memory performance than controls on a range of memory tests (Whyte et al 2005).

Working memory (WM) and attention have also been extensively examined in schizophrenia (Cohen et al 1999; Stone et al 1998; Barch et al 1998; Gold et al 1997; Goldberg et al 1998; Gooding and Tallent 2001; Park and Holzman 1992, 1993; Wexler et al 1998). Alterations in WM performance and performance correlations with abnormalities in PFC function are consistently found across laboratories (Andreasen et al 1992; Barch et al 2001, 2004; Berman et al 1986, 1992; Callicott et al 1998, 2003; Carter et al 1998; Manoach et al 1999; Menon et al 2001; Perlstein et al 2001; Volz et al 1997; Weinberger 1988) in unmedicated as well as medicated volunteers (Barch e al 2001; Berman et al 1986; Andreasen et al 1997).

For example, verbal and spatial WM alterations are pervasive in the illness and correlate with many of the other cognitive alterations in schizophrenia. Moreover, WM impairments have been correlated with negative symptoms and the language aspects of formal thought disorder (Weinberger et al 1986). Some consider WM alterations to be 'core' (66-68) and a putative marker for endophenotypes (Callicott et al 1998, 2003; Glahn et al 2003).

Arguments parallel to the ones above have been articulated for distinguishing the negative symptom domain from positive psychotic symptoms (Kirkpatrick et al 2001; Castellon et al 1994). Although other domains of dysfunction have been proposed, they have been less frequently considered, and are excluded from this discussion for simplicity of the model. It is the evidence supporting a role of the medial temporal lobe in the mediation of these symptom domains that is the focus of this paper. What does the hippocampal formation contribute to the domains of symptoms in schizophrenia?

We know that the hippocampal formation is affected in many brain diseases, particularly epilepsy and Alzheimer disease, and there is growing evidence that the hippocampus plays an important role in the pathophysiology of schizophrenia, perhaps in more than one domain. The overlapping symptom constellations of these diseases have led to a productive inquiry into how pathological lesions of this region might cause common symptoms. Here we review the normal and diseased hippocampal formation structure and function with what we know about the clinical syndromes of psychosis.

8.2 The anatomy of the hippocampal formation

The hippocampal formation is a small, finger-shaped, cortical structure resembling a seahorse's tail when cut sagittally. It is broadly connected both to many cortical regions and to other limbic structures in a distinctive fashion, receiving indirect input from most sensory areas of the brain. Information from primary and multimodal sensory cortices funnels into the parahippocampal gyrus (including the entorhinal, [EC] perirhinal, [PRc] and parahippocampal [PHc] cortices), which is positioned adjacent to the hippocampus proper in the medial temporal lobe. After cortical afferents converge onto the parahippocampal gyrus, their information is projected to specific subregions of the hippocampal formation (Insausti and Amaral 2004). The hippocampus proper is composed of four subfields, Cornu Ammonis (CA) 1-4, along with the dentate gyrus (DG). The dentate gyrus receives projections from EC and projects through the trisynaptic pathway to CA3 and CA3 projects to CA1; a second but lesser tract from EC, the temporoammonic, projects directly to CA3 and CA1. Both the DG and CA3 have widespread networks of recurrent collaterals by which neurons influence other cells within their own subregion (Lisman and Otmakhova 2001). Efferent connections from CA1 reach the subiculum, and from there most project to the limbic circuit by way of the fornix. Some of the subicular fibers project back to the EC, and from there project back to the same cortical areas that were the sources of input into this region (Insausti and Amaral 2004; Eichenbaum 2000; Sweatt 2003). One unusual feature of the hippocampal subfields is their lack of reciprocal projections backward (Insausti and Amaral 2004). Thus, the hippocampus is a broadly-connected structure, with a "one way" internal circuitry and indirect reciprocal connections with virtually all cortical regions (Fig. 8.1).

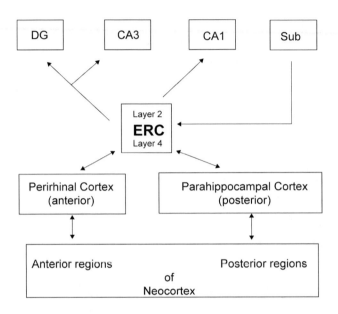

Fig. 8.1. Schematic diagram depicting connectivity of the hippocampus.

In addition to its subfield architecture, the anterior to posterior distribution of the medial temporal lobe roughly mirrors the topographical arrangement of the neocortex (Witter et al 2000; Moser and Moser 1998). For example, the anterior hippocampus formation has reciprocal connections with the anterior neocortex while the posterior hippocampus receives visual-spatial inputs from primary sensory areas in posterior neocortical areas via association, perirhinal and entorhinal areas (Moser and Moser 1998). These different connectivity patterns may have functional implications. For example, the anterior hippocampus is involved in the processing of visual object information while the posterior hippocampus plays a role in visuospatial information processing. This diverse and heterogeneous connectivity pattern may influence the regional development of the hippocampus.

8.3 Development of the Hippocampal Formation

While CNS "development" often covers only prenatal and early years, studies of human brain maturation using quantitative MRI reveal that brain development continues well into adolescence. Age related changes in grey and white matter components have been documented well into young adult years by several researchers (Kitano 2002; Shelton and Weinberger 1987; Gur and Pearlson 1993; Shenton et al 2001; Johnstone et al 1976; Lawrie et al 2002; Suddath et al 1990; Barta et al 1990). Development in cortical gray matter is non-linear, with a pre-adolescent increase followed by a post adolescent decrease. These changes in cortical gray matter are regionally specific, with different cortical regions showing developmental peaks at different ages (Johnstone et al 1976). Some of the more dramatic developmental morphological changes in the brain occur in the frontal and medial temporal (Johnstone et al 1976; Lawrie et al 2002; Suddath et al 1990; Barta et al 1990) gray matter regions. The period of gray matter maturation coincides temporally with post mortem findings of an increase in synaptic pruning that occurs during adolescence and early adulthood (Bogerts et al 1990; Breier et al 1992; Csernansky et al 2004). White matter changes are also seen in MTL; these show a protracted progression of myelination, especially in perihippocampal regions; in MTL this occurs between the 1st and 2nd decades of life (Shihabuddin et al 1998). Recently, a longitudinal MRI study of human hippocampal development between ages of 4 and 25 demonstrates that hippocampal subdivisions follow distinct developmental trajectories (Gogtay et al 2006). The anterior and posterior hippocampal subregions were found to exhibit distinct developmental patterns with the anterior hippocampus showing bilateral reduction in volume while the posterior subregion increases in size during development. The changes in brain morphology are likely related to maturing neurocognitive abilities during the same time period and reflect the dynamic and plastic nature of the developing brain. It is possible that genetic and environmental factors could influence specific regions of the hippocampus in a developmental stage-specific manner.

8.4 The Hippocampal Formation in Schizophrenia

There is considerable evidence implicating diverse brain regions in the pathophysiology of schizophrenia, with the strongest evidence implicating the prefrontal cortex and the medial temporal lobe in the mediation of its symptoms. In this discussion, however, we will focus on one brain region, the hippocampal formation, in the context of relevant neural circuits. This section is divided into three broad sections (i) in vivo imaging data, (ii) human post mortem data and (iii) the risk factors implicated in schizophrenia.

8.4.1 Human in vivo imaging data

Structural imaging studies

MRI studies report a reduction in overall brain size, an increase in ventricular size, and variable cortical wasting in schizophrenia (Shelton et al. 1987; Gur and Pearlson 1993; Shenton et al. 2001; Johnstone et al. 1976). These reports confirm and extend older literature using the computerized axial tomography (CAT) examination of schizophrenia that demonstrated ventricular enlargement (Johnstone et al. 1976) More recently, MRI studies have reported a volume reduction in medial temporal cortical structures (hippocampus, amygdala, and parahippocampal gyrus) (Lawrie et al. 2002; Suddath et al.; Barta et al. 1990; Bogerts et al. 1990; Breier et al. 1992). New analytic techniques for shape analysis show regional shape differences of hippocampus (Csernansky et al. 1998) in schizophrenia. The volume of the superior temporal gyrus may be reduced in schizophrenia, a change that correlates with the presence of hallucinations (Barta et al. 1990; Shenton et al. 1992) and with regional EEG changes. Proton Magnetic Resonance Spectroscopic imaging (^{1}H MRS) studies show decreases in N-acetyl aspartate (NAA) in the temporal lobe (Maier et al. 1995; Bertolino et al. 1996; Yurgeun-Todd et al. 1996; Deicken et al. 1999) while others show insignificant trends or no difference (Kegeles et al. 2000; Bartha et al. 1999). NAA, found exclusively in neurons, may reflect neuronal integrity. Phosphorous MRS (^{31}P MRS) data reflect the integrity of neuronal cell membranes. Decreased phosphomonoester resonance (precursors of membrane phospholipids) and increased phosphodiester resonances (breakdown products) in the temporal cortices in schizophrenia are postulated to reflect an increased turnover of membranes, possibly due to abnormalities in synaptic pruning (Reddy and Keshavan 2003).

Hippocampal size, measured in vivo, is reduced bilaterally in schizophrenia, especially in anterior areas (Barch et al. 1998; Barnes 1989; Benedict 1997; Nakazawa 2003). This reduction in size is seen as early as the first psychotic episode (Jaeger et al. 2003; Norman and O'Reilly 2003). It has been detected to a lesser degree in non-psychotic siblings of schizophrenia probands (Packard 1999), and in persons at risk for schizophrenia (Gur et al. 1999; Park and Holzman 1992). Moreover, studies of hippocampal shape have suggested regional abnormalities of contour in affected persons (Brewer et al. 1998) and in well siblings of schizophrenic probands (Packard 1999). Importantly, these regional shape abnormalities are often found in the head of the hippocampus, implicating the anterior subregion within hippocampus as abnormal (Brewer et al. 1998). While the degree to which shape abnormalities can reflect pathology in the underlying tissue is unclear, they are consistent with the presence of pathology.

Functional imaging

Initial neuroimaging studies of schizophrenia identified functional alterations in MTL regions as well as in PFC (Cornblatt and Erlenmeyer-Kimling 1985;

Eichenbaum and Cohen 2001; Eldridge et al. 2005; Jessen et al. 2003; O'Reilly and Rudy 2001). Even though these early studies indicated an abnormality in MTL, they typically lacked precise task paradigms and image resolution needed to address sophisticated questions regarding MTL function in the illness. Furthermore, the critical basic cognitive neuroscience models for examining human MTL function in a brain disease were not yet refined. More recently, with an expanded knowledge base and tasks that specifically target precise aspects of declarative memory processes, several studies have demonstrated abnormal MTL functioning in schizophrenia. These more recent studies have reported alterations in performance and function of the MTL in schizophrenia (Eldridge et al. 2005; Jessen et al. 2003; O'Reilly and Rudy 2001; Ranganath et al. 2005; Golby et al. 2001; Harris et al. 1996; Rissman et al. 2004; Gazzaley et al. 2005). Recently, Heckers et al. (1998) specifically identified the more complex relational learning as altered in schizophrenia. Such flexible use of declarative memory is particularly of interest because of the reduced ability of persons with schizophrenia to make transitive inferences (Heckers et al. 2004). Along similar lines, Keri reported a selective deficit in stimulus generalization in SV in the "acquired equivalence task", with a sparing in stimulus-response learning. Studies examining MTL interactions with PFC have consistently found alterations in functional interactions between these two structures in SV (Hicks and Marsh 1999; Henson et al. 1999), raising the issue of whether documented MTL changes in SV are primary or are secondary to alterations in PFC function.

Collectively, extant studies show a high degree of consistency in broadly implicating physiological failure of MTL activation in schizophrenia, consistent with pathology reported in postmortem tissue studies (Eichenbaum 2000). Nonetheless, these initial studies have not identified the critical MTL regions or broader circuits which are altered in SV; nor have they delineated the specific memory conditions that pose challenges for SV; nor have they addressed how these deficits respond to APD treatment and whether functional differences from normal remain to be 'normalized' (i.e., treated) in MTL. We anticipate that such knowledge will prove valuable for developing specific treatments for the cognitive alterations that stem from MTL dysfunction in schizophrenia.

8.5 Systems of Dysfunction in Schizophrenia

Previous research has supported the modulatory role of striato-thalamo-cortical circuits on limbic and neocortical function (Alexander et al. 1986, 1990). The anatomy of these circuits has been supported by evidence from non-human primate (Haber 2003; McFarland and Haber 2002), and human (Tamminga et al. 1989) studies. While most of the evidence to support the functional significance of these circuits has derived from motor studies, scientists have long speculated that parallel circuits also modulate cognitive and affective functions (Alexander et al. 1990; McFarland and Haber 2002). Here, we extend the concept and proposed anatomy of the striato-thalamo-cortical pathways to the cognitive behaviors of schizophrenia as they are organized into domains. These pathways transmit the modulatory influence of the basal ganglia and thalamus to prefrontal- and limbic-mediated cortical functions not only in domains of motor, but also for cognitive and affective functions.

Accumulating evidence suggests that the medial temporal lobe may be a component of circuits involved in schizophrenia. Most of this evidence has come from functional imaging studies that demonstrate the involvement of the MTL in various cognitive functions and an association with psychosis. Here we consider the evidence that support the role of the MTL in the domains of schizophrenia and propose neural circuits that may underlie each symptom domain.

8.5.1 Domain Networks

Psychosis neural network

Fluorodeoxyglucose (FDG) PET scans at rest show glucose utilization (rCMRglu) differences between normal and positive symptom-schizophrenia groups in the anterior cingulate cortex (ACC) and the hippocampus (HC) (Tamminga et al. 1992). A follow-up study of similar design tested the regional associations of positive symptoms and showed that as long as patient volunteers were medication-free, there was a significant association between rCMRglu in the limbic cortex and the magnitude of positive symptoms in the illness; this correlation was not obtained when patients were medicated or when other symptom domains were tested (Tamminga et al. 2000). The results of these studies allow us to speculate that it is the limbic cortex which is associated with the positive symptoms of the illness, while the PFC may support negative and/or cognitive symptoms. Blood flow changes in the anterior cingulate and adjacent medial frontal cortex also correlate with induction of positive symptoms with the NMDA antagonist, ketamine (Lahti et al 1995). PET studies in hallucinating schizophrenic persons is associated with activations in several brain regions including the medial prefrontal cortex, left superior temporal gyrus (STG), right medial temporal gyrus (MTG), left hippocampus/parahippocampal region, thalamus, putamen and cingulate (Silbersweig et al. 1995; Copolov et al. 2003). These studies provide evidence regarding the anatomic structures that may be involved in a 'psychosis neural circuit'. Limbic regions, in particular, are frequently implicated in in vivo psychosis studies.

Taking these data along with the basic knowledge that already exists about anatomic connections between these areas (Alexander et al 1986, 1990; Haber 2003; McFarland and Haber 2002; Tamminga et al. 1989; Heimer and Van Hoesen 2006), we postulate that a neural system for the psychosis cluster in schizophrenia consists of the anterior hippocampal formation, the anterior cingulate and medial PFC, the anterior thalamus and the basal ganglia (Fig. 8.2). On a more speculative level, we postulate that the core pathology of "psychosis" begins in the hippocampal formation, with this adversely affecting other regions in the network.

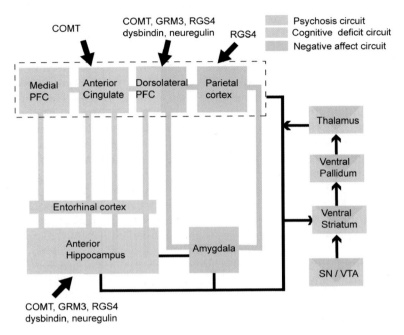

Fig. 8.2. Hypothetical neural circuits underlying the different domains of schizophrenia. Certain risk genes have been shown to influence structure, function or neurochemistry of brain region (blue arrows) important to one or more of these neural circuits.

Cognitive deficit neural network

Traditionally, processing of working memory is associated with the PFC, declarative memory with the MTL, and procedural memory with basal ganglia suggesting neural templates for memory systems in brain. It is likely that other cognitive systems are similarly distributed as well. In volunteers with schizophrenia, functional MRI studies using cognitively demanding tasks, such as working memory tasks, produce diverse results. Manoach et al. (1999, 2000) used the Sternberg Item Recognition working memory paradigm which required the subjects to remember either two or five digits. Unlike many PET studies of working memory, they found an increase instead of a decrease in prefrontal rCBF in the schizophrenic volunteers as compared to the normal controls. Research has suggested that although rCBF increases in prefrontal regions with greater working memory demands, if working memory capacity is exceeded, then the activation decreases (Callicott et al. 1999). Manoach suggests that the discrepant findings in schizophrenia may be explained by an overload of working memory in schizophrenic subjects for some tasks. In separate studies, disorganization was associated with flow in anterior cingulate and mediodorsal thalamus (Liddle et al. 1992) while apomorphine, a dopamine agonist that has antipsychotic properties, normalizes anterior cingulate blood flow of schizophrenic persons during verbal fluency task performance (Dolan et al. 1995).

Impaired performance and hippocampal activation during encoding and retrieval is seen in schizophrenia (Henson et al. 2003; Yonelinas 2002). Using novelty detection, a hippocampal-specific task, baseline rCBF is altered in schizophrenia (Henson et al. 1999; Eyler-Zorrilla et al. 2002). BOLD signal higher and reduced activation from baseline with novelty in schizophrenia. Heckers et al. (2004) studied transitive inference in control subjects and described hippocampal activation during discrimination of overlapping pair of stimuli, but not during non-overlapping pair. This is particularly interesting in schizophrenia, because of the reduced ability in persons with this illness to make transitive inferences (Heckers et al. 2004). Moreover, studies examining MTL interactions with PFC in schizophrenia, have consistently found alterations in the interactions between these two structures in schizophrenia (Wagner et al. 1998; Nakazawa et al. 2002). These studies show a high degree of consistency in implicating physiological failure of MTL activation in schizophrenia during cognitive tasks.

Taking these data together with basic anatomic information, we propose that a neural system for cognitive deficits involves the DLPFC, anterior hippocampus, anterior cingulate cortex, thalamus and basal ganglia (Fig. 8.2).

Negative symptom neural network

Brain activation patterns associated with negative symptoms have been studied. Hypoactivation of the frontal lobe is seen with increased negative symptoms in schizophrenia (Andreasen et al. 1992, 1994; Volkow et al. 1987; Wolkin et al. 1992; Schroeder et al. 1994). Decrease in regional cerebral blood flow is observed in the prefrontal and parietal cortex in patients exhibiting negative symptoms (Tamminga et al. 1992; Friston 1992; Lahti et al. 2001). It is interesting to note that the DLPFC and parietal cortex have dense reciprocal interconnections (Schwartz and Goldman-Rakic 1984). Lower activity was also noted in the thalamus (Tamminga et al. 1992), in particular the mediodorsal nucleus of the thalamus (Hazlett et al. 2004). The amygdala, a key component in the circuit of emotion, implicated in emotional processing in schizophrenia (Gur et al. 2002). Functional and structural imaging has also implicated the amygdala in schizophrenia (Kosaka et al. 2002; Hulshoff et al. 2001; Wright et al. 1999). Of interest, deficit symptom persons with schizophrenia exhibit greater impairment in cognitive performance (Buchanan et al. 1994, 1997) that may reflect overlapping systems. Considering these data, the neural system we propose for the negative symptom cluster includes the DLPFC, parietal cortex, amygdala, thalamus, ventral pallidum, striatum and SN/VTA (Fig. 8.2).

Functional imaging studies provide the most direct data on the anatomic substrates associated with symptom domains in schizophrenia. Also, neuroreceptor and other molecular imaging studies allow *in vivo* determination of specific receptor abnormalities and indirect measures of neurotransmitter release in specific brain regions. These studies in schizophrenic volunteers can provide *in vivo* data on molecular abnormalities in specific neural systems. In this section, we have proposed neural systems that may underlie each of the symptom clusters.

8.6 Other support for MTL involvement in Circuit domains

8.6.1 Post mortem Studies of the MTL

Postmortem studies have heretofore not been designed to identify complex system changes in schizophrenia. Neither have brain regions of neural systems been sampled and examined together in a comprehensive manner. Moreover, neurochemical changes are difficult to associate with domain symptoms due to the imprecision of clinical symptom data gathered in post mortem retrospective diagnostics. Therefore, little information is available from these studies that can be assigned to specific domains. It is likely that we will have to surmise the domain-specific neurochemical lesions by regional data and/or use clues from *in vivo* neurochemistry where symptom correlates are straightforward.

Although schizophrenia lacks an identifiable neuropathologic lesions such as occur in Parkinson's disease or Alzheimer's dementia, there are observable differences at the cellular and molecular levels in several brain regions believed to be important in the pathophysiology of schizophrenia; e.g., PFC, ACC, hippocampus, entorhinal cortex and thalamus. Here, we review what neurochemistry evidence exists to support the involvement of the MTL in schizophrenia.

Morphometric studies using semi quantitative or qualitative techniques have demonstrated aberrant positioning of clusters of neurons in the entorhinal cortex in schizophrenia (Jakob and Beckmann 1986; Arnold et al. 1991a; Falkai and Bogerts 1986) although repeated negative studies have been reported (Akil and Lewis 1997; Krimer et al. 1997; Bernstein et al. 1998). Similarly, abnormal orientation of pyramidal neurons has been reported in subregions of the hippocampus (Kovelman and Scheibel 1984; Altschuler et al. 1987; Conrad et al. 1991) but not consistently (Benes et al. 1991; Arnold et al. 1995; Zaidel et al. 1997). There are also reports that cite decreases in cortical thickness, abnormalities of cell size, cell number and packing density, area, neuronal organization, gross structure and neurochemistry (Bogerts 1993; Harrison 1999). These data, however, have not been consistently replicated across laboratories, making it difficult to build a consistent story across all of the findings. The inconsistencies may be due to differences in patient populations, stage of illness, although possible confounds of tissue artifact, agonal state, chronic drug treatment, lifelong-altered mental state, and relevant demographic factors must always be considered in evaluating postmortem brain tissue studies.

Neurochemical studies began in schizophrenia at a time when scientists were anticipating a single protein defect. While no single defect has been found, investigators interpret studies in a broader systems context. Nonetheless, findings are often still organized by neurotransmitter system.

Studies of biochemical markers of the dopamine system in schizophrenia were stimulated by the early pharmacologic observation that blockade of dopamine receptors in the brain reduces psychotic symptoms (Carlsson and Lindquist 1963). The hypothesis derived from this observation that dysfunction of the CNS dopaminergic system either in whole or in part accounts for psychosis in schizophrenia has been explored in all body fluids and in various conditions of rest and stimulation over the last half century (Elkashef et al. 1995; Davis et al. 1991) with little real support except for the recent imaging studies that show higher occupancy of D_2 receptors by dopamine in patients with schizophrenia (Abi-Dargham et al. 2000) and changes in DA release in acute illness phases.

More recently, because of its ubiquitous and prominent location in the CNS, and because the antiglutamatergic drugs phencyclidine (PCP) and ketamine cause a schizophrenic-like reaction in humans, the glutamate system has become a focus of study. Several studies have examined ionotropic glutamate receptor subtypes (Meador-Woodruff and Healy 2000). Most studies have focused on the mesial temporal lobe and, in general, report abnormalities in AMPA, KA and NMDA receptor expression at the mRNA, protein and ligand-binding level. Few studies have examined the metabotropic glutamate receptors (mGluRs) in schizophrenia. Group II (mGluR2 and 3) receptors are implicated in animal (Moghaddam and Adams 1998) and human (Krystal et al. 2005) studies. N-acetylaspartylglutamate (NAAG), an endogenous agonist of mGluR3 (Coyle 1997; Neale et al. 2000; Tsai et al. 1995) and its metabolic enzyme (Tsai et al. 1995; Ghose et al. 2004) are abnormal in schizophrenia. Additionally, mGluR3 may be a risk gene for schizophrenia (Egan et al. 2004).

Evidence of GABAergic involvement is found in reduced expression of presynaptic markers in subpopulations of interneurons in the hippocampal formation (Lewis et al. 2004; Benes and Berretta 2001). GABAergic neurons can be defined by the presence of one of three calcium binding proteins, namely parvalbumin, calretinin, and calbindin. The most characteristic morphological types of neurons that express parvalbumin are the large basket and chandelier cells (Lewis and Lund 1990). GAT1 is decreased in the parvalbumin-expressing prefrontal interneurons (Lewis et al. 2005) and to a lesser extent in the hippocampus (Konopaske et al 2006). An increase in expression of GABA-A receptors on cells receiving input from hippocampal interneurons is increased in schizophrenia has been reported (Benes et al. 1996, 1997). There are also studies demonstrating Glutamic acid decarboxylase 65 (GAD65) (193) and GAD67-kd mRNA expression is reported to be unchanged in the hippocampus in schizophrenia (Heckers et al. 2002).

The affinity of newer antipsychotic drugs for serotonergic receptors has raised speculation over the role of this neurotransmitter system in the treatment and perhaps in the pathophysiology of the illness. Years ago, serotonin was hypothesized to be central to the pathophysiology of schizophrenia, because of the psychotomimetic actions of serotonergic drugs, like LSD (Freedman 1975). Postmortem studies have failed to find consistent change in measures of the serotonin system in schizophrenia, including in receptors (in vivo and postmortem) or in metabolites (Harrison 1999). Since serotonin has been shown to modify dopamine release in striatum (Marcus et al. 2000; Kuroki et al. 2003), the augmented antipsychotic action of the new drugs may be mediated through modulation of dopamine release into the synapse. Indeed, drugs without any dopamine receptor affinity, but with only 5 HT_{2a} receptor antagonism, do behave as antipsychotic drugs in animal models and show antipsychotic activity in humans (de Paulis 2001). Since the serotonin system has diverse receptors and functions, it is not surprising that this aspect is not yet fully explicated.

Cholinergic neurotransmission, integral to cognition and memory, may be dysfunctional in schizophrenia. Clinically, it is well known that schizophrenic patients have a much higher incidence of cigarette smoking (Hughes et al. 1986). Although 'control' smokers exhibit an upregulation in nicotinic receptors (Benwell et al. 1988; Wonnacott 1990), decreased levels of nicotinic and muscarinic receptors are reported in the hippocampus and other brain regions (Hyde and Crook 2001).

Synaptic proteins have been assessed in schizophrenia postmortem brain tissue and compared with normal tissue chiefly in hippocampus and in prefrontal cortex. However, the existing literature is divergent, often contradictory, and difficult to interpret. In hippocampus, Eastwood first reported decreased synaptophysin expression and protein (Eastwood et al. 1995). Even earlier, Browning et al. (1999) evaluated synapsin I, IIb, and synaptophysin proteins and noted high variability in hippocampus. Harrison and Eastwood (1998) reported alterations in complexin I and II mRNA (reductions), but no change in protein, and analyzed the same proteins with similar results in the cerebellum. Webster et al. (2001) also found decreased synaptophysin message (but no change in GAP-43). Vawter et al. (2002) found no synaptophysin protein decreases, but did find a synapsin protein decrease in hippocampus. Eastwood and Harrison (2000) found decreased complexin I and II RNA in schizophrenia and in bipolar samples. Blennow et al. (2000) reported decreases in Rab_3 protein in several regions in schizophrenia, including the hippocampus. Fatemi measured SNAP-25 protein in hippocampus and found decreases in schizophrenia and bipolar, but increases in depression (Fatemi et al. 2001). In the ACC, Eastwood and Harrison (2001) reported decreased synaptophysin protein, but the biggest decrease in bipolar tissue, smaller in depression tissue, and smallest in schizophrenia.

Each of these studies of synaptic proteins in schizophrenia took a fragmented look at these proteins, without a systematic screen. Synaptophysin was most often reported to show low RNA or protein in hippocampus, ACC, or prefrontal cortex. Where synaptic proteins are reported to be abnormal in schizophrenia, it is usually a decrease, but with inconsistent results between mRNA and protein. The decreases in GAP-43 mRNA, complexin I and II mRNA and protein, SNAP-25 protein, and Rab3A protein are the most frequently reported. However, several studies showed lack of specificity for the diagnosis of schizophrenia. In addition, individual presynaptic trafficking proteins do not exist alone but rather as a dynamic system. A comprehensive examination of these proteins would likely provide a clue to their true contribution in schizophrenia.

In summary, *in vivo* studies implicate the hippocampus as a brain region important to the pathophysiology of schizophrenia and human post mortem findings confirm the hippocampus as a brain region of molecular dysfunction. There are regional abnormalities within the hippocampus as well as along the axis of this structure (Ghose et al. 2006), perhaps providing clues to which neural circuits may be compromised in this disease.

8.6.2 Risk Factors

Genetic variations

In recent years, several genes have been associated with the illness, including neuregulin (NRG1) (Stefansson et al. 2002), dysbindin (DTNBP1) (Schwabb et al. 2003), G72 (Chumakov et al 2002), D-amino-acid oxidase (DAAO), regulator of G protein-signaling-4 (RGS4), (Chowdari et al. 2002) proline dehydrogenase (ProDH) (Liu et al. 2002), catecol-O-methyl transferase (COMT) (Egan et al. 2001; Shifman et al. 2002), metabotropic glutamate receptor 3 (mGluR3) (Lawrie et al. 2002). Each of these genes codes for a protein that has been speculatively linked with a purported illness mechanism (Harrison and Owen 2003; Harrison and Weinberger 2005) but no clear disease pathophysiology has yet emerged. The effects of these risk genes on specific brain regions is currently being pursued by several research groups; one goal of these studies would be to determine if risk genes are associated with specific domains of schizophrenia. The findings so far suggest that this may be the case (Fig. 8.2). We review the main risk genes identified to date and their influence on the MTL in schizophrenia.

Dystrobrevin binding protein 1 (DTNBP1 or Dysbindin)

The association between dysbindin and schizophrenia has since been shown in several other studies (Straub et al. 2002; Schwab et al. 2002; Williams et al. 2004; Kirov et al. 2004; Numakawa et al. 2004; Funke et al. 2004) although not in all (Van Den et al. 2003; Morris et al. 2003). The studies that implicate dysbindin differ in the risk alleles and haplotypes involved. Nonetheless, the evidence in favor of dysbindin as a susceptibility gene for schizophrenia is strong. The function of dysbindin in the human brain has not been determined. It is a 40- to 50-kDa protein that binds α- and β-dystrobrevins, components of the dystrophin glycoprotein complex. This complex is believed to play a role in stabilization of the post synaptic membrane, cytoskeletal rearrangement, and signal transduction (Adams et al. 2000; Grady et al. 2000). β-dystrobrevin is expressed in neurons, where it is associated primarily with postsynaptic densities. A possible functional role of dysbindin in schizophrenia is supported by the finding of decreased protein in the human post mortem hippocampus from schizophrenic donors (Talbot et al. 2004). This decrease was most marked in glutamatergic neurons in the CA2, CA3 and dentate gyrus. These authors also found an inverse relationship with vesicular glutamate transporter 1 (vGluT1) suggesting perhaps, a role for dysbindin in glutamatergic neurotransmission (Talbot et al. 2004).

Neuregulin 1 (NRG1)

Substantial evidence exists for neuregulin to be considered a susceptibility gene for schizophrenia (Stefansson et al. 2002, 2003; Corvin et al. 2004; Williams et al. 2003; Yang et al. 2003; Tang et al. 2004; Zhao et al. 2004) although there are also negative findings that have also been reported (Iwata et al. 2004; Thiselton et al. 2004). Three studies (Stefansson et al. 2002, 2003 Williams et al. 2003) have implicated the specific core haplotype identified in the initial study. NRG peptides are involved in a host of physiologic processes including neuronal migration, axon guidance, synaptogenesis, glial differentiation, myelination, neurotransmission and synaptic plasticity. A recent meta-analysis of 13 population-based and family-based association studies in schizophrenia (Li et al. 2006) found strong positive associations for six polymorphisms which included SNP8NRG221132 and SNP8NRG243177. These two SNPs are among the five identified in the original core at-risk haplotype of neuregulin identified by Stefansson et al. (2002). Both SNPs are associated with gene expression profiles in the human post mortem brain. Specifically, SNP8NRG221132 significantly influences neuregulin 1 mRNA in the hippocampus in schizophrenia. SNP8NRG243177, on the other hand, was found to be associated with a novel isoform of neuregulin in the human brain. SNP8NRG243177 is also reported to alter putative binding sites for three transcription factors in the neuregulin promoter region (Law et al 2006). In human functional imaging studies, this SNP is found to be associated with decreased activation of frontal and temporal lobe regions and with the development of psychotic symptoms (Hall et al 2006). These data suggest that neuregulin may have biologic significance in the hippocampus in schizophrenia.

Catechol-O-methyl transferase (COMT)

There have been numerous association studies examining the COMT val/met polymorphism in schizophrenia with inconsistent results. The first meta-analysis conducted by Glatt et al. (2003), suggested that the val allele may be confer a small risk in European cases of schizophrenia. Subsequent meta-analyses found no significant effect (Fan et al. 2005) of the COMT val/met genotype while the most recent report found significance for the val allele when all studies were included but no association when studies that deviated from Hardy-Weinberg equilibrium were removed (Munafo et al. 2005). There is stronger evidence, however, that the COMT val158met polymorphism influences prefrontal (PFC) and hippocampal function in normal controls, schizophrenics and their siblings (Bilder et al. 2002; Egan et al. 2001; Goldberg et al. 2003; de Frias et al. 2004; Malhotra et al. 2002; Joober et al. 2002). Carriers of the val/val genotype show poorer episodic memory than do carriers of the less active met allele. Consistent with this, tolcapone, a COMT inhibitor, enhances episodic memory (Iudicello et al. 2004). It has been postulated that the COMT val158met polymorphism impacts tonic and phasic dopamine neurotransmission to modulate corticolimbic reactivity (Bilder et al. 2004).

Regulator of G protein signaling 4 (RGS 4)

RGS-4 is a gene that maps to chromosome 1q21-22, a region implicated in linkage studies. Decreased levels of this gene were found in human post mortem microarray studies in schizophrenia (Mirnics et al. 2001). Associations were found with four SNPs of varying haplotypes in RGS 4 in three different populations of subjects (Chowdari et al. 2002). Later studies report positive associations with RGS4 and schizophrenia (Williams et al. 2004; Chen et al. 2004; Morris et al. 2004; Fallin et al. 2005; Zhang et al. 2005) but as is reminiscent for the other genes, there is inconsistency in the alleles and haplotypes implicated. Studies with no association between RGS4 and schizophrenia have also been reported (Sobell et al. 2005). RGS proteins are a family of about 30 proteins that are GTPase activating proteins serving to negatively regulate G protein coupled receptors (GPCR). RGS4 is abundantly expressed in the brain and can regulate multiple GPCRs including dopaminergic and metabotropic glutamate receptors. Allelic variations in RGS4 are associated with psychosis and blood flow in the frontoparietal and frontotemporal brain regions (Buckholtz et al. 2007). While several studies have demonstrated reduced RGS4 message in the prefrontal cortex (Mirnics et al. 2001; Chowdari et al. 2002; Williams et al. 2004), the hippocampus has gone largely unexplored. In the adjacent superior temporal gyrus, there is reduced expression of RSG4 (Bowden et al. 2007). Since RGS4 protein function to reduce the effect of G protein coupled receptor activation such as dopamine and metabotropic glutamate receptors, reduced RGS4 expression could potentially disrupt neuronal communication. It is also interesting to note that hippocampal RGS4 mRNA has been shown to be influenced by the COMT Val158Met polymorphism in schizophrenia supporting the idea that multiple risk genes could have a converging influence on target proteins (Lipska et al. 2006).

Disrupted in Schizophrenia 1 and 2 (DISC1 and 2)

A balanced translocation in chromosomes 1 and 11 (1;11)(q42.1;q14.3) linked to the major mental disorders- schizophrenia, depression, and mania- was found to disrupt 2 genes, called DISC1 and DISC2 (St Clair et al 1990). Linkage analysis indicated that 1q42 as a possible locus for schizophrenia and the strongest signal was found in a marker in the DISC1 gene (Ekelund et al. 2004). Positive associations have been found in several later studies (Hennah et al. 2005; Hodgkinson et al. 2004; Zhang et al. 2006). There is one negative study that examined four SNPs in DISC1 (Devon et al. 2001). The positive studies implicate DISC1 although the precise patterns of association differ. A SNP, Ser704Cys, is associated with schizophrenia and, in control subjects, this SNP is found to be associated with hippocampal structure and function including reduced gray matter in the hippocampus (Callicott et al. 2005). A significant association is reported between a DISC1 haplotype containing this SNP and severity of positive symptoms

(Derosse et al. 2007). DISC1 interacts with components of the cytoskeletal system such as NUDEL, and the centromere proteins to impair neurite growth and development of the cerebral cortex (Kamiya et al. 2005). DISC1 is predominantly expressed in the hippocampus. While levels of DISC1 mRNA itself is reportedly unchanged in schizophrenia, several DISC 1 partners have been reported to be expressed at lower levels in the hippocampus in schizophrenia (Lipska et al. 2006).

Metabotropic glutamate receptor 3 (GRM3)

GRM3 may be a risk gene for schizophrenia (Lawrie et al. 2002; Marti et al. 2002; Fuji et al 2003). Egan et al (2004) reported functional effects of GRM3 variants on prefrontal and hippocampal-dependent tasks in both control and schizophrenic volunteers. An endogenous peptide, N-acetylaspartateglutamate (NAAG) is a highly selective agonist at mGluR3 receptors (Neale et al. 2000). Altered levels of NAAG and the enzyme that metabolizes it, glutamate carboxypeptidase II (GCP II), are abnormally expressed in the hippocampus in schizophrenia (Tsai et al. 1995; Ghose et al. 2004).

Environmental factors

Even though several risk genes have been implicated, the precise variations in the genes are inconsistent and until specific mutations are identified, it will be difficult to determine the biologic effect of each risk gene. Another level of complexity to factor in is the interaction between risk genes and also the interaction between the risk genes and environmental factors.

The role of environmental factors in the development of schizophrenia is evident from the fact that monozygotic twins have less than 100% concordance rates of schizophrenia. Observational studies in humans suggest that prenatal stress, such as infections, exposure to toxins, nutritional deficiencies, severe maternal duress, and obstetric complications are risk factors for schizophrenia (Sullivan 2005). In animal models, prenatal immune activation leads to a psychosis-related phenotype later in life. This phenotype is associated with abnormal morphological development and neurochemical abnormalities in the hippocampus and entorhinal cortex. Similarly, prolonged social stress, a factor believed to be important in the development of schizophrenia (Bebbington et al. 1993; Hirsch et al. 1996), is associated with changes in the rodent hippocampal neurochemistry (Pickering et al. 2006; Melendez et al. 2004). Data such as these suggest that the hippocampus is vulnerable to the same stressors important to schizophrenia.

The hippocampus is believed to develop later than the neocortex with regards to neuronal migration and maturation (Nowakowski and Rakic 1981) with myelination in the hippocampus especially in perihippocampal regions occurring through the 1st and 2nd decades of life (Kornack and Rakic 1999; Benes et al. 2004). The period of gray matter maturation coincides temporally with post mortem findings of an increase in synaptic pruning that occurs during adolescence and early adulthood (Huttenlocher 1994; Bourgeois et al 1994; Rakic 1996). The prolonged developmental course of the hippocampus may make it a brain region that is more susceptible to environmental stressors. In addition, the spatiotemporal pattern of hippocampal development with the anterior and posterior hippocampi showing specific developmental trajectories (Gogtay et al. 2006) suggest that these two regions may have different susceptible periods.

8.7 Summary: the role of the hippocampal formation in mediating selective dysfunctions in schizophrenia, its anatomy, neurochemistry, and genetics

Our understanding of the pathophysiology of schizophrenia is still obscure and it is becoming evident that schizophrenia may not be one disease but possibly a syndrome. The identification of candidate risk genes, specific developmental environmental risk factors and the effects of gene-environmental interactions speak to the heterogeneity we can expect to find in schizophrenia. There is early evidence that some of the candidate risk genes of schizophrenia influence function of the hippocampus. In addition, the development of the hippocampus well into early adulthood provides an opportunity for specific stressors of childhood and adolescence to influence the ongoing maturation of this brain region and its connections, thereby influencing neural circuit function adulthood. Functional imaging studies have repeatedly found hippocampal function to be abnormal in the psychosis and cognitive domains of schizophrenia. Consistent with this, neurochemical abnormalities are found in human post mortem MTL tissue of cases with schizophrenia supporting the idea that this may be one substrate important in the pathophysiology of this illness.

Researchers have focused on finding specific molecular targets that could explain schizophrenia. So far, this has proved unsuccessful. A shift in conceptual framework from defects in one protein in schizophrenia to defects in neural networks of schizophrenia may be a useful paradigm to facilitate the study of disease mechanism. This will require identification of relevant neural systems and an understanding of the dynamics of the systems. As an initial formulation, we propose neural networks for each symptom domains (Fig. 8.2). This formulation proposes that the broadly connected hippocampus is positioned to influence multiple neural networks involving distinct cortical regions and subcortical structures.

Whether the symptom domains are manifestations of a single disease pathophysiology perturbing the neural networks or independent disease construct remains unknown. An integral part of determining function of affected neural systems understands the molecular composition of the connections in the network. Identifying the pivotal molecular targets for schizophrenia rests on the application of modern concepts and techniques to clearly diagnosed and characterized populations of persons with schizophrenia.

References

Abi-Dargham A, Rodenhiser J, Printz D, Zea-Ponce Y, Gil R, Kegeles LS, Weiss R, Cooper TB, Mann JJ, Van Heertum RL, Gorman JM, Laruelle M (2000) Increased baseline occupancy of D2 receptors by dopamine in schizophrenia. Proc Natl Acad Sci USA 97:8104-8109.

Adams ME, Kramarcy N, Krall SP, Rossi SG, Rotundo RL, Sealock R, Froehner SC (2000) Absence of alpha-syntrophin leads to structurally aberrant neuromuscular synapses deficient in utrophin. J Cell Biol 150:1385-1398.

Akil M & Lewis DA (1997) Cytoarchitecture of the entorhinal cortex in schizophrenia. Am J Psychiat 154:1010-1012.

Alexander GE, Crutcher MD, DeLong MR (1990) Basal ganglia-thalamocortical circuits: parallel substrates for motor, oculomotor, "prefrontal" and "limbic" functions. Prog Brain Res 85:119-146.

Alexander GE, DeLong MR, Strick PL (1986) Parallel organization of functionally segregated circuits linking basal ganglia and cortex. Annu Rev Neurosci 9:357-381.

Altshuler LL, Conrad A, Kovelman JA, et al (1987) Hippocampal pyramidal cell orientation in schizophrenia. Arch Gen Psychiat 44:1094-1098.

Andreasen NC, Arndt S, Alliger RJ, Miller D & Flaum M (1995) Symptoms of schizophrenia. Methods, meanings, and mechanisms. Arch Gen Psychiatry 52:341-351.

Andreasen NC, Flashman L, Flaum M, Arndt S, Swayze V, O'Leary DS, Ehrhardt JC, Yuh WT (1994) Regional brain abnormalities in schizophrenia measured with magnetic resonance imaging. JAMA 272:1763-1769.

Andreasen NC, O'Leary DS, Flaum M, Nopoulos P, Watkins GL, Boles Ponto LL & Hichwa RD (1997) Hypofrontality in schizophrenia: Distributed dysfunctional circuits in neuroleptic-naive patients. Lancet 349:1730-1734.

Andreasen NC, Paradiso S & O'Leary D (1998) "Cognitive dysmetria" as an integrative theory of schizophrenia: a dysfunction in cortical-subcortical-cerebellar circuitry? Schiz Bull 24:203-218.

Andreasen NC, Rezai K, Alliger RJ, Swayze VW, Flaum M, Kirchner P, Cohen G & O'Leary DS (1992) Hypofrontality in neuroleptic-naive patients and in patients with chronic schizophrenia. Arch Gen Psychiat 49.943-958.

Arndt S, Alliger RJ & Andreasen NC (1991) The distinction of positive and negative symptoms. The failure of a two-dimensional model. Br J Psychiatry 158:317-322.

Arnold SE, Franz BR, Gur RC, Gur RE, Shapiro RM, Moberg PJ, Trojanowski JQ (1995) Smaller neuron size in schizophrenia in hippocampal subfields that mediate cortical-hippocampal interactions. Am J Psychiat 152:738-748.

Arnold SE, Hyman BT, Van Hoesen GW, et al (1991) Some cytoarchitectural abnormalities of the entorhinal cortex in schizophrenia. Arch Gen Psychiat 48:625–632.

Barch DM, Braver TS, Cohen JD, Servan-Schreiber D (1998) Context processing deficits in schizophrenia. Arch Gen Psychiat 55:187-188.

Barch DM, Carter CS, Braver TS, Sabb FW, MacDonald AW, Noll DC & Cohen JD (2001) Selective deficits in prefrontal cortex function in medication-naive patients with schizophrenia. Arch Gen Psychiat 58:280-288.

Barch DM, Mitropoulou V, Harvey PD, New AS, Silverman JM & Siever LJ (2004) Context-processing deficits in schizotypal personality disorder. J Abnorm Psychol 113:556-568.

Barnes TR & Liddle PF (1990) Evidence for the validity of negative symptoms Pharmacopsychiatry 24:43-72.

Barnes TR (1989) A rating scale for drug-induced akathisia. Br J Pharmacol 154:672-676.

Barta PE, Pearlson GD, Powers RE, Richards SS & Tune LE (1990) Auditory hallucinations and smaller superior temporal gyral volume in schizophrenia. Am J Psychiat 147:1457-1462.

Bartha R, al-Semaan YM, Williamson PC, Drost DJ, Malla AK, Carr TJ, Densmore M, Canaran G, Neufeld RW (1999) A short echo proton magnetic resonance spectroscopy study of the left mesial-temporal lobe in first-onset schizophrenic patients. Biol Psychiat 45:1403-1411.

Bebbington P, Wilkins S, Jones P, Foerster A, Murray R, Toone B, Lewis S (1993) Life events and psychosis. Initial results from the Camberwell Collaborative Psychosis Study. Br J Psychiat 162:72-79.

Benedict RH (1997) Brief Visuospatial Memory Test – Revised. Odessa, Florida, Psychological Assessment Resources, Inc.

Benes FM, Berretta S (2001) GABAergic interneurons: implications for understanding schizophrenia and bipolar disorder. Neuropsychopharmacol 25:1-27.

Benes FM, Sorensen I, Bird ED (1991) Reduced neuronal size in posterior hippocampus of schizophrenic patients. Schiz Bull 17:597-608.

Benes FM, Vincent SL, Marie A, Khan Y (1996) Up-regulation of GABAA receptor binding on neurons of the prefrontal cortex in schizophrenic subjects. Neuroscience 75:1021-1031.

Benes FM, Wickramasinghe R, Vincent SL, Khan Y, Todtenkopf M (1997) Uncoupling of GABA(A) and benzodiazepine receptor binding activity in the hippocampal formation of schizophrenic brain. Brain Res 755:121-129.

Benwell MEM, Balfour DJK, Anderson JM (1988) Evidence that tobacco smoking increase the density of $(-)$-[^{3}H]nicotine binding sites in human brain. J Neurochem 50:1243-1247.

Berman KF, Torrey EF, Daniel DG & Weinberger DR (1992) Regional cerebral blood flow in monozygotic twins discordant and concordant for schizophrenia. Arch Gen Psychiat 49:927-934.

Berman KF, Zec RF & Weinberger DR (1986) Physiologic dysfunction of dorsolateral prefrontal cortex in schizophrenia. II. Role of neuroleptic treatment, attention, and mental effort. Arch Gen Psychiat 43:126-135.

Bernstein HG, Krell D, Baumann B, Danos P, Falkai P, Diekmann S, Henning H, Bogerts B (1998) Morphometric studies of the entorhinal cortex in neuropsychiatric patients and controls: clusters of heterotopically displaced lamina II neurons are not indicative of schizophrenia. Schiz Res 33:125-132.

Bertolino A, Nawroz S, Mattay VS, Barnett AS, Duyn JH, Moonen CT, Frank JA, Tedeschi G, Weinberger DR (1996) Regionally specific pattern of neurochemical pathology in schizophrenia as assessed by multislice proton magnetic resonance spectroscopic imaging. Am J Psychiat153:1554-1563.

Bilder RM, Bogerts B, Ashtari M, Wu H, Alvir JM, Jody D, Reiter G, Bell L & Lieberman JA (1995) Anterior hippocampal volume reductions predict frontal lobe dysfunction in first episode schizophrenia. Schiz Res 17:47-58.

Bilder RM, Volavka J, Czobor P, Malhotra AK, Kennedy JL, Ni X, Goldman RS, Hoptman MJ, Sheitman B, Lindenmayer JP, Citrome L, McEvoy JP, Kunz M, Chakos M, Cooper TB, Lieberman JA (2002) Neurocognitive correlates of the COMT Val(158)Met polymorphism in chronic schizophrenia. Biol Psychiat 52:701-707.

Bilder RM, Volavka J, Lachman HM, Grace AA (2004) The catechol-O-methyltransferase polymorphism: relations to the tonic-phasic dopamine hypothesis and neuropsychiatric phenotypes. Neuropsychopharmacol 29:1943-1961.

Blennow K, Bogdanovic N, Heilig M, Grenfeldt B, Karlsson I, Davidsson P (2000) Reduction in the synaptic protein rab3a in the thalamus and connecting brain regions in postmortem schizophrenic brains. J Neural Transm 107:1085-1097.

Bogerts B, Ashtari M, Degreef G, Alvir JM, Bilder RM & Lieberman JA (1990) Reduced temporal limbic structure volumes on magnetic resonance images in first episode schizophrenia. Psychiat Res 35:1-13.

Bogerts B (1993) Recent advances in the neuropathology of schizophrenia Schiz Bull 19:431-445.

Bourgeois JP, Goldman-Rakic PS & Rakic P (1994) Cereb Cortex 4:78–96.

Bowden NA, Scott RJ, Tooney PA (2007) Altered expression of regulator of G-protein signalling 4 (RGS4) mRNA in the superior temporal gyrus in schizophrenia. Schiz Res 89:165-168.

Braff DL (1993) Information processing and attention dysfunctions in schizophrenia. Schiz Bull 19:233-259.

Breier A, Buchanan RW, Elkashef A, Munson RC, Kirkpatrick B & Gellad F (1992) Brain morphology and schizophrenia. A magnetic resonance imaging study of limbic, prefrontal cortex, and caudate structures. Arch Gen Psychiat 49:921-926.

Brewer JB, Zhao Z, Desmond JE, Glover GH, Gabrieli JD (1998) Making Memories: Brain Activity that Predicts How Well Visual Experience Will Be Remembered. Science 281:1185-1187.

Brickman AM, Buchsbaum MS, Bloom R, Bokhoven P, Paul-Odouard R, Haznedar MM, Dahlman KL, Hazlett E, Aronowitz J, Heath D, Shihabuddin L (2004) Neuropsychological functioning in first-break, never-medicated adolescents with psychosis. J Nerv Men Dis 192:615-622.

Bruder GE, Wexler BE, Sage MM, Gil RB, Gorman JM (2004) Verbal memory in schizophrenia: additional evidence of subtypes having different cognitive deficits. Schiz Res 68:137-147.

Buchanan RW, Strauss ME, Breier A, Kirkpatrick B, Carpenter WT (1997) Attentional impairments in deficit and nondeficit forms of schizophrenia. Am J Psychiat 154:363-370.

Buchanan RW, Strauss ME, Kirkpatrick B, Holstein C, Breier A, Carpenter WT (1994) Neuropsychological impairments in deficit vs nondeficit forms of schizophrenia. Arch Gen Psychiat 51:804-811.

Buckholtz JW, Meyer-Lindenberg A, Honea RA, Straub RE, Pezawas L, Egan MF, Vakkalanka R, Kolachana B, Verchinski BA, Sust S, Mattay VS, Weinberger DR, Callicott JH (2007) Allelic variation in RGS4 impacts functional and structural connectivity in the human brain. J Neurosci 27:1584-1593.

Callicott JH, Egan MF, Mattay VS, Bertolino A, Bone AD, Verchinksi B & Weinberger DR (2003) Abnormal fMRI response of the dorsolateral prefrontal cortex in cognitively intact siblings of patients with schizophrenia. Am J Psychiat 160:709-719.

Callicott JH, Mattay VS, Bertolino A, Finn K, Coppola R, Frank JA (1999) Physiological characteristics of capacity constraints in working memory as revealed by functional MRI. Cereb Cortex 9:20-26.

Callicott JH, Ramsey NF, Tallent K, Bertolino A, Knable MB, Coppola R, Goldberg T, van Gelderen P, Mattay VS, Frank JA, Moonen CT & Weinberger DR (1998) Functional magnetic resonance imaging brain mapping in psychiatry: methodological issues illustrated in a study of working memory in schizophrenia. Neuropsychopharmacol 18:186-196.

Callicott JH, Straub RE, Pezawas L, Egan MF, Mattay VS, Hariri AR, Verchinski BA, Meyer-Lindenberg A, Balkissoon R, Kolachana B, Goldberg TE, Weinberger DR (2005) Variation in DISC1 affects hippocampal structure and function and increases risk for schizophrenia. Proc Natl Acad Sci USA 102:8627-8632.

Cannon TD, Huttunen MO, Lonnqvist J, Tuulio-Henriksson A, Pirkola T, Glahn D, Finkelstein J, Hietanen M, Kaprio J & Koskenvuo M (2000) The inheritance of neuropsychological dysfunction in twins discordant for schizophrenia. Am J Hum Genet 67:369-382.

Carlsson A, Lindquist M (1963) Effect of chlorpromazine and haloperidol of formation of 3- methoxytyramine and normetanephrine in mouse brain. Acta Pharmacol Toxicol 20:140-144.

Carpenter WT & Buchanan RW (1989) Domains of psychopathology relevant to the study of etiology and treatment in schizophrenia. In: Schulz SC & Tamminga CA (Eds). Schizophrenia: Scientific Progress. Oxford University Press: New York, NY, 13-22.

Carter CS, Braver TS, Barch DM, Botvinick MM, Noll D & Cohen JD (1998) Anterior cingulate cortex, error detection, and the on-line monitoring of performance. Science 280:747-749.

Castellon SA, Asarnow RF, Goldstein MJ & Marder SR (1994) Persisting negative symptoms and information-processing deficits in schizophrenia: implications for subtyping. Psychiat Res. 54:59-69.

Censits DM, Daniel Ragland J, Gur RC, Gur RE (1997) Neuropsychological evidence supporting a neurodevelopmental model of schizophrenia: a longitudinal study. Schiz Res 24:289-298.

Chen X, Dunham C, Kendler S, Wang X, O'Neill FA, Walsh D, Kendler KS (2004) Regulator of G-protein signaling 4 (RGS4) gene is associated with schizophrenia in Irish high density families. Am J Med Genet B Neuropsychiat Genet 129:23-26.

Chowdari KV, Mirnics K, Semwal P, Wood J, Lawrence E, Bhatia T et al (2002) Association and linkage analyses of RGS4 polymorphisms in schizophrenia. Hum Mol Genet 11:1373-1380.

Chumakov I, Blumenfeld M, Guerassimenko O, Cavarec L, Palicio M, Abderrahim H et al. (2002) Genetic and physiological data implicating the new human gene G72 and the gene for D-amino acid oxidase in schizophrenia. Proc Natl Acad Sci USA 99:13675-13680.

Cohen RA, Kaplan RF, Zuffante P, Moser DJ, Jenkins MA, Salloway S & Wilkinson H (1999) Alteration of intention and self-initiated action associated with bBilateral anterior cingulotomy. J Neuropsychiat Clin Neurosci 11:444-453.

Conrad AJ, Abebe T, Austin R, Forsythe S, Scheibel AB (1991) Hippocampal pyramidal cell disarray in schizophrenia as a bilateral phenomenon. Arch Gen Psychiat 48:413.

Copolov DL, Seal ML, Maruff P, Ulusoy R, Wong MT, Tochon-Danguy HJ, Egan GF (2003) Cortical activation associated with the experience of auditory hallucinations and perception of human speech in schizophrenia: a PET correlation study. Psychiat Res 122:139-152.

Cornblatt BA & Erlenmeyer-Kimling L (1985) Global attentional deviance as a marker of risk for schizophrenia: specificity and predictive validity. J Abnormal Psychol 94:470-486.

Corvin AP, Morris DW, McGhee K, Schwaiger S, Scully P, Quinn J, Meagher D, Clair DS, Waddington JL, Gill M (2004) Confirmation and refinement of an 'at-risk' haplotype for schizophrenia suggests the EST cluster, Hs.97362, as a potential susceptibility gene at the Neuregulin-1 locus. Mol Psychiat 9:208-213.

Coyle JT (1997) The nagging question of the function of N-acetylaspartylglutamate. Neurobiol Dis 4:231-238.

Csernansky JG, Joshi S, Wang L, Haller JW, Gado M, Miller JP (1998) Hippocampal morphometry in schizophrenia by high dimensional brain mapping. Proc Natl Acad Sci USA 95:11406-11411.

Csernansky JG, Schindler MK, Splinter NR, Wang L, Gado M, Selemon LD, Rastogi-Cruz D, Posener JA, Thompson PA & Miller MI (2004) Abnormalities of thalamic volume and shape in schizophrenia. Am J Psychiat 161:896-902.

Davis KL, Kahn RS, Ko G, Davidson M (1991) Dopamine in schizophrenia: a review and reconceptualization. Am J Psychiat 148:1474-1486.

Deicken RF, Pegues M, Amend D (1999) Reduced hippocampal N-acetylaspartate without volume loss in schizophrenia. Schiz Res 37:217-223.

de Frias CM, Annerbrink K, Westberg L, Eriksson E, Adolfsson R, Nilsson LG (2004) COMT gene polymorphism is associated with declarative memory in adulthood and old age. Behav Genet 34:533-539.

de Paulis T (2001) M-100907 (Aventis). Curr Opin Investig Drugs 2:123-132.

Derosse P, Hodgkinson CA, Lencz T, Burdick KE, Kane JM, Goldman D, Malhotra AK (2007) Disrupted in schizophrenia 1 genotype and positive symptoms in schizophrenia. Biol Psychiat 61:1208-1210.

Devon RS, Anderson S, Teague PW, Burgess P, Kipari TM, Semple CA, Millar JK, Muir WJ, Murray V, Pelosi AJ, Blackwood DH, Porteous DJ (2001) Identification of polymorphisms within Disrupted in Schizophrenia 1 and Disrupted in Schizophrenia 2, and an investigation of their association with schizophrenia and bipolar affective disorder. Psychiat Genet 11:71-78.

Dickinson D, Iannone VN, Wilk CM & Gold JM (2004) General and specific cognitive deficits in schizophrenia. Biol Psychiat 55:826-833.

Dolan RJ, Fletcher P, Frith CD, Friston KJ, Frackowiak RSJ, Grasby PM (1995) Dopaminergic modulation of impaired cognitive activation in the anterior cingulate cortex in schizophrenia. Nature 378:180-182.

Eastwood SL, Harrison PJ (1995) Decreased synaptophysin in the medial temporal lobe in schizophrenia demonstrated using immunoautoradiography. Neuroscience 69:339-343.

Eastwood SL, Harrison PJ (2000) Hippocampal synaptic pathology in schizophrenia, bipolar disorder and major depression: a study of complexin mRNAs. Mol Psychiat 5:425-432.

Eastwood SL, Harrison PJ (2001) Synaptic pathology in the anterior cingulate cortex in schizophrenia and mood disorders. A review and a western blot study of synaptophysin, GAP-43 and the complexins. Brain Res Bull 55:569-578.

Egan MF, Goldberg TE, Kolachana BS, Callicott JH, Mazzanti CM, Straub RE et al. (2001) Effect of COMT Val108-158 Met genotype on frontal lobe function and risk for schizophrenia. Proc Natl Acad Sci USA 98:6917-6922.

Egan MF, Straub RE, Goldberg TE, Yakub I, Callicott JH, Hariri AR, Mattay VS, Bertolino A, Hyde TM, Shannon-Weickert C, Akil M, Crook J, Vakkalanka RK, Balkissoon R, Gibbs RA, Kleinman JE, Weinberger DR (2004) Variation in GRM3 affects cognition, prefrontal glutamate, and risk for schizophrenia. Proc Natl Acad Sci USA 101:12604-12609.

Eichenbaum H & Cohen NJ (2001) From conditioning to conscious recollection: memory systems of the brain. New York, Oxford University Press.

Eichenbaum H (2000) A cortical-hippocampal system for declarative memory. Neuroscience 1:41-50.

Ekelund J, Hennah W, Hiekkalinna T, Parker A, Meyer J, Lonnqvist J, Peltonen L (2004) Replication of 1q42 linkage in Finnish schizophrenia pedigrees. Mol Psychiat 9:1037-1041.

Eldridge LL, Engel SA, Zeineh MM, Bookheimer SY, Knowlton BJ (2005) A dissociation of encoding and retrieval processes in the human hippocampus. J Neurosci 25:3280-3286.

Elkashef AM, Issa F, Wyatt RJ (1995) The biochemical basis of schizophrenia. In: Shriqui CL, Nasrallah HA (eds). Contemporary Issues in the treatment of schizophrenia. American Psychiatric Press: Washington, DC, pp 863.

Erdely HA, Tamminga CA, Roberts RC, Vogel MW (2006) Regional alterations in RGS4 protein in schizophrenia. Synapse 59:472-479.

Eyler-Zorrilla LT, Jeste DV, Paulus MP, Brown G (2002) Functional abnormalities of medial temporal cortex during novel picture learning among patients with chronic schizophrenia. Schiz Res 59:187-198.

Falkai P & Bogerts B (1986) Cell loss in the hippocampus of schizophrenics. Eur Arch Psychiat Neurol Sci 236:154-161.

Fallin MD, Lasseter VK, Avramopoulos D, Nicodemus KK, Wolyniec PS, McGrath JA, Steel G, Nestadt G, Liang KY, Huganir RL, Valle D, Pulver AE (2005) Bipolar I disorder and schizophrenia: a 440-single-nucleotide polymorphism screen of 64 candidate genes among Ashkenazi Jewish case-parent trios. Am J Hum Genet 77:918-936.

Fan JB, Zhang CS, Gu NF, Li XW, Sun WW, Wang HY, Feng GY, St Clair D, He L (2005) Catechol-O-methyltransferase gene Val/Met functional polymorphism and risk of schizophrenia: a large-scale association study plus meta-analysis. Biol Psychiat 57:139-144.

Fatemi SH, Earle JA, Stary JM, Lee S, Sedgewick J (2001) Altered levels of the synaptosomal associated protein SNAP-25 in hippocampus of subjects with mood disorders and schizophrenia. Clin Neurosci Neuropathol 12:3257-3262.

Flashman LA & Green MF (2004) Review of cognition and brain structure in schizophrenia: profiles, longitudinal course, and effects of treatment. Psychiat Clin.North Am 27:1-18.

Freedman DX (1975) LSD, psychotogenic procedures, and brain neurohumors. Psychopharmacol Bull 11:42-43.

Friston KJ (1992) The dorsolateral prefrontal cortex, schizophrenia and PET. J Neur Transm Suppl 37:79-93.

Fujii Y, Shibata H, Kikuta R, Makino C, Tani A, Hirata N, Shibata A, Ninomiya H, Tashiro N, Fukumaki Y (2003) Positive associations of polymorphisms in the metabotropic glutamate receptor type 3 gene (GRM3) with schizophrenia. Psychiat Genet 13:71-76.

Funke B, Finn CT, Plocik AM, Lake S, DeRosse P, Kane JM, Kucherlapati R, Malhotra AK (2004) Association of the DTNBP1 locus with schizophrenia in a U.S. population. Am J Hum Genet 75:891-898.

Gazzaley A, Cooney JW, McEvoy K, Knight RT, D'Esposito M (2005) Top-down enhancement and suppression of the magnitude and speed of neural activity. J Cogn Neurosci 17:507-517.

Ghose S, Weickert CS, Colvin SM, Coyle JT, Herman MM, Hyde TM, Kleinman JE (2004) Glutamate carboxypeptidase II gene expression in the human frontal and temporal lobe in schizophrenia.Neuropsychopharmacol 29:117-125.

Ghose S., Chin R., Gao X.M., Stan A., Lewis-Amezcua K., Frost D., Roberts R., Tamminga C.A (2006) Distinct gene expression correlations in schizophrenia between NR1 and GAD67 with BDNF in the anterior hippocampus. Abstract, 36th Annual Meeting of the Society for Neuroscience, Atlanta, GA.

Glahn DC, Therman S, Manninen M, Huttunen M, Kaprio J, Lonnqvist J & Cannon TD (2003) Spatial working memory as an endophenotype for schizophrenia. Biol Psychiat 53:626.

Glatt SJ, Faraone SV, Tsuang MT (2003) Association between a functional catechol O-methyltransferase gene polymorphism and schizophrenia: Meta-analysis of case-control and family-based studies. Am J Psychiat 160:469-476.

Gogtay N, Nugent TF, Herman DH, Ordonez A, Greenstein D, Hayashi KM, Clasen L, Toga AW, Giedd JN, Rapoport JL & Thompson PM (2006) Dynamic mapping of normal human hippocampal development. Hippocampus16:664-672.

Golby AJ, Poldrak RA, Brewer JB, Spencer D, Desmond JE, Aron AP, Gabrieli JD (2001) Material-specific lateralization in the medial temporal lobe and prefrontal cortex during memory encoding. Brain 134:1841-1854.

Gold JM, Carpenter C, Randolph C, Goldberg TE & Weinberger DR (1997) Auditory working memory and Wisconsin Card Sorting Test performance in schizophrenia. Arch Gen Psychiat 54:159-165.

Goldberg TE, Berman KF, Fleming K, Ostrem J, Van Horn JD, Esposito G, Mattay VS, Gold JM & Weinberger DR (1998) Uncoupling cognitive workload and prefrontal cortical physiology: a PET rCBF study. Neuroimage 7:296-303.

Goldberg TE, Egan MF, Gscheidle T, Coppola R, Weickert T, Kolachana BS, Goldman D, Weinberger DR (2003) Executive subprocesses in working memory: relationship to catechol-O-methyltransferase Val[158]Met genotype and schizophrenia. Arch Gen Psychiat 60:889-896.

Goldberg TE & Weinberger DR (1996) Effects of neuroleptic medications on the cognition of patients with schizophrenia: a review of recent studies. J Clin Psychiat 57:62-65.

Goldman RS, Axelrod BN, Taylor SF (1996) Neuropsychological aspects of schizophrenia in Neuropsychological assessment of neuropsychiatric disorders. Grant I, Adams K (Eds).

Goldman-Rakic PS (1994) Working memory dysfunction in schizophrenia. J Neuropsychiat Clin Neurosci 6:348-357.

Gooding D & Tallent K (2001) The association between antisaccade task and working memory task performance in schizophrenia and bipolar disorder. J Nerv Mental Dis 189:8-16.

Grady RM, Zhou H, Cunningham JM, Henry MD, Campbell KP, Sanes JR (2000) Maturation and maintenance of the neuromuscular synapse: genetic evidence for roles of the dystrophin--glycoprotein complex. Neuron 25:279-293.

Green MF, Kern RS, Braff DL, Mintz J (2000) Neurocognitive deficits and functional outcome in schizophrenia: are we measuring the "right stuff"? Schiz Bull 26:119-136.

Gur RC, Ragland JD, Moberg PJ, Bilker WB, Kohler C, Siegel SJ & Gur RE (2001) Computerized neurocognitive scanning: II. The profile of schizophrenia. Neuropsychopharmacology 25:777-788.

Gur RE, McGrath C, Chan RM, Schroeder L, Turner T, Turetsky BI, Kohler C, Alsop D, Maldjian J, Ragland JD, Gur RC (2002) An fMRI study of facial emotion processing in patients with schizophrenia. Am J Psychiat 159:1992-1999.

Gur RE & Pearlson GD (1993) Neuroimaging in schizophrenia research. Schiz Bull 19:337-353.

Gur RE, Turetsky BI, Bilker WB & Gur RC (1999) Reduced gray matter volume in schizophrenia. Arch Gen Psychiat 56:905-911.

Haber SN (2003) The primate basal ganglia: parallel and integrative networks. J Chem Neuroanat 26:317-330.

Hall J, Whalley HC, Job DE, Baig BJ, McIntosh AM, Evans KL, Thomson PA, Porteous DJ, Cunningham-Owens DG, Johnstone EC, Lawrie SM (2006) A neuregulin 1 variant associated with abnormal cortical function and psychotic symptoms. Nat Neurosci 9:1477-1478.

Harris JG, Adler LE, Young DA, Cullum CM, Rilling LM, Cicerello A, Intemann PM, Freedman R (1996) Neuropsychological dysfunction in parents of schizophrenics. Schiz Res 20:253-260.

Harrison PJ & Eastwood SL (1998) Preferential involvement of excitatory neurons in medial temporal lobe in schizophrenia. Lancet 352:1669-1673.

Harrison PJ & Eastwood SL (2001) Neuropathological studies of synaptic connectivity in the hippocampal formation in schizophrenia. Hippocampus 11:508-519.

Harrison PJ & Owen M (2003) Genes for schizophrenia? Recent findings and their pathophysiological implications. Lancet 361:417-419.

Harrison PJ & Weinberger D (2005) Schizophrenia genes, gene expression, and neuropathology: on the matter of their convergence. Mol Psychiat 10:40-68.

Harrison PJ (1999) The neuropathology of schizophrenia. A critical review of the data and their interpretation. Brain 122:593-624.

Harrison PJ (1999) Neurochemical alterations in schizophrenia affecting the putative receptor targets of atypical antipsychotics. Focus on dopamine (D1, D3, D4) and 5-HT2a receptors. Br J Psychiatry Suppl 38:12-22.

Hazlett EA, Buchsbaum MS, Kemether E, Bloom R, Platholi J, Brickman AM, Shihabuddin L, Tang C, Byne W (2004) Abnormal glucose metabolism in the mediodorsal nucleus of the thalamus in schizophrenia. Am J Psychiat 161:305-314.

Heckers S, Rauch SL, Goff D, Savage CR, Schacter DL, Alpert NM (1998) Impaired recruitment of the hippocampus during conscious recollection in schizophrenia. Nature 1:318-323.

Heckers S, Stone D, Walsh J, Shick J, Koul P, Benes FM (2002) Differential hippocampal expression of glutamic acid decarboxylase 65 and 67 messenger RNA in bipolar disorder and schizophrenia. Arch Gen Psychiat 59:521-529.

Heckers S, Zalesak M, Weiss A, Ditman T, Titone D (2004) Hippocampal Activation During Transitive Inference in Humans. Hippocampus 14:153-162.

Heimer L & Van Hoesen GW (2006) The limbic lobe and its output channels: implications for emotional functions and adaptive behavior. Neurosci Biobehav Rev 30:126-147.

Heinrichs RW & Zakzanis KK (1998) Neurocognitive deficit in schizophrenia: a quantitative review of the evidence. Neuropsychology 12:426-445.

Hennah W, Tuulio-Henriksson A, Paunio T, Ekelund J, Varilo T, Partonen T, Cannon TD, Lonnqvist J, Peltonen L (2005) A haplotype within the DISC1 gene is associated with visual memory functions in families with a high density of schizophrenia. Mol Psychiat 10:1097-1103.

Henson RN, Cansino S, Herron JE, Robb WG, Rugg MD (2003) A familiarity signal in human anterior medial temporal cortex? Hippocampus 13:301-304.

Henson RN, Rugg MD, Shallice T, Josephs O, Dolan RJ (1999) Recollection and Familiarity in Recognition Memory: An Event-Related Functional Magnetic Resonance Imaging Study. J Neurosci 19:3962-3972.

Hicks JL & Marsh RL (1999) Remember-know judgements can depend on how memory is tested. Psychol Bull Rev 6:117-122.

Hill SK, Ragland JD, Gur RC & Gur RE (2002) Neuropsychological profiles delineate distinct profiles of schizophrenia, an interaction between memory and executive function, and uneven distribution of clinical subtypes. J Clin Exp Neuropsychol 24:765-780.

Hirsch S, Bowen J, Emami J, Cramer P, Jolley A, Haw C, Dickinson M (1996) A one year prospective study of the effect of life events and medication in the aetiology of schizophrenic relapse. Br J Psychiat 168:49-56.

Hodgkinson CA, Goldman D, Jaeger J, Persaud S, Kane JM, Lipsky RH, Malhotra AK (2004) Disrupted in schizophrenia 1 (DISC1): association with schizophrenia, schizoaffective disorder, and bipolar disorder. Am J Hum Genet 75:862-872.

Hughes JR, D.K. Hatsukami DK, J.E. Mitchell JE, Dahlgren LA (1986) Prevalence of smoking among psychiatric outpatients. Am J Psychiat 143:993-997.

Hulshoff Pol HE, Schnack HG, Mandl RC, van Haren NE, Koning H, Collins DL, Evans AC, Kahn RS (2001) Focal gray matter density changes in schizophrenia. Arch Gen Psychiat 58:1118-1125.

Huttenlocher PR (1994) in Human Behavior and the Developing Brain, eds. Dawson, G. & Fischer, K. (Guilford, New York), pp. 137-152;

Hyde TM, Crook JM (2001) Cholinergic systems and schizophrenia: primary pathology or epiphenomena? J Chem Neuroanat 22:53-63.

Insausti R & Amaral DG (2004) Hippocampal Formation. In: The Human Nervous System Paxinos GJM (Ed.). San Diego, Elsevier.

Iudicello JE, Apud JA, Egan MF, Straub RE, Goldberg TE & Weinberger DR (2004) The Effects of the COMT-Inhibitor Tolcapone on Cognition in Healthy Human Subjects: Modulation by the COMT Val Met Polymorphism. ACNP 43rd Annual Meeting , San Juan, Puerto Rico, Neuropsychopharmacology.

Iwata N, Suzuki T, Ikeda M, Kitajima T, Yamanouchi Y, Inada T, Ozaki N (2004) No association with the neuregulin 1 haplotype to Japanese schizophrenia. Mol Psychiat 9:126-127.

Jaeger J, Czobor P, Berns SM (2003) Basic neuropsychological dimensions in schizophrenia. Schiz Res 65:105-116.

Jakob H & Beckmann H (1986) Prenatal development disturbances in the limbic allocortex in schizophrenics. J Neural Transm 65:303-326.

Jessen F, Scheef L, Germeshausen L, Tawo Y, Kockler M, Kuhn K-U, Maier W, Schild HH, Heun R (2003) Reduced hippocampal activation during encoding and recognition of words in schizophrenia patients. Am J Psychiat 160:1305-1312.

Johnstone EC, Crow TJ, Frith CD, Husband J & Kreel L (1976) Cerebral ventricular size and cognitive impairment in chronic schizophrenia. Lancet 2:924-926.

Joober R, Gauthier J, Lal S, Bloom D, Lalonde P, Rouleau G, Benkelfat C, Labelle A (2002) Catechol-O-methyltransferase Val-108/158-Met gene variants associated with performance on the Wisconsin Card Sorting Test. Arch Gen Psychiat 59:662-663.

Kamiya A, Kubo K, Tomoda T, Takaki M, Youn R, Ozeki Y, Sawamura N, Park U, Kudo C, Okawa M, Ross CA, Hatten ME, Nakajima K, Sawa A (2005) A schizophrenia-associated mutation of DISC1 perturbs cerebral cortex development. Nat Cell Biol 7:1067-1078.

Kay SR & Sevy S (1990) Pyramidical model of schizophrenia. Schiz Bull 16:537-545.

Keefe RS, Silverman JM, Mohs RC, Siever LJ, Harvey PD, Friedman L, Roitman SE, DuPre RL, Smith CJ, Schmeidler J, Davis KL (2003) Eye tracking, attention, and schizotypal symptoms in nonpsychotic relatives of patients with schizophrenia. Arch Gen Psychiat 54:169-176.

Kegeles LS, Shungu DC, Anjilvel S, Chan S, Ellis SP, Xanthopoulos E, Malaspina D, Gorman JM, Mann JJ, Laruelle M, Kaufmann CA (2000) Hippocampal pathology in schizophrenia: magnetic resonance imaging and spectroscopy studies. Psychiatry Res 98:163-175.

Kirkpatrick B, Buchanan RW, Ross DE & Carpenter WT Jr (2001) A separate disease within the syndrome of schizophrenia. Arch Gen Psychiat 58:165-171.

Kirov G, Ivanov D, Williams NM, Preece A, Nikolov I, Milev R, Koleva S, Dimitrova A, Toncheva D, O'Donovan MC, Owen MJ (2004) Strong evidence for association between the dystrobrevin binding protein 1 gene (DTNBP1) and schizophrenia in 488 parent-offspring trios from Bulgaria. Biol Psychiat 55:971-975.

Kitano H (2002) Computational systems biology. Nature 420:206-210.

Konopaske GT, Sweet RA, Wu Q, Sampson A, Lewis DA (2006) Regional specificity of chandelier neuron axon terminal alterations in schizophrenia. Neuroscience 138:189-196.

Kornack DR, Rakic P (1999) Continuation of neurogenesis in the hippocampus of the adult macaque monkey. Proc Natl Acad Sci USA 96:5768-5773.

Kosaka H, Omori M, Murata T, Iidaka T, Yamada H, Okada T, Takahashi T, Sadato N, Itoh H, Yonekura Y, Wada Y (2002) Differential amygdala response during facial recognition in patients with schizophrenia: an fMRI study. Schiz Res 57:87-95.

Kovelman JA & Scheibel AB (1984) A neurohistological correlate of schizophrenia. Biol Psychiat 19:1601-1621.

Kravariti E, Morris RG, Rabe-Hesketh S, Murray RM, Frangou S (2003) The Maudsley Early-Onset Schizophrenia Study: cognitive function in adolescent-onset schizophrenia. Schiz Res 65:95-103.

Krimer LS, Hyde TM, Herman MM, Saunders RC (1997) The entorhinal cortex: an examination of cyto- and myeloarchitectonic organization in humans. Cereb Cortex 7:722-731.

Krystal JH, Abi-Saab W, Perry E, D'Souza DC, Liu N, Gueorguieva R, McDougall L, Hunsberger T, Belger A, Levine L, Breier A (2005) Preliminary evidence of attenuation of the disruptive effects of the NMDA glutamate receptor antagonist, ketamine, on working memory by pretreatment with the group II metabotropic glutamate receptor agonist, LY354740, in healthy human subjects. Psychopharmacol (Berl) 179:303-309.

Kuroki T, Meltzer HY, Ichikawa J (2003) 5-HT 2A receptor stimulation by DOI, a 5-HT 2A2C receptor agonist, potentiates amphetamine-induced dopamine release in rat medial prefrontal cortex and nucleus accumbens. Brain Res 972:216-221.

Kurtz MM, Ragland JD, Bilker W, Gur RC, Gur RE (2001) Comparison of the continuous performance test with and without working memory demands in healthy controls and patients with schizophrenia. Schiz.Res 48:307-316.

Lahti AC, Holcomb HH, Medoff DR, Tamminga CA (1995) Ketamine activates psychosis and alters limbic blood flow in schizophrenia. Neuroreport 6:869-872.

Lahti AC, Holcomb HH, Medoff DR, Weiler MA, Tamminga CA, Carpenter WT (2001) Abnormal patterns of regional cerebral blood flow in schizophrenia with primary negative symptoms during an effortful auditory recognition task. Am J Psychiat 158:1797-1808.

Law AJ, Lipska BK, Weickert CS, Hyde TM, Straub RE, Hashimoto R, Harrison PJ, Kleinman JE, Weinberger DR (2006) Neuregulin 1 transcripts are differentially expressed in schizophrenia and regulated by 5' SNPs associated with the disease. Proc Natl Acad Sci USA103:6747-6752.

Lawrie SM, Whalley HC, Abukmeil SS, Kestelman JN, Miller P, Best JJ, Owens DG & Johnstone EC (2002) Temporal lobe volume changes in people at high risk of schizophrenia with psychotic symptoms.Br J Psychiat 181:138-143.

Lenzenweger MF, Dworkin RH & Wethington E (1991) Examining the underlying structure of schizophrenic phenomenology: evidence for a three-process model. Schiz Bull 17:515-524.

Lewis DA, Hashimoto T, Volk DW (2005) Cortical inhibitory neurons and schizophrenia. Nat Rev Neurosci 6:312-324.

Lewis DA & Lund JS (1990) Heterogeneity of chandelier neurons in monkey neocortex: corticotropin-releasing factor- and parvalbumin-immunoreactive populations. J Comp Neurol 293:599-615.

Lewis DA, Volk DW, Hashimoto T (2004) Selective alterations in prefrontal cortical GABA neurotransmission in schizophrenia: a novel target for the treatment of working memory dysfunction. Psychopharmacol (Berl) 174:143-150.

Li D, Collier DA, He L (2006) Meta-analysis shows strong positive association of the neuregulin 1 gene with schizophrenia. Hum Mol Genet 15:1995-2002.

Liddle PF (1987) The symptoms of chronic schizophrenia: a re-examination of the positive-negative dichotomy. Br J Psychiat 151:145-151.

Liddle PF, Friston KJ, Frith CD, Hirsch SR, Jones T, Frackowiak RS (1992) Patterns of cerebral blood flow in schizophrenia. Br J Psychiat 160:179-186.

Lipska BK, Mitkus S, Caruso M, Hyde TM, Chen J, Vakkalanka R, Straub RE, Weinberger DR, Kleinman JE (2006) RGS4 mRNA expression in postmortem human cortex is associated with COMT Val[158]Met genotype and COMT enzyme activity. Hum Mol Genet 15:2804-2812.

Lisman J & Otmakhova N (2001) Storage, recall, and novelty detection of sequences by the hippocampus: elaborating on the socratic model to account for normal and aberrant effects of dopamine. Hippocampus 11:551-568.

Liu H, Heath SC, Sobin C, Roos JL, Galke BL, Blundell ML et al. (2002) Genetic variation at the 22q11 PRODH2DGCR6 locus presents an unusual pattern and increases susceptibility to schizophrenia. Proc Natl Acad Sci USA 99:3717-3722.

Maier M, Ron MA, Barker GJ, Tofts PS (1995) Proton magnetic resonance spectroscopy: an in vivo method of estimating hippocampal neuronal depletion in schizophrenia. Psychol Med 25:1201-1209.

Malhotra AK, Kestler LJ, Mazzanti CM, Bates JA, Goldberg T, Goldman D (2002) A functional polymorphism in the COMT gene and performance on a test of prefrontal cognition. Am J Psychiat 159:652-654.

Manoach D (2003) Prefrontal cortex dysfunction during working memory performance in schizophrenia: reconciling discrepant findings. Schiz Res 60:285-298.

Manoach DS, Press DZ, Thangaraj V, Searl MM, Goff DC, Halpern E, Saper CB & Warach S (1999) Schizophrenic subjects activate dorsolateral prefrontal cortex during a working memory task, as measured by fMRI. Biol Psychiat 45:1128-1137

Manoach DS, Gollub RL, Benson ES, Searl MM, Goff DC, Halpern E (2000) Schizophrenic subjects show aberrant fMRI activation of dorsolateral prefrontal cortex and basal ganglia during working memory performance. Biol Psychiat 48:99-109.

Marcus MM, Nomikos GG, Svensson TH (2000) Effects of atypical antipsychotic drugs on dopamine output in the shell and core of the nucleus accumbens: role of 5-HT (2A) and alpha(1)-adrenoceptor antagonism. Eur Neuropsychopharmacol 10:245-253.

Marti SB, Cichon S, Propping P, Nothen M (2002) Metabotropic glutamate receptor 3 (GRM3) gene variation is not associated with schizophrenia or bipolar affective disorder in the German population. Am J Med Genet 114:46-50.

Meador-Woodruff JH, Healy DJ (2000) Glutamate receptor expression in schizophrenic brain. Brain Res Rev 31:288-294.

Melendez RI, Gregory ML, Bardo MT, Kalivas PW (2004) Impoverished rearing environment alters metabotropic glutamate receptor expression and function in the prefrontal cortex. Neuropsychopharmacol 29:1980-1987.

Menon V, Anagnoson RT, Mathalon DH, Glover GH & Pfefferbaum A (2001) Functional neuroanatomy of auditory working memory in schizophrenia: Relation to positive and negative symptoms. NeuroImage 13:433-446.

McCarley RW, Shenton ME, O'Donnell BF, Faux SF, Kikinis R, Nestor PG, Jolesz FA (1993) Auditory P300 abnormalities and left posterior superior temporal gyrus volume reduction in schizophrenia. Arch Gen Psychiat 50:190-197.

McFarland NR & Haber SN (2002) Thalamic relay nuclei of the basal ganglia form both reciprocal and nonreciprocal cortical connections, linking multiple frontal cortical areas. J Neurosci 22:8117-8132.

Mirnics K, Middleton FA, Marquez A, Lewis DA, Levitt P (2000) Molecular characterization of schizophrenia viewed by microarray analysis of gene expression in prefrontal cortex. Neuron 28:53-67.

Mirnics K, Middleton FA, Stanwood GD, Lewis DA, Levitt P (2001) Disease-specific changes in regulator of G-protein signaling 4 (RGS4) expression in schizophrenia. Mol Psychiat 6:293-301.

Moghaddam B, Adams BW (1998) Reversal of phencyclidine effects by a group II metabotropic glutamate receptor agonist in rats. Science 281:1349-1352.

Morris DW, McGhee KA, Schwaiger S, Scully P, Quinn J, Meagher D, Waddington JL, Gill M, Corvin AP (2003) No evidence for association of the dysbindin gene [DTNBP1] with schizophrenia in an Irish population-based study. Schiz Res 60:167-172.

Morris DW, Rodgers A, McGhee KA, Schwaiger S, Scully P, Quinn J, Meagher D, Waddington JL, Gill M, Corvin AP (2004) Confirming RGS4 as a susceptibility gene for schizophrenia. Am J Med Genet B Neuropsychiatr Genet 125:50-53.

Moser MB & Moser EI (1998) Functional differentiation in the hippocampus. Hippocampus 8:608-619.

Munafo MR, Bowes L, Clark TG, Flint J (2005) Lack of association of the COMT (Val158/108 Met) gene and schizophrenia: a meta-analysis of case-control studies. Mol Psychiat 10:765-770.

Nakazawa K (2003) Hippocampal CA3 NMDA receptors are crucial for memory acquisition of one-time experience. Neuron 38:305-315.

Nakazawa K, Quirk MC, Chitwood RA, Watanabe M, Yechel MF, Sun LD, Kato A, Carr CA, Johnston D, Wilson MA, Tonegawa S (2002) Requirement for hippocampal CA3 NMDA receptors in associative memory recall. Science 297:211-218.

Neale JH, Bzdega T, Wroblewska B (2000) N-Acetylaspartylglutamate: the most abundant peptide neurotransmitter in the mammalian central nervous system. J Neurochem 75:443-452.

Niendam TA, Bearden CE, Rosso IM, Sanchez LE, Hadley T, Nuechterlein KH, Cannon TD (2003) A prospective study of childhood neurocognitive functioning in schizophrenic patients and their siblings. Am J Psychiat 160:2060-2062.

Norman KA & O'Reilly RC (2003) Modeling hippocampal and neocortical contributions to recognition memory: A complementary-learning-systems approach. Psychol Rev 110:611-646.

Nowakowski RS, Rakic P (1981) The site of origin and route and rate of migration of neurons to the hippocampal region of the rhesus monkey. J Comp Neurol 196:129-154.

Nuechterlein K, Barch D, Gold J, Goldberg T, Green M, Heaton R (2004) Identification of separate cognitive factors in schizophrenia. Schiz Res 72:29-39.

Numakawa T, Yagasaki Y, Ishimoto T, Okada T, Suzuki T, Iwata N, Ozaki N, Taguchi T, Tatsumi M, Kamijima K, Straub RE, Weinberger DR, Kunugi H, Hashimoto R (2004) Evidence of novel neuronal functions of dysbindin, a susceptibility gene for schizophrenia. Hum Mol Genet 13:2699-2708.

O'Reilly RC & Rudy JW (2001) Conjunctive representations in learning and memory: principles of cortical and hippocampal function. Psychol Rev 108:311-345.

Packard MG (1999) Glutamate infused posttraining into the hippocampus or caudate-putamen differentially strengthens place and response learning. Proc Natl Acad Sci USA 96:12881-12886.

Park S & Holzman PS (1992) Schizophrenics show spatial working memory deficits. Arch Gen Psychiat 49:975-982.

Park S & Holzman PS (1993) Association of working memory deficit and eye tracking dysfunction in schizophrenia. Schiz Res 11:55-61.

Perlstein WH, Carter CS, Noll DC & Cohen JD (2001) Relation of prefrontal cortex dysfunction to working memory and symptoms in schizophrenia. Am J Psychiat 158:1105-1113.

Perlstein WM, Dixit NK, Carter CS, Noll DC & Cohen JD (2003) Prefrontal cortex dysfunction mediates deficits in working memory and prepotent responding in schizophrenia. Biol Psychiat 53:25-38.

Pickering C, Gustafsson L, Cebere A, Nylander I, Liljequist S (2006) Repeated maternal separation of male Wistar rats alters glutamate receptor expression in the hippocampus but not the prefrontal cortex. Brain Res 1099:101-108.

Potkin SG, Alva G, Fleming K, Anand R, Keator D, Carreon D, Doo M, Jin Y, Wu JC, Fallon JH (2002) A PET study of the pathophysiology of negative symptoms in schizophrenia. Am J Psychiat 159:227-237.

Rakic P (1996) In Child and Adolescent Psychiatry, ed. Lewis, M. (Williams and Wilkins, Baltimore), pp. 9-30.

Ranganath C, Heller A, Cohen MX, Brozinsky CJ, Rissman J (2005) Functional connectivity with the hippocampus during successful memory formation. Hippocampus 15:997-1005.

Reddy R & Keshavan MS (2003) Phosphorus magnetic resonance spectroscopy: its utility in examining the membrane hypothesis of schizophrenia. Prostaglandins Leukot Essent Fatty Acids 69:401-405.

Rissman J, Gazzaley A, D'Esposito M (2004) Measuring functional connectivity during distinct stages of a cognitive task. NeuroImage 23:752-763.

Sawada K, Barr AM, Nakamura M, Arima K, Young CE, Dwork AJ, Falkai P, Phillips AG, Honer WG (2005) Hippocampal Complexin Proteins and Cognitive Dysfunction in Schizophrenia. Arch Gen Psychiat 62:263-272.

Saykin AJ, Shtasel DL, Gur RE, Kester DB, Mozley LH, Stafiniak P, Gur RC (1994) Neuropsychological deficits in neuroleptic naive patients with first-episode schizophrenia. Arch Gen Psychiat 51:124-131.

Schroeder J, Buchsbaum MS, Siegel BV, Geider FJ, Haier RJ, Lohr J, Wu J, Potkin S (1994) Patterns of cortical activity in schizophrenia. Psychol Med 24:947-955.

Schwab SG, Knapp M, Mondabon S, Hallmayer J, Borrmann-Hassenbach M, Albus M et al. (2003) Support for association of schizophrenia with genetic variation in the 6p22.3 gene, dysbindin, in sib-pair families with linkage and in an additional sample of triad families. Am J Hum Genet 72:185-190.

Schwartz ML & Goldman-Rakic PS (1984) Callosal and intrahemispheric connectivity of the prefrontal association cortex in rhesus monkey: relation between intraparietal and principal sulcal cortex. J Comp Neurol 226:403-420.

Shelton RC & Weinberger DR (1987) Brain morphology in schizophrenia. In: Meltzer HY (ed). Psychopharmacology: The Third Generation of Progress. Raven Press: New York, NY, pp 773-781.

Shenton ME, Dickey CC, Frumin M & McCarley RW (2001) A review of MRI findings in schizophrenia. Schiz Res 49:1-52.

Shenton ME, Kikinis R, Jolesz FA, Pollak SD, LeMay M, Wible CG (1992) Abnormalities of the left temporal lobe and thought disorder in schizophrenia. A quantitative magnetic resonance imaging study. N Engl J Med 327:604-612.

Shifman S, Bronstein M, Sternfeld M, Pisante-Shalom A, Lev-Lehman E, Weizman A et al. (2002) A highly significant association between a COMT haplotype and schizophrenia. Am J Hum Genet 71:1296-1302.

Shihabuddin L, Buchsbaum MS, Hazlett EA, Haznedar MM, Harvey PD, Newman A, Schnur DB, Spiegel-Cohen J, Wei T, Machac J, Knesaurek K, Vallabhajosula S, Biren MA, Ciravolo TM & Luu-Hsia C (1998) Dorsal striatal size, shape, and metabolic rate in never-medicated and previously medicated schizophrenics performing a verbal learning task. Arch Gen Psychiat 55:235-243.

Silver H, Feldman P, Bilker W & Gur RC (2003) Working memory deficit as a core neuropsychological dysfunction in schizophrenia. Am J Psychiat 160:1809-1816.

Silbersweig DA, Stern E, Frith C, Cahill C, Holmes A, Grootoonk S (1995) A functional neuroanatomy of hallucinations in schizophrenia. Nature 378:176-179.

Sobell JL, Richard C, Wirshing DA, Heston LL (2005) Failure to confirm association between RGS4 haplotypes and schizophrenia in Caucasians. Am J Med Genet B Neuropsychiat Genet 139:23-27.

St Clair D, Blackwood D, Muir W, Carothers A, Walker M, Spowart G, Gosden C, Evans HJ (1990) Association within a family of a balanced autosomal translocation with major mental illness. Lancet 336:13-16.

Stefansson H, Sigurdsson E, Steinthorsdottir V, Bjornsdottir S, Sigmundsson T, Ghosh S, Brynjolfsson J, Gunnarsdottir S, Ivarsson O, Chou TT, Hjaltason O, Birgisdottir B, Jonsson H, Gudnadottir VG, Gudmundsdottir E, Bjornsson A, Ingvarsson B, Ingason A, Sigfusson S, Hardardottir H, Harvey RP, Lai D, Zhou M, Brunner D, Mutel V, Gonzalo A, Lemke G, Sainz J, Johannesson G, Andresson T, Gudbjartsson D, Manolescu A, Frigge ML, Gurney ME, Kong A, Gulcher JR, Petursson H, Stefansson K (2002) Neuregulin 1 and susceptibility to schizophrenia. Am J Hum Genet 71:877-892.

Stefansson H, Sarginson J, Kong A, Yates P, Steinthorsdottir V, Gudfinnsson E, Gunnarsdottir S, Walker N, Petursson H, Crombie C, Ingason A, Gulcher JR, Stefansson K, St Clair D (2003) Association of neuregulin 1 with schizophrenia confirmed in a Scottish population. Am J Hum Genet 72:83-87.

Stone M, Gabrieli JD, Stebbins GT, Sullivan EV (1998) Working and strategic memory deficits in schizophrenia. Neuropsychology 12:278-288.

Straub RE, Jiang Y, MacLean CJ, Ma Y, Webb BT, Myakishev MV, Harris-Kerr C, Wormley B, Sadek H, Kadambi B, Cesare AJ, Gibberman A, Wang X, O'Neill FA, Walsh D, Kendler KS (2002) Genetic variation in the 6p22.3 gene DTNBP1, the human ortholog of the mouse dysbindin gene, is associated with schizophrenia. Am J Hum Genet 71:337-348.

Suddath RL, Christison GW, Torrey EF, Casanova MF & Weinberger DR (1990) Anatomical abnormalities in the brains of monozygotic twins discordant for schizophrenia. N Engl J Med 322:789-794.

Sullivan PE (2005) The genetics of schizophrenia. PLoS Med 2:e212.

Sweatt J (2003) Mechanisms of Memory. Academic Press.

Takahashi H, Koeda M, Oda K, Matsuda T, Matsushima E, Matsuura M, Asai K, Okubo Y (2004) An fMRI study of differential neural response to affective pictures in schizophrenia. Neuroimage 22:1247-1254.

Talbot K, Eidem WL, Tinsley CL, Benson MA, Thompson EW, Smith RJ, Hahn CG, Siegel SJ, Trojanowski JQ, Gur RE, Blake DJ, Arnold SE (2004) Dysbindin-1 is reduced in intrinsic, glutamatergic terminals of the hippocampal formation in schizophrenia. J Clin Invest 113:1353-1363.

Tamminga, CA, Burrows, GH, Chase, TN, Alphs, LA, Thaker, GK (1989) Dopamine neuronal tracts in schizophrenia: Their pharmacology and In Vivo glucose metabolism. NY Acad Sci 443-450.

Tamminga CA & Holcomb HH (2004) Phenotype of schizophrenia: a review and formulation. Mol Psychiat 10:27-39.

Tamminga CA, Thaker GK, Buchanan R, Kirkpatrick B, Alphs LD, Chase TN (1992) Limbic system abnormalities identified in schizophrenia using positron emission tomography with fluorodeoxyglucose and neocortical alterations with deficit syndrome. Arch Gen Psychiat 49:522-530.

Tamminga CA, Vogel M, Gao X, Lahti AC, Holcomb HH (2000) The limbic cortex in schizophrenia: focus on the anterior cingulate. Brain Res Rev 31:364-370.

Tang JX, Chen WY, He G, Zhou J, Gu NF, Feng GY, He L (2004) Polymorphisms within 5' end of the Neuregulin 1 gene are genetically associated with schizophrenia in the Chinese population. Mol Psychiat 9:11-12.

Thiselton DL, Webb BT, Neale BM, Ribble RC, O'Neill FA, Walsh D, Riley BP, Kendler K (2004) No evidence for linkage or association of neuregulin-1 (NRG1) with disease in the Irish study of high-density schizophrenia families (ISHDSF). Mol Psychiat 9:777-783.

Todtenkopf MS & Benes FM (1998) Distribution of glutamate decarboxylase65 immunoreactive puncta on pyramidal and nonpyramidal neurons in hippocampus of schizophrenic brain. Synapse 29:323-332.

Tsai G, Passani LA, Slusher BS, Carter R, Baer L, Kleinman JE, Coyle JT (1995) Abnormal excitatory neurotransmitter metabolism in schizophrenic brains. Arch Gen Psychiat 52:829-836.

Van Den BA, Schumacher J, Schulze TG, Otte AC, Ohlraun S, Kovalenko S, Becker T, Freudenberg J, Jonsson EG, Mattila-Evenden M, Sedvall GC, Czerski PM, Kapelski P, Hauser J, Maier W, Rietschel M, Propping P, Nothen MM, Cichon S (2003) The DTNBP1 (dysbindin) gene contributes to schizophrenia, depending on family history of the disease. Am J Hum Genet 73:1438-1443.

Vawter MP, Thatcher L, Usen N, Hyde TM, Kleinman JE, Freed WJ (2002) Reduction of synapsin in the hippocampus of patients with bipolar disorder and schizophrenia. Mol Psychiat 7:571-578.

Volkow ND, Wolf AP, Van Gelder P, Brodie JD, Overall JE, Cancro R, Gomez-Mont F (1998) Phenomenological correlates of metabolic activity in 18 patients with chronic schizophrenia. Am J Psychiat 144:151-158.

Volz HP, Gaser C, Hager F, Rzanny R, Mentzel HJ, Kreitschmann-Andermahr I, Kaiser WA & Sauer H (1997) Brain activation during cognitive stimulation with the Wisconsin Card Sorting Test – a functional MRI study on healthy volunteers and schizophrenics. Psychiatry Res 75:145-157.

Wagner AD, Schacter DL, Rotte M, Koutstaal W, Maril A, Dale AM, Rosen BR, Buckner RL (1998) Building memories: Remembering and forgetting of verbal experiences as predicted by brain activity. Science 281:1188-1191.

Webster MJ, Weickert CS, Herman MM, Hyde TM, Kleinman JE (2001) Synaptophysin and GAP-43 mRNA levels in the hippocampus of subjects with schizophrenia. Schiz Res 49:89-98.

Weinberger D (1988) Schizophrenia and the frontal lobe. Trends in Neuroscience 11:367-370.

Weinberger DR, Berman KF & Zec RF (1986) Physiologic dysfunction of dorsolateral prefrontal cortex in schizophrenia. I. Regional cerebral blood flow evidence. Arch Gen Psychiat 43:114-124.

Weiss A, Zalesak M, DeWitt I, Goff D, Kunkel L & Heckers S (2004) Impaired hippocampal function during the detection of novel words in schizophrenia. Biol Psychiat 55:668-675.

Wexler BE, Stevens AA, Bowers AA, Sernyak MJ & Goldman-Rakic PS (1998) Word and tone working memory deficits in schizophrenia. Arch Gen Psychiat 55:1093-1096.

Williams NM, Preece A, Morris DW, Spurlock G, Bray NJ, Stephens M, Norton N, Williams H, Clement M, Dwyer S, Curran C, Wilkinson J, Moskvina V, Waddington JL, Gill M, Corvin AP, Zammit S, Kirov G, Owen MJ, O'Donovan MC (2004) Identification in 2 independent samples of a novel schizophrenia risk haplotype of the dystro-brevin binding protein gene (DTNBP1). Arch Gen Psychiat 61:336-344.

Williams NM, Preece A, Spurlock G, Norton N, Williams HJ, Zammit S, O'Donovan MC, Owen MJ (2003) Support for genetic variation in neuregulin 1 and susceptibility to schizophrenia. Mol Psychiat 8:485-487.

Whyte MC, McIntosh AM, Johnstone EC & Lawrie SM (2005) Declarative memory in un-affected adult relatives of patients with schizophrenia: a systematic review and meta-analysis. Schiz Res 78:13-26.

Witter MP, Wouterlood FG, Naber PA & Van Haeften T (2000) Anatomical organization of the parahippocampal-hippocampal network. Ann NY Acad Sci 9:1-24.

Wolkin A, Sanfilipo M, Wolf AP, Angrist B, Brodie JD, Rotrosen J (1992) Negative symp-toms and hypofrontality in chronic schizophrenia. Arch Gen Psychiat 49:959-965.

Wonnacott S (1990) The paradox of nicotinic acetylcholine receptor up-regulation by nico-tine. Trends Pharmacol Sci 11:216-219.

Wright IC, Ellison ZR, Sharma T, Friston KJ, Murray RM, McGuire PK (1999) Mapping of grey matter changes in schizophrenia. Schiz Res 35:1-14.

Yang JZ, Si TM, Ruan Y, Ling YS, Han YH, Wang XL, Zhou M, Zhang HY, Kong QM, Liu C, Zhang DR, Yu YQ, Liu SZ, Ju GZ, Shu L, Ma DL, Zhang D (2003) Association study of neuregulin 1 gene with schizophrenia. Mol Psychiat 8:706-709.

Yonelinas AP (2002) The nature of recollection and familiarity: A review of 30 years of re-search. J Memory & Language 46:441-517.

Yurgelun-Todd DA, Renshaw PF, Gruber SA, Ed M, Waternaux C, Cohen BM (1996) Pro-ton magnetic resonance spectroscopy of the temporal lobes in schizophrenics and nor-mal controls. Schiz Res 19:55-59.

Zaidel DW, Esiri MM, Harrison PJ (1997) Size, shape, and orientation of neurons in the left and right hippocampus: investigation of normal asymmetries and alterations in schizophrenia. Am J Psychiat 154:812-818.

Zhang F, Sarginson J, Crombie C, Walker N, Stclair D, Shaw D (2006) Genetic association between schizophrenia and the DISC1 gene in the Scottish population. Am J Med Genet B Neuropsychiatr Genet 141:155-159.

Zhang F, St Clair D, Liu X, Sun X, Sham PC, Crombie C, Ma X, Wang Q, Meng H, Deng W, Yates P, Hu X, Walker N, Murray RM, Collier DA, Li T (2005) Association analy-sis of the RGS4 gene in Han Chinese and Scottish populations with schizophrenia. Genes Brain Behav 4:444-448.

Zhao X, Shi Y, Tang J, Tang R, Yu L, Gu N, Feng G, Zhu S, Liu H, Xing Y, Zhao S, Sang H, Guan Y, St Clair D, He L (2004) A case control and family based association study of the neuregulin1 gene and schizophrenia. J Med Genet 41:31-34.

Chapter 9

Identifying Novel Target Genes in Psychotic Disorder Network Association Analyses

Francine M. Benes, Benjamin Lim, David Matzilevich, John Walsh

Abstract. Many believe that unique genes for schizophrenia (SZ) and bipolar disorder (BD) are associated with the susceptibility for the respective illnesses. To explore this idea, gene expression profiling (GEP) studies of "whole" hippocampus (HIPP) have been performed on a cohort consisting of SZs, BDs and normal controls and the data have been interrogated using a combination of a global functional analysis of biopathways/clusters and network association analysis. Overall, it appears that BDs may show a down-regulation of genes associated with growth pathways and their receptors. Such changes may mitigate against cell survival and potentially even facilitate apoptotic injury in subjects with BD.

9.1 Introduction

After many years of intensive study (Kopnisky et al. 2002), there is a consensus view that both genetic and environmental factors are needed to express the clinical phenotype for schizophrenia (SZ) (Barondes et al. 1997; Bray and Owen 2001). A two-factor model affords the most pragmatic explanation for the findings that have been reported from linkage studies of monozygotic and dizygotic twin pairs discordant for schizophrenia and their first degree relatives (Kety et al. 1968; Sullivan et al. 2003). Indeed, there may be sets of genes independently regulated by environmental influences that exist in parallel with heritable genes that impart susceptibility to individuals "at risk" for schizophrenia and bipolar disorder (Gottesman and Shields 1971; Gottesman and Gould 2003; Sullivan et al. 2003). The identification of susceptibility genes for SZ and BD represents a great challenge to the field of psychiatry (McDonald and Murphy 2003). Although linkage and association analyses have been commonly used strategies, many believe that they are not sensitive enough to detect relevant genes, unless very large numbers of subjects are collected (Barondes et al. 1997). More recently, gene expression profiling

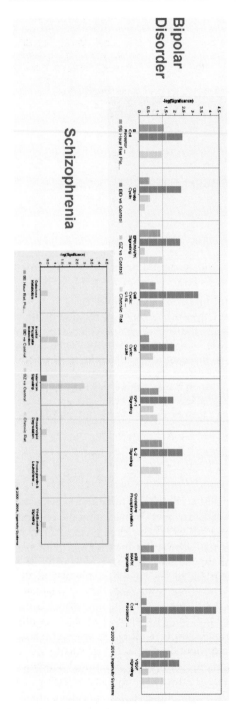

(GEP) has become a rich source linking genomic information with susceptibility genes (Glanzer et al. 2004). Using expression analysis, we recently reported that there may be significant and fundamental differences in the profiles of gene expression in SZs when compared to BDs, with the latter showing activation of the apoptosis cascade (Benes et al. 2006). These data contrast with those showing a down-regulation of GAD expression in both SZ and BD (Benes et al. 2000; Guidotti et al. 2000).

As a strategy for identifying HIPP genes that may be regulated, at least in part, by environmental stress, we have searched GEP results from the "whole" HIPP to find genes that might show changes that are similar in SZs and BDs. It can be inferred that similar patterns of gene expression within groups of individual genes common to both the rodent model and in SZ and BD may be related, either directly or indirectly to acute or chronic environmental stress. Conversely, changes in the expression of genes found in SZ or BD, but not in both, may be related directly or indirectly to genes associated with the heritability for one or the other disorders. In the discussion that follows, similarities and differences in the expression of genes have been sought using a combination of a global functional analysis and network association analysis to facilitate this process.

Fig. 9.1. A Global functional analysis as a preliminary step in identifying focus genes-of-interest. Four separate databases for gene expression profiling studies of HIPP were included in the post hoc analysis, including those from SZs vs. normal controls (CONs), BDs vs. CONs, as well as acute and chronic picrotoxin infusion in the basolateral amygdala of rats. Using the canonical pathways programmed into Ingenuity Pathways Analysis, the significance levels of individual functional pathways was determined and plotted separately for each dataset. For BDs (dark blue), several pathways log significance levels that exceeded the alpha level (red line). For SZ, there were fewer pathways and less significance, but the interferon signaling pathway.

9.2 Focus Genes Unique to Schizophrenia or Bipolar Disorder

Using the Ingenuity Pathways Knowledge Base (IPKB) for, a number of global functional pathways (GFPs) relevant to BD or SZ, but not both, were identified (Figure 1; upper). In the BD group, but not the SZ group, there were preferential changes in expression in several different signaling pathways, including those for B and T cell receptors, ERK/MAPK, p38/MAPK, insulin-like growth factor (IGF-1), interleukin-2 (IL2) and vascular-endothelial growth factor (VEGF). In addition, two pathways associated with cell cycle regulation, G1/S and G2/M, also showed preferential changes in BDs, but not SZs. For the SZs, the only global functional pathway showing unique changes and exceeding the alpha level of significance was interferon signaling (Figure 9.1; lower). Interferon deficiencies have been implicated in the pathophysiology of SZ (Leszek 1992), although both increases (Cazzullo et al. 2002) and decreases (Hornberg et al. 1995) of gamma-interferon and other cytokines have been reported. Of interest is the fact that interferons are associated with increased repair of DNA (Zasukhina et al. 1990), a mechanism that has been invoked to explain the marked reduction of single-stranded DNA breaks in patients with this disorder (Benes et al. 2003). Other pathways showing unique, but non-significant changes in SZ included galactose metabolism, phospholipid degradation and prostaglandin metabolism.

Networks associated with each of the global functional pathways described above were identified and evaluated for the presence of specific *focus genes* that made numerous associations with other genes within the pathways showing significant changes in BDs or SZs, but not in both. In the BD group, Ingenuity Networks #9 (Figure 9.2) was found to have frequent associations with the global functional pathways for the BD group. In Network #9, the dopamine D4 receptor showed an upregulation, but this gene did not appear in any of the networks derived for SZ. Additionally, as noted below, both groups showed an equivalently robust upregulation when qRT-PCR was employed (Fig. 9.3). Another potential focus gene was GRB2 (growth factor receptor bound protein 2) that was found to Fig. 9.2. Network associations for GRB2 derived from global functional pathways (shown to the right) obtained through Ingenuity Pathways analysis of a database in which subjects with bipolar disorder (N = 7) were compared with normal controls (N = 10) from a previous study (Konradi et al. 2003). The global functional pathways are those showing unique and significant changes in expression in the BDs. One of the genes associated with GRB2 is that for the D4 dopamine receptor. To the right, human chromosome 17 shows a linkage site for bipolar disorder and the gene for GRB2 maps to this site have the most associations with several of the signaling pathways that emerged from the global functional analysis (Fig 9.1), including VEGF, IL-2 and IGF-1. In addition, GRB2 was also found to be associated with the insulin, integrin and epithelial growth factor (EGF-1) signaling pathways (Fig. 9.5). The receptors for of these growth factors and cytokines all showed increased expression in BDs (Table 9.1). GRB2 regulates these growth-associated receptors by forming heterodimeric and heterotrimeric complexes with other regulatory proteins, such as "son of sevenless" (SOS) (Watanabe et al., 2000) and sarcoma viral oncogene homologue (SHC) (Jiang and Sorkin 2002), both known to influence EGF signaling (Jiang and Sorkin 2002), probably through internalization of its receptors (Jiang et al. 2003) or subsequent activation of RAS. Although these latter two genes did not show changes in regulation in the BD group, GRB2 was found to be down-regulated (Figs. 9.4 and 9.5), such a change would tend to dampen the overall activation of the growth-associated signaling pathways showing changes in the HIPP of BDs. However, GRB2 also showed a significant down-regulation in SZs (Fig. 9.4), suggesting that this gene is probably not playing a specific role in either disorder. For several of the growth factor/cytokine receptor signaling pathways, GRB2 acts as a key regulatory element by forming a complex with activated receptors and the RAS-specific guanine nucleotide exchange factor, SOS (Fig. 9.5).

Another potential *focus gene* in the BD group was kRAS2. Activation of the RAS signaling cascade is a common downstream mechanism related to GRB2 and its function as an adaptor protein (Sorkin et al. 2000). Like GRB2, kRAS2 also showed decreased expression in BDs (Figs. 9.4 and 9.5). In this case, however, the down-regulation of kRAS2 occurred selectively in BDs, but not SZs (Fig. 9.1). KRAS2 is part of a complex set of "switching" mechanisms that employ RAF, MEK and ERK to transmit signals to the nucleus where a broad array of transcription factors (Kolch 2000), including Bcl-6, EGR-1, ELK1, STAT1, c-myc, c-jun and CREB, are regulated (see Fig. 9.4).

A third focus gene-of-interest identified with the network association analysis of functional pathways was phosphoinositol-3-kinase regulatory protein (PIK3R). This protein helps to regulate the spatial and temporal components of chemotaxic cell responses (Huang et al. 2003) and is found in many of the same pathways showing unique changes in regulation in BDs and in which GRB2 and kRAS2 are also key components, including B and T cell, integrin, insulin, IL-2, IGF-1 IL-2, ERK/MAPK and VEGF signaling pathways. Like GRB2 and kRAS2, PI3KR was also down-regulated in the HIPP of BDs. In SZ, PI3KR also showed significant changes, but in the opposite direction (Fig. 1), suggesting that its regulation is uniquely different in the two disorders. Like GRB2 and kRAS2, PI3KR plays an important role in EGF signaling (Gogg and Smith 2002) and other pro-survival mechanisms, such as the NF-kB signaling pathway (Cuni et al. 2004). The down-regulation of these three closely aligned regulatory factors in BDs could potentially contribute to our observed increase in pro-apoptotic signaling potentially leading to increases in cell death.

BD vs Control: BDvsCONp=0.25FoldChange.xls
Network 9

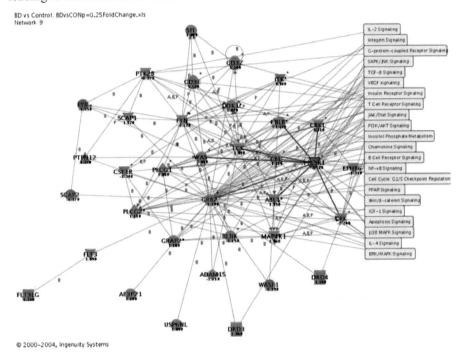

© 2000–2004, Ingenuity Systems

Fig. 9.2. Network associations for GRB2 derived from global functional pathways (right) obtained through Ingenuity Pathways analysis of a database in which subjects with bipolar disorder (n=7) were compared with normal controls (n=10) from a previous study (Konradi et al. 2003). The global functional pathways are those showing unique and significant changes in expression in the BDs. One of the genes associated with GRB2 is that for the D4 dopamine receptor. To the right, human chromosome 17 shows a linkage site for bipolar disorder and the gene for GRB2 maps to this site.

In the SZ vs. CON DB, the global functional analysis suggested that Ingenuity Networks might contain genes-of-interest. No focus genes appeared to be uniquely related to SZ. For one of the networks, GSK-3β, a gene that plays a pivotal role in regulating cell survival via the Wnt pathway (Inestrosa et al. 2002; De Ferrari et al. 2003) initially emerged as a potential focus gene for this disorder; however, it also showed a similar change in expression in BDs (Fig. 9.3) and was not considered any further. Casein kinase 2 (CK2) or CSNK2A1), another key regulatory component of the Wnt/catenin signaling pathway, was also detected in one of the SZ networks; however, like GSK-3β, this gene also showed a change in regulation in SZ similar to that seen in BDs when qRT-PCR was employed (Fig. 9.1) and was also not considered a focus gene-of-interest.

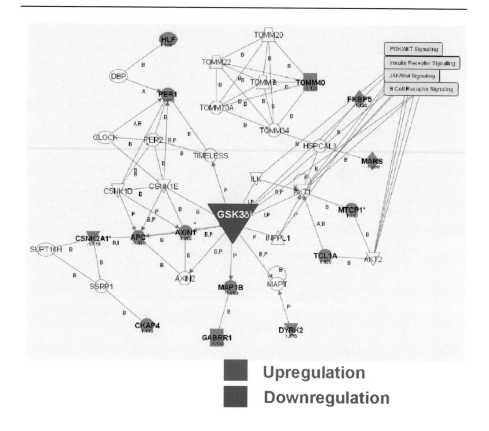

Upregulation

Downregulation

Schizophrenia

Fig. 9.3. Network associations for GRB2 derived from global functional pathways (right) obtained through Ingenuity Pathways analysis of a database in which subjects with schizophrenia (n=7) were compared with normal controls (n=10) from a previous study (Konradi et al. 2003). The global functional pathways are those showing unique and significant changes in expression in the BDs.

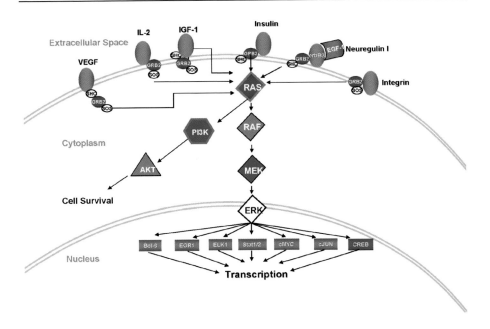

Fig. 9.4. Quantitative RT-PCR validation of changes in gene expression in the hippocampus of SZs and BDs. The genes analyzed include the dopamine D4 receptor, GRB2, kRAS2, PI3K, CK-2 and GSK-3β.

9.3 Discussion

The results of this series of post hoc analyses have demonstrated that there are similarities and differences in the regulation of some functional pathways and genes in SZ and BD. When the databases for the human postmortem cases are analyzed simultaneously with the DBs for rats treated acutely and chronically to model of the effects of amygdalar activation of the hippocampus, changes in gene expression that were similar in BDs and SZs were also identified in the acute and/ or chronic rat models (not shown). Accordingly, these post hoc analyses suggest that there may be discrete clusters of hippocampal genes that may be regulated through environmental influences processed through the basolateral amygdala. On the other hand, there were also some pathways and genes that showed changes in expression that appeared to be unique to BD, but not in SZ. Such focus genes could theoretically be associated with the susceptibility for either BD or SZ.

Quantitative RT-PCR Validation of Gene Expression Profiling Results

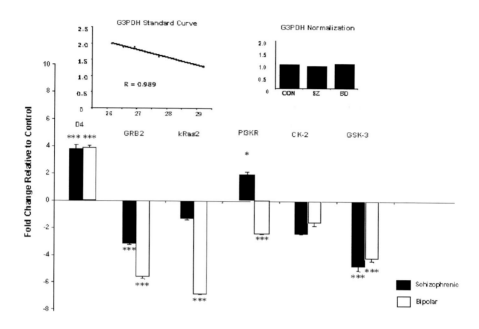

Fig. 9.5. Schematic diagram depicting focus genes detected in pro-survival pathways. GRB2 forming dimers and trimers with SOS and SHC in several different growth-associated pathways that use VEGF, IL-2, IGF-1, insulin, integrins and neuregulins as ligands their respective receptors. The RAS signaling cascade is activated through these interactions. In BDs, although the various growth-associated receptors are all upregulated, GRB2 and kRAS2 are both down-regulated. Since RAS signaling is a final common pathway for all these factors, a reduced expression of GRB2 and kRAS2 could be associated with diminished pro-survival potential in HIPP cells of BDs. Decreased expression of PI3KR would also diminish the potential for cell survival through an AKT-mediated pathway.

Using the network association analysis, a reduction in the expression of GSK-3β and CK2 were observed in both SZs and BDs and this finding was confirmed using qRT-PCR. In studies from other laboratories, GSK-3β was found to be decreased in the prefrontal cortex of both SZs and BDs (Beasley et al. 2001, 2002). These results suggest that the expression of both genes may be influenced by a non-specific factor, such as stress. GSK-3β and CK2 are known to play an important role in cell survival (Litchfield et al. 2001) by helping to regulate cell cycle via the Wnt and NF kappa B pathways (Landesman-Bollag et al. 2001). Both genes also regulate the rate of proteosomal degradation by either directly or indirectly influencing the phosphorylation of β-catenin, (Nusse 2002) a pivotal element in the Wnt signaling pathway. Overall, it seems reasonable to conclude that decreased expression of GSK-3β and CK2 probably play a non-specific role in defining the endophenotype of SZ and BD.

Table 9.1. Expression Data for Growth-Associated Receptor Systems in Bipolars vs Controls and Schizophrenics vs Controls

	GenBank Accession	BD Fold Changes	BD P-Value	SZ GenBank Accession	SZ Fold Change	SZ P-Value
VEGF						
VEGF (FLT1)S Fms-related tyrosine kinase 1 (ascular endo-thelial growth factor /vascular permeability	X51602	1.15	0.036			
factor receptor)	U01134	1.13	0.039		1.08	0.205
IL-2						
IL2RA	X01057	1.12	0.139			
IL2RB	M26062	1.14	0.018			
IGF-1						
IGF1R	AF020763	1.08	0.077		1.05	0.243
Insulin-like growth factor binding protein 5	AF055033	-1.16	0.224			
Human insulin-like growth factor binding protein 5 (IGFBP5) mRNA				L27560	1.25	0.220
Insulin Receptor						
INSR (Insulin Receptor)	M10051	1.13	0.003		1.08	0.003
Neuregulin						
EGF (epidermal growth factor -- beta-urogastrone)	X04571	1.06	0.201			
EGFR (epidermal growth factor receptor)	X00588	1.26	0.073			
ERBB3	AA584202	1.12	0.063			
Human erbB3 binding protein EBP1	AI817548	1.13	0.087			
ITGBL1 (integrin, beta-like 1 with EGF-like re-peat domains)				AB008375	-1.21	0.081
Epidormal growth faotor receptor pathway sub-strate 8				U12535	1.12	0.236
Activated leukocyte cell adhesion molecule	Y10183	-1.23	0.155		-1.24	0.022
Integrin						

ITGB1 (Beta1 Integrin Rreceptor)	M34189	1.11	0.105		
ITGB3 (Integrin Beta 3)	AF052141	1.19	0.054		
ITGB6 (Integrin Beta 6)	U88834	1.13	0.108		
ITGA3 (Integrin Alpha 3)	M59911	1.07	0.154		
ITGA6 (Integrin Alpha 6)	X53586	1.18	0.187	1.17	0.146

Expression data shown are from RNA extracts of whole human hippocampus that was hybridized to the Affymetrix HG-U95A chip. In both BDs and SZs, most growth-associated genes showing increased expression. There are twice as many genes in BDs showing changes when the low stringency criterion of p = 0.25 is used.

Other genes showed significant changes in expression in BDs, but not SZs and included kRASb, a component of the RAS signaling pathway. There was a down-regulation of kRAS2 and MEK1/2, changes that would tend to increase the cell death potential of the apoptosis cascade. These results are consistent with another recent study from this laboratory demonstrating that there is a marked upregulation of pro-apoptotic genes in the HIPP of BDs, but not in SZs (Benes et al. 2006). The other potential focus gene, GRB2, is known to be a pivotal factor in the activation of several different pro-survival pathways, including VEGF (Takahashi and Shibuya 2001), IL-2 (Beck et al. 2005), insulin, IGF-1 (Chesnel et al. 2003; Khawaja et al. 2004), neuregulin I (Eilam et al. 1998), integrin (Gary and Mattson 2001; Lin et al. 2003) and EGF-1(Jiang et al. 2003) signaling cascades, were all upregulated-regulated in BDs. All of these factors and in the case of neuregulin, its receptor erbb3, were upregulated in BDs. The down-regulation of GRB2, however, would tend to work against these pro-survival pathways and potentially result in a facilitation of apoptotic cell death.

Another gene showing unique changes in expression is PI3KR; however, this was in opposite directions with an upregulation found in SZs and a down-regulation found in BDs. Like GRB2 and kRAS2, this protein was also downregulated in BDs and because it facilitates growth associated signaling pathways and is pro-survival in its influence on cells (Cuni et al. 2004), this change would also tend to facilitate apoptosis in the HIPP of BDs. In SZs, however, PI3KR was upregulated and this change would tend to promote cell survival. Overall, this finding is consistent with a study showing a marked reduction of DNA breaks in SZs. Consistent with these current findings, a combination of cell counting (Benes et al. 1991), in situ hybridization (Heckers et al. 2002; Woo et al. 2004) and in situ end-labeling of DNA breaks (Benes et al. 2003; Buttner et al. 2007) has suggested that the susceptibility for cell death may be decreased in SZ, but increased in BD. Indeed, the results reported here are consistent with these hypotheses and may help to elucidate underlying molecular mechanisms.

Taking together all of these results, it seems likely that the specific endophenotypes for SZ and BD, i.e. quantifiable components that make genetic and biological studies of etiologies for such diseases more manageable (Gould and Gottesman 2006), will probably involve many different transduction, signaling and metabolic pathways (Manji et al. 2003; Hasler et al. 2006) operating simultaneously and in parallel to create non-adaptive changes in neuronal and non-neuronal cell populations. The current study was undertaken in RNA extracts taken from the "whole" hippocampus of the SZ, BD and CON subjects. A more recent analysis using laser microdissection to "deconstruct" the hippocampus into component sectors and layers has demonstrated that the phenotype of GABAergic interneurons may be regulated by fundamentally different clusters of genes in SZ vs BD (Benes et al. 2007). The use of "whole" homogenates of the hippocampus is inherently limited by the heterogeneous nature of cellular constituents contributing to RNA extracts. Nevertheless, as discussed above, such material has provided additional new insights into possible genes associated with the susceptibility for metabolic injury to neurons in subjects with BD.

Acknowledgements

This work was supported by grants from the National Institutes of Health (MH42261, MH62822, MH/NS31862 and MH60450).

References

Barondes SH, Alberts BM, Andreasen NC, Bargmann C, Benes F, Goldman-Rakic P, Gottesman I, Heinemann SF, Jones EG, Kirschner M, Lewis D, Raff M, Roses A, Rubenstein J, Snyder S, Watson SJ, Weinberger DR, Yolken RH (1997) Workshop on schizophrenia. Proc Natl Acad Sci USA 94:1612-1614.

Beasley C, Cotter D, Everall I (2002) An investigation of the Wnt-signalling pathway in the prefrontal cortex in schizophrenia, bipolar disorder and major depressive disorder. Schizophr Res 58:63-67.

Beasley C, Cotter D, Khan N, Pollard C, Sheppard P, Varndell I, Lovestone S, Anderton B, Everall I (2001) Glycogen synthase kinase-3beta immunoreactivity is reduced in the prefrontal cortex in schizophrenia. Neurosci Lett 302:117-120.

Beck RD, Jr., King MA, Ha GK, Cushman JD, Huang Z, Petitto JM (2005) IL-2 deficiency results in altered septal and hippocampal cytoarchitecture: relation to development and neurotrophins. J Neuroimmunol 160:146-153.

Benes FM, Lim B, Matzilevich D, Walsh JP, Subbaraju S, Minns S (2007) Regulation of the GABA cell phenotype in hippocampus of schizophrenics and bipolars. Proc Natl Acad Sci USA 104:10164-10169.

Benes FM, Matzilevich D, Burke RE, Walsh J (2006) The expression of proapoptosis genes is increased in bipolar disorder, but not in schizophrenia. Mol Psychiatry 11:241-251.

Benes FM, McSparren J, Bird ED, SanGiovanni JP, Vincent SL (1991) Deficits in small interneurons in prefrontal and cingulate cortices of schizophrenic and schizoaffective patients. Arch Gen Psychiatry 48:996-1001.

Benes FM, Walsh J, Bhattacharyya S, Sheth A, Berretta S (2003) DNA fragmentation decreased in schizophrenia but not bipolar disorder. Arch Gen Psychiatry 60:359-364.

Bray NJ, Owen MJ (2001) Searching for schizophrenia genes. Trends Mol Med 7:169-174.

Buttner N, Bhattacharyya S, Walsh J, Benes GM (2007) DNA fragmentation is increased in non-GABAergic neurons in bipolar disorder but not in schizophrenia.

Cazzullo CL, Sacchetti E, Galluzzo A, Panariello A, Adorni A, Pegoraro M, Bosis S, Co-lombo F, Trabattoni D, Zagliani A, Clerici M (2002) Cytokine profiles in schizophrenic patients treated with risperidone: a 3-month follow-up study. Prog Neuropsychophar-macol Biol Psychiatry 26:33-39.

Chesnel F, Heligon C, Richard-Parpaillon L, Boujard D (2003) Molecular cloning and characterization of an adaptor protein Shc isoform from Xenopus laevis oocytes. Biol Cell 95:311-320.

Cuni S, Perez-Aciego P, Perez-Chacon G, Vargas JA, Sanchez A, Martin-Saavedra FM, Ballester S, Garcia-Marco J, Jorda J, Durantez A (2004) A sustained activation of PI3K/NF-kappaB pathway is critical for the survival of chronic lymphocytic leukemia B cells. Leukemia 18:1391-1400.

De Ferrari GV, Chacon MA, Barria MI, Garrido JL, Godoy JA, Olivares G, Reyes AE, Alvarez A, Bronfman M, Inestrosa NC (2003) Activation of Wnt signaling rescues neurodegeneration and behavioral impairments induced by beta-amyloid fibrils. Mol Psychiatry 8:195-208.

Eilam R, Pinkas-Kramarski R, Ratzkin BJ, Segal M, Yarden Y (1998) Activity-dependent regulation of Neu differentiation factor/neuregulin expression in rat brain. Proc Natl Acad Sci USA 95:1888-1893.

Gary DS, Mattson MP (2001) Integrin signaling via the PI3-kinase-Akt pathway increases neuronal resistance to glutamate-induced apoptosis. J Neurochem 76:1485-1496.

Glanzer JG, Haydon PG, Eberwine JH (2004) Expression profile analysis of neurodegen-erative disease: advances in specificity and resolution. Neurochem Res 29:1161-1168.

Gogg S, Smith U (2002) Epidermal growth factor and transforming growth factor alpha mimic the effects of insulin in human fat cells and augment downstream signaling in in-sulin resistance. J Biol Chem 277:36045-36051.

Gottesman II, Shields J (1971) Schizophrenia: geneticism and environmentalism. Hum He-red 21:517-522.

Gottesman II, Gould TD (2003) The endophenotype concept in psychiatry: etymology and strategic intentions. Am J Psychiatry 160:636-645.

Gould TD, Gottesman, II (2006) Psychiatric endophenotypes and the development of valid animal models. Genes Brain Behav 5:113-119.

Hasler G, Drevets WC, Gould TD, Gottesman, II, Manji HK (2006) Toward Constructing an Endophenotype Strategy for Bipolar Disorders. Biol Psychiatry.

Heckers S, Stone D, Walsh J, Shick J, Koul P, Benes FM (2002) Differential hippocampal expression of glutamic acid decarboxylase 65 and 67 messenger RNA in bipolar disor-der and schizophrenia. Arch Gen Psychiatry 59:521-529.

Hornberg M, Arolt V, Wilke I, Kruse A, Kirchner H (1995) Production of interferons and lymphokines in leukocyte cultures of patients with schizophrenia. Schizophr Res 15:237-242.

Huang YE, Iijima M, Parent CA, Funamoto S, Firtel RA, Devreotes P (2003) Receptor-mediated regulation of PI3Ks confines PI(3,4,5)P3 to the leading edge of chemotaxing cells. Mol Biol Cell 14:1913-1922.

Inestrosa N, De Ferrari GV, Garrido JL, Alvarez A, Olivares GH, Barria MI, Bronfman M, Chacon MA (2002) Wnt signaling involvement in beta-amyloid-dependent neurode-generation. Neurochem Int 41:341-344.

Jiang X, Sorkin A (2002) Coordinated traffic of Grb2 and Ras during epidermal growth factor receptor endocytosis visualized in living cells. Mol Biol Cell 13:1522-1535.

Jiang X, Huang F, Marusyk A, Sorkin A (2003) Grb2 regulates internalization of EGF re-ceptors through clathrin-coated pits. Mol Biol Cell 14:858-870.

Kety S, Rosenthal D, Wender P, Schulsinger F (1968) The type and prevalence of mental illness in the biological and adoptive families, of adopted schizophrenics. In: The

Transmission of Schizophrenia (Rosenthal D, Kety S, eds), pp 25-37. Oxford: Pregamon Press.

Khawaja X, Xu J, Liang JJ, Barrett JE (2004) Proteomic analysis of protein changes developing in rat hippocampus after chronic antidepressant treatment: Implications for depressive disorders and future therapies. J Neurosci Res 75:451-460.

Kolch W (2000) Meaningful relationships: the regulation of the Ras/Raf/MEK/ERK pathway by protein interactions. Biochem J 351 Pt 2:289-305.

Kopnisky KL, Cowan WM, Hyman SE (2002) Levels of analysis in psychiatric research. Dev Psychopathol 14:437-461.

Landesman-Bollag E, Song DH, Romieu-Mourez R, Sussman DJ, Cardiff RD, Sonenshein GE, Seldin DC (2001) Protein kinase CK2: signaling and tumorigenesis in the mammary gland. Mol Cell Biochem 227:153-165.

Leszek J (1992) [Interferon theory of schizophrenia]. Psychiatr Pol 26:381-387.

Lin B, Arai AC, Lynch G, Gall CM (2003) Integrins regulate NMDA receptor-mediated synaptic currents. J Neurophysiol 89:2874-2878.

Litchfield DW, Bosc DG, Canton DA, Saulnier RB, Vilk G, Zhang C (2001) Functional specialization of CK2 isoforms and characterization of isoform-specific binding partners. Mol Cell Biochem 227:21-29.

Manji HK, Gottesman, II, Gould TD (2003) Signal transduction and genes-to-behaviors pathways in psychiatric diseases. Sci STKE 2003:pe49.

McDonald C, Murphy KC (2003) The new genetics of schizophrenia. Psychiatr Clin North Am 26:41-63.

Nusse R (2002) The Wnt gene Homepage. http://wwwstanfordedu/-musse/wntwindowhtml.

Sorkin A, McClure M, Huang F, Carter R (2000) Interaction of EGF receptor and grb2 in living cells visualized by fluorescence resonance energy transfer (FRET) microscopy. Curr Biol 10:1395-1398.

Sullivan PF, Kendler KS, Neale MC (2003) Schizophrenia as a complex trait: evidence from a meta-analysis of twin studies. Arch Gen Psychiatry 60:1187-1192.

Takahashi T, Shibuya M (2001) The overexpression of PKCdelta is involved in vascular endothelial growth factor-resistant apoptosis in cultured primary sinusoidal endothelial cells. Biochem Biophys Res Commun 280:415-420.

Watanabe T, Shinohara N, Moriya K, Sazawa A, Kobayashi Y, Ogiso Y, Takiguchi M, Yasuda J, Koyanagi T, Kuzumaki N, Hashimoto A (2000) Significance of the Grb2 and son of sevenless (Sos) proteins in human bladder cancer cell lines. IUBMB Life 49:317-320.

Woo TU, Walsh JP, Benes FM (2004) Density of glutamic acid decarboxylase 67 messenger RNA-containing neurons that express the N-methyl-D-aspartate receptor subunit NR2A in the anterior cingulate cortex in schizophrenia and bipolar disorder. Arch Gen Psychiatry 61:649-657.

Zasukhina GD, Lvova GN, Sinelshchikova TA (1990) DNA repair and human pathology. Acta Biol Hung 41:267-274.

Chapter 10

Alterations in Cortical GABA Neurotransmission in Schizophrenia: Causes and Consequences

David A. Lewis, Takanori Hashimoto, Jaime G. Maldonado-Aviles and Harvey Morris

10.1 Introduction

Schizophrenia is frequently noted to be a heterogeneous disorder, at both the levels of etiological risk factors and clinical manifestations. The extent to which this heterogeneity reflects variability (variance of a particular parameter within a class) and diversity (the existence of different classes within a population) remains to be determined. As demonstrated by the other volumes in this book, multiple molecular, cellular and circuitry candidates could contribute to this heterogeneity.

Of the reported brain alterations in postmortem studies of schizophrenia, a deficit in expression of the mRNA for the 67kD form of glutamic acid decarboxylase (GAD_{67}), which is principally responsible for the synthesis of GABA, appears to be the most consistent and widely-replicated finding in the illness (Akbarian and Huang 2006; Torrey et al. 2005). The expression of GAD_{67} mRNA has been most widely studied in the dorsolateral prefrontal cortex (DLPFC) in schizophrenia. Because the activity of GABA neurons in the DLPFC is essential for normal working memory function in non-human primates (Rao et al. 2000; Sawaguchi et al. 1989), reduced GAD_{67} mRNA expression and hence altered GABA neurotransmission, in the DLPFC could contribute to the impairments in working memory present in schizophrenia (Lewis et al. 2005).

Although GAD_{67} mRNA is undetectable in ~25-30% of DLPFC GABA neurons (Akbarian et al. 1995; Volk et al. 2000), the remaining GABA neurons have normal levels of GAD_{67} mRNA expression (Volk et al. 2000). The affected GABA neurons include those that express the calcium-binding protein parvalbumin (PV), which is found in ~25% of GABA neurons in the primate DLPFC, whereas the ~50% of GABA neurons that express the calcium-binding protein calretinin appear to be unaffected (Hashimoto et al. 2003). These findings suggest that the synthesis of GABA is reduced in the PV-containing subset of DLPFC neurons in schizophrenia.

However, understanding how lower levels of GAD_{67} mRNA actually contribute to the clinical features of schizophrenia requires knowledge of both the potential causes and consequences of this impairment in gene expression, and evaluation of concurrent changes in expression of other molecules mediating GABA neurotransmission. Consequently, in this chapter we consider both of these issues, and place them in the larger context of cortical circuitry disturbances in schizophrenia.

10.2 Potential Causes

Recently, allelic variants in the gene ($GAD1$) for GAD_{67} have been reported to be associated with increased risk for schizophrenia (Addington et al. 2005; Straub et al. 2007), although other studies have failed to find such an association (De Luca et al. 2004; Lundorf et al. 2005; Ikeda et al. 2007). Interestingly, a SNP in the 5' untranslated region, which is predicted to be in the promoter, was associated with reduced GAD_{67} mRNA in the DLPFC of individuals with schizophrenia (Straub et al. 2007). However, it remains unclear how such a genetic liability would be associated with cell type-specific alterations in GAD_{67} mRNA expression.

Reductions in mRNA and protein levels for reelin in association with deficits in GAD_{67} mRNA expression have been reported in subjects with schizophrenia (Impagnatiello et al. 1998), although the levels of reelin and GAD_{67} mRNAs were not correlated in subjects with schizophrenia (Guidotti et al. 2000). Furthermore, reelin is reportedly expressed in ~50% of GABA neurons in the adult rodent cortex (Pesold et al. 1998), and heterozygote reeler mice, which express only 50% of the normal levels of reelin, also show a decrement in GAD_{67} mRNA expression (Liu et al. 2001). However, reelin is apparently not expressed in PV-containing cortical GABA neurons (Pesold et al. 1999); thus, the relationship between altered reelin expression and the preferential reduction of GAD_{67} mRNA expression in PV-containing neurons in schizophrenia remains unclear.

Because the expression of GAD_{67} mRNA is activity-dependent, the reduction observed in schizophrenia could reflect diminished activity of excitatory circuits in the DLPFC (Jones 1997). Several lines of evidence suggest that PV-containing GABA neurons are particularly sensitive to reductions in excitatory transmission through NMDA receptors (Lewis and Moghaddam 2006). First, PV-containing cells receive a much larger number of excitatory inputs than do calretinin-containing GABA neurons (Gulyás et al. 1999; Melchitzky and Lewis 2003). Second, PV-containing neurons may be particularly sensitive to changes in excitatory signaling via NMDA receptors. For example, in rodents NMDA receptor antagonists produce a decreased density of PV-immunoreactive neurons in the hippocampus (Keilhoff et al. 2004) [but no change in calretinin neurons (Rujescu et al. 2006)] and decreased PV mRNA expression in the prefrontal cortex (Cochran et al. 2003). In the latter study, the density of PV mRNA-positive neurons was unchanged, but the expression level of PV mRNA per neuron was decreased by 25%, a pattern strikingly similar to that present in the DLPFC cortex of subjects with schizophrenia (Hashimoto et al. 2003). Similarly, in mouse cortical neuron cultures, ketamine reduced both PV and GAD_{67} immunoreactivity in PV-containing interneurons, an effect that appeared to be mediated by NR2A-containing NMDA receptors (Kinney et al. 2006). Finally, in living slice preparations from mouse entorhinal cortex, both the genetically engineered reduction in lysophosphatidic acid-1 receptor (LPA-1) and the acute blockade of NMDA receptors produced a layer-specific decrease in induced gamma oscillations, and in the LPA-1 deficient animals, these physiological changes were associated with a ~40% reduction in the number of GABA- and PV-containing neurons, without a change in the number of calretinin-containing neurons (Cunningham et al. 2006). Thus, reduced signaling through NMDA receptors could be an upstream event that selectively affects PV-containing GABA neurons. Consistent with this interpretation, expression of the NR2A subunit of the NMDA receptor was reported to be decreased in a subset of cortical GABA neurons in schizophrenia (Woo et al. 2004).

Alterations in signaling mediated by neuregulin-1 (NRG1) and its receptor ErbB4, both of which are encoded by candidate risk genes for schizophrenia (Fischbach 2007), could also contribute to the disturbances in GABA neurotransmission in schizophrenia by altering the expression of $GABA_A$ receptor subunits or the strength of GABA-induced currents (Corfas et al. 2004). Indeed, ErbB4 receptors are localized to GABA terminals in the prefrontal cortex, and inhibition or mutation of ErbB4 impairs the NRG1-mediated regulation of GABA neurotransmission (Woo et al. 2007). Interestingly, ErbB4 is expressed in nearly 90% of PV-containing neurons, but in only 15% of calretinin-containing neurons in the adult rat neocortex (Yau et al. 2003), suggesting that any effects of altered NRG1-ErbB4 signaling in schizophrenia might be most prominent in the PV-containing subclass of GABA neurons. In addition, NRG1-ErbB4 signaling is required for the maintenance of synaptic NMDA currents (Li et al. 2007). Thus, in concert with the evidence cited above, these findings converge on the hypothesis that the selective alterations in GABA neurotransmission in PV-containing neurons in schizophrenia are a downstream consequence of impaired NMDA receptor-mediated excitatory neurotransmission and that genetically-driven changes in NRG1-ErbB4 signaling could affect PV neurons both directly and through NMDA-mediated mechanisms.

Signaling by the neurotrophin, brain-derived neurotrophic factor (BDNF), through its receptor tyrosine kinase (TrkB) promotes the development of GABA neurons and induces the expression of GAD_{67} mRNA (Marty et al. 2000; Yamada et al. 2002). In addition, these effects may be relatively selective for PV-containing cortical GABA neurons because TrkB is predominantly expressed by PV-containing and not by calretinin-containing GABA neurons (Cellerino et al. 1996). Interestingly, the mRNA and protein levels for both BDNF and TrkB are reduced in the DLPFC of subjects with schizophrenia (Weickert et al. 2003;

Hashimoto et al. 2005; Weickert et al. 2005), and the changes in TrkB and GAD_{67} mRNA expression levels were strongly correlated (r = 0.74, P < 0.001) in the same subjects as were those in BDNF and GAD_{67} mRNA expression levels (r = 0.52, P = 0.007). Interestingly, the correlation was significantly (P = 0.043) stronger between TrkB and GAD_{67} mRNAs than between BDNF and GAD_{67} mRNAs, suggesting that altered TrkB might be a pathogenetic mechanism driving the reduced GABA-related gene expression in schizophrenia. This interpretation was supported by the findings that in mice genetically engineered to have reduced TrkB mRNA expression, expression levels of GAD_{67} and PV mRNAs in the prefrontal cortex were significantly decreased in a TrkB gene dose related fashion (Hashimoto et al. 2005). Furthermore, the cellular pattern of reduced GAD_{67} mRNA expression in these mice precisely paralleled that seen in schizophrenia (Volk et al. 2002). That is, the density of neurons with detectable levels of GAD_{67} mRNA was significantly reduced, but the level of GAD_{67} mRNA expression per neuron was unchanged (Hashimoto et al. 2005). Furthermore, consistent with the selective vulnerability of a GABA neuron subpopulation in schizophrenia, TrkB genotype had no effect on the expression of calretinin mRNA. Thus, the alterations in GABA-related gene expression in TrkB hypomorphic mice replicate those found in subjects with schizophrenia at both the tissue and cellular levels.

Thus, a number of potential causes, many of which are directly or indirectly related to putative risk genes for schizophrenia, could underlie the molecular alterations in a subset of cortical GABA neurons in the illness. Whether these factors are additive or synergistic within a given individual, or represent distinct pathways to a common disturbance, remains to be determined.

The meta-analytic findings illustrate two points *vis à vis* convergence. First, that statistical genetics alone is unlikely to provide an unequivocal and unambiguous answer; it is necessary, but it is not sufficient. Thus, complementary evidence, particularly plausible molecular and biological mechanisms, will also be required, in order to provide convergent evidence to establish the candidacy of a gene. The same will apply to the search for gene-gene and gene-environment interactions. Second, the two NRG1 meta-analyses do agree on one important point: the specific variant(s) associated with schizophrenia has not been identified. What explains the consistent but unfocused association signals originating from the NRG1 locus? The same question pertains to all other putative schizophrenia genes; the issue is whether it has the same answer across genes, in which case a convergent molecular basis for association may exist. With regard to this possibility, two issues need consideration: allelic heterogeneity, and identification of the genuine risk variants.

10.3 Potential Consequences

The affected PV-containing GABA neurons in schizophrenia include the chandelier subclass whose axons give rise to a linear array of terminals (termed cartridges) that synapse exclusively on the axon initial segments of pyramidal neurons. In the

DLPFC of subjects with schizophrenia, the expression of the mRNA for the GABA membrane transporter (GAT1) is reduced in a subset of GABA neurons (Volk et al. 2001) and GAT1 immunoreactivity is selectively reduced in chandelier axon cartridges (Woo et al. 1998). In contrast, in the postsynaptic axon initial segments of pyramidal neurons immunoreactivity for the $GABA_A$ α_2 subunit is markedly increased (Volk et al. 2002). All of these changes appear to be specific to the disease process of schizophrenia since they are not found in subjects with other psychiatric disorders or in monkeys exposed chronically to antipsychotic medications (Hashimoto et al. 2003; Volk et al. 2000, 2001, 2002). Together, these findings suggest that reduced expression of GAD_{67} mRNA in PV-containing chandelier neurons results in decreased GABA synthesis and release, and induces compensatory responses including decreased pre-synaptic GABA re-uptake and upregulated post-synaptic $GABA_A$ receptors (Lewis et al. 2005).

Reduced GABA signaling from chandelier cells to pyramidal neurons could contribute to the pathophysiology of working memory dysfunction. Networks of PV-containing, GABA neurons, formed by both chemical and electrical synapses, give rise to oscillatory activity in the gamma band range, the synchronized firing of a neuronal population at 30-80 Hz (Whittington and Traub 2003). Gamma band oscillations in the human DLPFC increase in proportion to working memory load (Howard et al. 2003), and in subjects with schizophrenia, induced DLPFC gamma band oscillations are reduced during a cognitive control task (Cho et al. 2006). Thus, a deficit in the synchronization of pyramidal cells, resulting from impaired regulation by PV-containing GABA neurons, could contribute to reduced gamma band oscillations, and consequently to working memory dysfunction, in subjects with schizophrenia (Lewis et al. 2005).

Exactly how the apparent alterations in chandelier cells could contribute to working memory impairments requires further consideration. Traditionally, chandelier neurons have been considered to be powerful inhibitors of pyramidal cell output, exercising "veto power" by virtue of the close proximity their synaptic inputs to the site of action potential generation in pyramidal neurons. Like other cortical GABA neurons, the effect of GABA released from chandelier neuron axon terminals is mediated by binding to post-synaptic $GABA_A$ receptors which results in the opening of chloride ion channels. In the adult brain, high expression of the potassium-chloride co-transporter (KCC2) results in the extrusion of chloride from the cell (Stein and Nicol 2003). Thus, when $GABA_A$ receptors are activated, chloride ions flow into the cell along a concentration gradient, resulting in hyperpolarization of the membrane, and a reduced probability of cell firing. However, a recent study (Szabadics et al. 2006) reported that KCC2, while readily detectable in the cell body of adult pyramidal neurons, was apparently absent in the axon initial segment. Consistent with this observation, Szabadics and colleagues found that the release of GABA from chandelier neuron axon terminals resulted in depolarization of pyramidal cells in an in vitro slice preparation. In fact, the chandelier cell-mediated depolarization was so powerful that in ~50% of the cases in which a single chandelier cell was stimulated, the postsynaptic pyramidal cell was depolarized to the point of firing an action potential. If replicated, these findings would suggest that an alternative explanation for the physiological consequences of alterations in chandelier neurons in schizophrenia is needed. For example, reduced synthesis of GABA in chandelier neurons could result in a marked decrease in excitatory output of pyramidal neurons that project to other cortical regions or subcortical sites.

10.4. A broader view of alterations in cortical GABA neurotransmission in schizophrenia

Whatever the causes and functional consequences of their alterations, abnormalities in PV neurons alone can not completely account for the deficits in expression of GAD_{67} mRNA since such changes were also observed in cortical layers 2 and 5 where relatively few PV-containing GABA neurons are located (Condé et al. 1994; Hashimoto et al. 2003), and where no changes in PV mRNA expression were found (Hashimoto et al. 2003). In addition, further understanding of the significance of altered GABA neurotransmission in dysfunction of the DLPFC in schizophrenia requires the systematic analysis of the expression of other GABA-related transcripts whose protein products are involved in different aspects of GABA neurotransmission. In order to address these questions, we utilized a customized DNA microarray containing probes for GABA-related transcripts that included transcripts 1) selectively expressed in subsets of GABA neurons; 2) whose protein products are involved in the synthesis, release, uptake or degradation of GABA; 3) encoding GABA receptor subunits and related proteins; and 4) whose protein products regulate the function of GABA neurons (Hashimoto et al. 2007). In addition to replicating the deficits in GAD_{67} and GAT1 mRNAs, we also found decreased expression of the transcripts encoding the neuropeptides, somatostatin (SST), neuropeptide Y (NPY) and cholecystokinin (CCK), each of which is expressed and used as a neuromodulator by a subset of GABA neurons. These alterations were verified by real-time quantitative PCR in the same subjects and by in situ hybridization in a larger cohort of subjects.

The changes in SST and NPY mRNA expression in the subjects with schizophrenia were highly correlated with those for GAD_{67} mRNA (GAD_{67} and SST, r = 0.71, P < 0.005; GAD_{67} and NPY, r = 0.72, P < 0.004), suggesting that GAD_{67} mRNA expression is also decreased in the subset of GABA neurons that express both SST and NPY. In the cortex, SST is expressed by the majority of calbindin-containing GABA neurons, a separate population from those that express PV or calretinin (Condé et al. 1994; Kubota et al. 1994; González-Albo et al. 2001), and a subset of SST-containing neurons largely overlaps with the majority of NPY-containing neurons (Hendry et al. 1984; Kubota et al. 1994). Because SST- and NPY-containing neurons are predominantly found in layers 2 and 5 (Hendry et al. 1984; Kubota et al. 1994), they may account for the deficits in GAD_{67} mRNA expression in these layers which could not be explained by the expression deficits in PV-containing neurons.

Interneurons that contain both SST and NPY include the Martinotti cells (Kawaguchi and Kubota 1997; Gibson et al. 1999; Reyes et al. 1998; Ma et al. 2006). The axons of Martinotti cells project to layer 1 where they synapse on the apical dendrites of pyramidal neurons and in rodent and monkey neocortex, SST interneurons predominately innervate the dendrites of pyramidal neurons (Hendry et al. 1984; DeLima and Morrison 1989; Kawaguchi and Kubota 1996; Melchitzky and Lewis 2005). What might be the consequences of these disturbances in the SST/NPY-containing Martinotti cells for the function of the DLPFC and working memory performance in individuals with schizophrenia?

In computational modeling of cortical microcircuits and working memory, GABA neurons that target the dendritic domain of pyramidal neurons provide resistance against distracting stimuli by sending enhanced inhibition to dendrites of nearby pyramidal neurons that are selective for other stimuli (Wang et al. 2004). In addition, high frequency trains from pyramidal neurons produce facilitating excitatory inputs to Martinotti cells that, via synapses onto the dendrites of neighboring pyramidal neurons, cause disynaptic inhibition (Silberberg and Markram 2007), providing a neural basis for sensory gating within the cortex. Interestingly, in individuals with schizophrenia, disturbances in sensory-gating have been correlated with reductions in working memory performance (Silver and Feldman 2005), suggesting that the inability to filter distracting stimuli disrupts working memory. Therefore, alterations in GABA neurotransmission by SST/NPY-containing Martinotti cells appear to affect disynaptic inhibition of neighboring pyramidal neurons selective for other stimuli, resulting in a reduced capacity to filter distractors, and, consequently, impaired working memory performance in schizophrenia.

In addition, low threshold-spiking, SST-containing, Martinotti cells are extensively electrically coupled into networks that robustly synchronize their spiking activity (Gibson et al. 1999; Gibson et al. 2005) in the theta range (4-7 Hz) and produce corresponding synchronized inhibitory post-synaptic potentials in neighboring pyramidal neurons (Beierlein et al. 2000). EEG studies in humans and monkeys have demonstrated that theta band oscillations increase in power during working memory tasks (Raghavachari et al. 2001; Lee et al. 2005; Krause et al. 2000). Furthermore, subjects with schizophrenia demonstrate altered frontal theta oscillations during working memory tasks (Schmiedt et al. 2005). Thus, disturbances in SST/NPY-containing, low threshold-spiking Martinotti cells might contribute to alterations in theta oscillations and ultimately to impaired working memory performance in subjects with schizophrenia.

The changes in expression of GAD_{67} and CCK mRNAs in schizophrenia were also highly correlated ($r = 0.84$, $P < 0.001$), suggesting that GABA synthesis is also decreased in this subset of GABA neurons. CCK is heavily expressed in GABA neurons that do not contain PV or SST (Lund and Lewis 1993; Kawaguchi and Kondo 2002). Interestingly, the axon terminals of CCK-containing large basket neurons, which target selectively pyramidal neuron cell bodies, contain type I cannabinoid receptors and the activation of these receptors by exogenous cannabinoids suppresses GABA release (Katona et al. 1999; Bodor et al. 2005; Eggan and Lewis 2007). Thus, exposure to cannabis could exacerbate an intrinsic deficit in

GABA synthesis in these neurons, providing a mechanism by which cannabis use increases the risk for, and the severity of, schizophrenia (Fergusson et al. 2006).

In addition to these alterations in markers of GABA neurons, our microarray studies revealed alterations in the expression of several subunits of GABA$_A$ receptors, including α1, γ2, and δ subunits (Hashimoto et al. 2007). GABA$_A$ receptors containing the α1 and γ2 subunits are enriched in postsynaptic sites where they mediate phasic inhibition (Nusser et al. 1998; Mangan et al. 2005). In contrast, GABA$_A$ receptors containing the δ subunit, which is often coassembled with the α4 subunit in the forebrain, are selectively localized to extrasynaptic sites (Nusser et al. 1998; Wei et al. 2003; Mangan et al. 2005). These extrasynaptic receptors, which have a high sensitivity to GABA and thus can be activated by ambient GABA molecules in extracellular space, mediate tonic inhibition which reduces the effects of synaptic inputs over time (Wei et al. 2003; Farrant and Nusser 2005). Given the predominant localization of the α1, γ2, and δ subunits to dendrites (Hendry et al. 1994; Fritschy and Mohler 1995), the highly correlated expression deficits for these transcripts (α1 and γ2, r=0.84, P<0.001; α1 and δ, r=0.81, P<0.001; and γ2 and δ r=0.85, P<0.001) suggest coordinated downregulation of GABA$_A$ receptors mediating phasic and tonic inhibition in the dendritic domain of DLPFC pyramidal neurons in schizophrenia.

10.5. Summary

In concert, the findings reviewed above suggest that the GABA neurotransmission is altered in the DLPFC of subjects with schizophrenia in a manner that is cell type-specific, but not restricted to a single cell type, and that involves both pre- and post-synaptic mechanisms. As shown in Figure 9.1, mRNA expression deficits are present in PV-containing, but not in calretinin-containing, GABA neurons in the DLPFC of subjects with schizophrenia (Volk et al. 2000; Volk et al. 2001; Hashimoto et al. 2003). These changes are associated with the downregulation of GAT1 in the presynaptic terminals of PV-containing chandelier neurons (Pierri et al. 1999) and the upregulation of GABA$_A$ receptor α2 subunit in the postsynaptic axon initial segments of pyramidal neurons (Figure 10.1, lower enlarged square) (Volk et al. 2002). Alterations in GABA neurotransmission also appear to be present in the SST- and NPY-containing Martinotti neurons and CCK-containing basket neurons which predominately target the distal dendrites and cell bodies of pyramidal neurons, respectively. Furthermore, gene expression deficits for α1 and γ2 GABA$_A$ receptor subunits and for the δ subunit suggest decreased synaptic (phasic) and extrasynaptic (tonic) inhibition, respectively, in pyramidal neuron dendrites (Figure 10.1, upper enlarged square). GABA-mediated regulation at the dendritic domain of pyramidal neurons is important for the selection of excitatory inputs from different cortical and subcortical areas, whereas GABA inputs at the perisomatic domain, including the axon initial segment and cell body, are critical for control of the timing and synchronization of pyramidal neuron firing. Therefore, altered GABA-mediated regulation of both inputs to and outputs from DLPFC pyramidal neurons are likely to disrupt information processing in DLPFC circuitry and contribute to the working memory impairments present in schizophrenia.

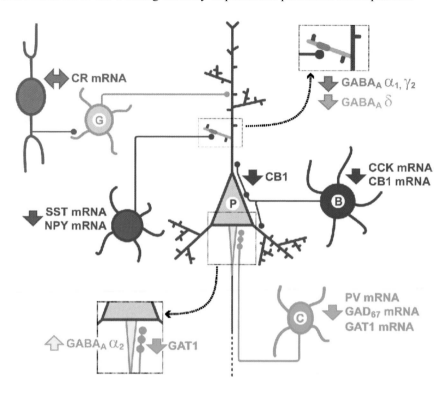

Fig. 10.1. Schematic summary of alterations in GABA-mediated circuitry in the DLPFC of subjects with schizophrenia. Altered GABA neurotransmission by PV-containing neurons (green) is indicated by gene expression deficits in these neurons and associated changes in their synapses, a decrease in GAT1 immunoreactivity in their terminals and an upregulation of $GABA_A$ receptor $\alpha 2$ subunit at the axon initial segments of pyramidal neurons (lower enlarged square). Decreased expression of both SST and NPY mRNAs indicates alterations in SST and/or NPY-containing neurons (blue) that target the distal dendrites of pyramidal neurons. These changes appear to be accompanied by a downregulation of $GABA_A$ receptor subunits, including the $\alpha 1$ and $\gamma 2$ subunits present in receptors that mediate synaptic (phasic) inhibition and the δ subunit present in receptors that mediate extrasynaptic (tonic) inhibition (upper enlarged square), in dendrites of pyramidal neurons. Decreased CCK mRNA levels indicate an alteration of CCK-containing large basket neurons (purple) that represent a separate source of perisomatic inhibition from PV-containing neurons. Gene expression in CR-containing GABA neurons (red) does not seem to be altered. Other neurons, such as PV-containing basket neurons, are not shown because the nature of their involvement in schizophrenia is unclear. G, generic GABA neuron; P, pyramidal neuron. Figure modified from (Hashimoto et al. 2007).

Acknowledgements

Work by the authors cited in this manuscript was supported by National Institutes of Health grants MH 045156, MH 051234 and MH 043784.

References

Addington AM, Gornick M, Duckworth J, Sporn A, Gogtay N, Bobb A, Greenstein D, Lenane M, Gochman P, Baker N, Balkissoon R, Vakkalanka RK, Weinberger DR, Rapoport JL, Straub RE (2005) GAD1 (2q31.1), which encodes glutamic acid decarboxylase (GAD(67)), is associated with childhood-onset schizophrenia and cortical gray matter volume loss. Mol Psychiatry 10:581-588

Akbarian S, Huang HS (2006) Molecular and cellular mechanisms of altered GAD1/GAD67 expression in schizophrenia and related disorders. Brain Res Rev 52:293-304

Akbarian S, Kim JJ, Potkin SG, Hagman JO, Tafazzoli A, Bunney Jr WE, Jones EG (1995) Gene expression for glutamic acid decarboxylase is reduced without loss of neurons in prefrontal cortex of schizophrenics. Arch Gen Psychiatry 52:258-266

Beierlein M, Gibson JR, Connors BW (2000) A network of electrically coupled interneurons drives synchronized inhibition in neocortex. Nat Neurosci 3:904-910

Bodor AL, Katona I, Nyiri G, Mackie K, Ledent C, Hajos N, Freund TF (2005) Endocannabinoid signaling in rat somatosensory cortex: laminar differences and involvement of specific interneuron types. J Neurosci 25:6845-6856

Cellerino A, Maffei L, Domenici L (1996) The distribution of brain-derived neurotrophic factor and its receptor trkB in parvalbumin-containing neurons of the rat visual cortex. Eur J Neurosci 8:1190-1197

Cho RY, Konecky RO, Carter CS (2006) Impairments in frontal cortical gamma synchrony and cognitive control in schizophrenia. Proc Natl Acad Sci USA 103:19878-19883

Cochran SM, Kennedy M, McKerchar CE, Steward LJ, Pratt JA, Morris BJ (2003) Induction of metabolic hypofunction and neurochemical deficits after chronic intermittent exposure to phencyclidine: differential modulation by antipsychotic drugs. Neuropsychopharm 28:265-275

Condé F, Lund JS, Jacobowitz DM, Baimbridge KG, Lewis DA (1994) Local circuit neurons immunoreactive for calretinin, calbindin D-28k, or parvalbumin in monkey prefrontal cortex: Distribution and morphology. J Comp Neurol 341:95-116

Corfas G, Roy K, Buxbaum JD (2004) Neuregulin 1-erbB signaling and the molecular/cellular basis of schizophrenia. Nat Neurosci 7:575-580

Cunningham MO, Hunt J, Middleton S, LeBeau FE, Gillies MJ, Davies CH, Maycox PR, Whittington MA, Racca C (2006) Region-specific reduction in entorhinal gamma oscillations and parvalbumin-immunoreactive neurons in animal models of psychiatric illness. J Neurosci 26:2767-2776

De Luca V, Muglia P, Masellis M, Jane DE, Wong GW, Kennedy JL (2004) Polymorphisms in glutamate decarboxylase genes: analysis in schizophrenia. Psychiatr Genet 14:39-42

DeLima AD, Morrison JH (1989) Ultrastructural analysis of somatostatin-immunoreactive neurons and synapses in the temporal and occipital cortex of the macaque monkey. J Comp Neurol 283:212-227

Eggan SM, Lewis DA (2007) Immunocytochemical distribution of the cannabinoid CB1 receptor in the primate neocortex: a regional and laminar analysis. Cereb Cortex 17:175-191

Farrant M, Nusser Z (2005) Variations on an inhibitory theme: phasic and tonic activation of GABA(A) receptors. Nat Rev Neurosci 6:215-229

Fergusson DM, Poulton R, Smith PF, Boden JM (2006) Cannabis and psychosis. BMJ 332:172-175

Fischbach GD (2007) NRG1 and synaptic function in the CNS. Neuron 54:495-497

Fritschy J-M, Mohler H (1995) GABA$_A$-receptor heterogeneity in the adult rat brain: Differential regional and cellular distribution of seven major subunits. J Comp Neurol 359:154-194

Gibson JR, Beierlein M, Connors BW (1999) Two networks of electrically coupled inhibitory neurons in neocortex. Nature 402:75-79

Gibson JR, Beierlein M, Connors BW (2005) Functional properties of electrical synapses between inhibitory interneurons of neocortical layer 4. J Neurophysiol 93:467-480

González-Albo MC, Elston GN, DeFelipe J (2001) The human temporal cortex: Characterization of neurons expressing nitric oxide synthase, neuropeptides and calcium-binding proteins, and their glutamate receptor subunit profiles. Cereb Cortex 11:1170-1181

Guidotti A, Auta J, Davis JM, Gerevini VD, Dwivedi Y, Grayson DR, Impagnatiello F, Pandey G, Pesold C, Sharma R, Uzunov D, Costa E (2000) Decrease in reelin and glutamic acid decarboxylase$_{67}$ (GAD$_{67}$) expression in schizophrenia and bipolar disorder. Arch Gen Psychiatry 57:1061-1069

Gulyás AI, Megías M, Emri Z, Freund TF (1999) Total number and ratio of excitatory and inhibitory synapses converging onto single interneurons of different types in the CA1 area of the rat hippocampus. J Neurosci 19:10082-10097

Hashimoto T, Arion D, Volk DW, Mirnics K, Lewis D (2007) Alterations in GABA-related transcriptome in the dorsolateral prefrontal cortex of subjects with schizophrenia. Mol Psychiatry Epub ahead of print:1-15

Hashimoto T, Bergen SE, Nguyen QL, Xu B, Monteggia LM, Pierri JN, Sun Z, Sampson AR, Lewis DA (2005) Relationship of brain-derived neurotrophic factor and its receptor TrkB to altered inhibitory prefrontal circuitry in schizophrenia. J Neurosci 25:372-383

Hashimoto T, Volk DW, Eggan SM, Mirnics K, Pierri JN, Sun Z, Sampson AR, Lewis DA (2003) Gene expression deficits in a subclass of GABA neurons in the prefrontal cortex of subjects with schizophrenia. J Neurosci 23:6315-6326

Hendry SHC, Huntsman MM, Viñuela A, Mohler H, de Blas AL, Jones EG (1994) GABA$_A$ receptor subunit immunoreactivity in primate visual cortex: Distribution in macaques and humans and regulation by visual input in adulthood. J Neurosci 14:2383-2401

Hendry SHC, Jones EG, Emson PC (1984) Morphology, distribution, and synaptic relations of somatostatin- and neuropeptide Y-immunoreactive neurons in rat and monkey neocortex. J Neurosci 4:2497-2517

Howard MW, Rizzuto DS, Caplan JB, Madsen JR, Lisman J, Aschenbrenner-Scheibe R, Schulze-Bonhage A, Kahana MJ (2003) Gamma oscillations correlate with working memory load in humans. Cereb Cortex 13:1369-1374

Ikeda M, Ozaki N, Yamanouchi Y, Suzuki T, Kitajima T, Kinoshita Y, Inada T, Iwata N (2007) No association between the glutamate decarboxylase 67 gene (GAD1) and schizophrenia in the Japanese population. Schizophr Res 91:22-26

Impagnatiello F, Guidotti AR, Pesold C, Dwivedi Y, Caruncho H, Pisu MG, Uzunov DP, Smalheiser NR, Davis JM, Pandey GN, Pappas GD, Teuting P, Sharma RP, Costa E (1998) A decrease of reelin expression as a putative vulnerability factor in schizophrenia. Proc Natl Acad Sci USA 95:15718-15723

Jones EG (1997) Cortical development and thalamic pathology in schizophrenia. Schizophr Bull 23:483-501

Katona I, Sperlagh B, Sik A, Kafalvi A, Vizi ES, Mackie K, Freund TF (1999) Presynaptically located CB1 cannabinoid receptors regulate GABA release from axon terminals of specific hippocampal interneurons. J Neurosci 19:4544-4558

Kawaguchi Y, Kondo S (2002) Parvalbumin, somatostatin and cholecystokinin as chemical markers for specific GABAergic interneuron types in the rat frontal cortex. J Neurocytol 31:277-287

Kawaguchi Y, Kubota Y (1996) Physiological and morphological identification of somatostatin- or vasoactive intestinal polypeptide-containing cells among GABAergic cell subtypes in rat frontal cortex. J Neurosci 16:2701-2715

Kawaguchi Y, Kubota Y (1997) GABAergic cell subtypes and their synaptic connections in rat frontal cortex. Cereb Cortex 7:476-486

Keilhoff G, Becker A, Grecksch G, Wolf G, Bernstein HG (2004) Repeated application of ketamine to rats induces changes in the hippocampal expression of parvalbumin, neuronal nitric oxide synthase and cFOS similar to those found in human schizophrenia. Neuroscience 126:591-598

Kinney JW, Davis CN, Tabarean I, Conti B, Bartfai T, Behrens MM (2006) A specific role for NR2A-containing NMDA receptors in the maintenance of parvalbumin and GAD67 immunoreactivity in cultured interneurons. J Neurosci 26:1604-1615

Krause CM, Sillanmaki L, Koivisto M, Saarela C, Haggqvist A, Laine M, Hamalainen H (2000) The effects of memory load on event-related EEG desynchronization and synchronization. Clin Neurophysiol 111:2071-2078

Kubota Y, Hattori R, Yui Y (1994) Three distinct subpopulations of GABAergic neurons in rat frontal agranular cortex. Brain Res 649:159-173

Lee H, Simpson GV, Logothetis NK, Rainer G (2005) Phase locking of single neuron activity to theta oscillations during working memory in monkey extrastriate visual cortex. Neuron 45:147-156

Lewis DA, Hashimoto T, Volk DW (2005) Cortical inhibitory neurons and schizophrenia. Nat Rev Neurosci 6:312-324

Lewis DA, Moghaddam B (2006) Cognitive dysfunction in schizophrenia: Convergence of GABA and glutamate alterations. Arch Neurol 63:1372-1376

Li B, Woo RS, Mei L, Malinow R (2007) The neuregulin-1 receptor erbB4 controls glutamatergic synapse maturation and plasticity. Neuron 54:583-597

Liu WS, Pesold C, Rodriguez MA, Carboni G, Auta J, Lacor P, Larson J, Condie BG, Guidotti A, Costa E (2001) Down-regulation of dendritic spine and glutamic acid decarboxylase 67 expressions in the reelin haploinsufficient heterozygous reeler mouse. Proc Natl Acad Sci USA 98:3477-3482

Lund JS, Lewis DA (1993) Local circuit neurons of developing and mature macaque prefrontal cortex: Golgi and immunocytochemical characteristics. J Comp Neurol 328:282-312

Lundorf MD, Buttenschon HN, Foldager L, Blackwood DH, Muir WJ, Murray V, Pelosi AJ, Kruse TA, Ewald H, Mors O (2005) Mutational screening and association study of glutamate decarboxylase 1 as a candidate susceptibility gene for bipolar affective disorder and schizophrenia. Am J Med Genet B Neuropsychiatr Genet 135:94-101

Ma Y, Hu H, Berrebi AS, Mathers PH, Agmon A (2006) Distinct subtypes of somatostatin-containing neocortical interneurons revealed in transgenic mice. J Neurosci 26:5069-5082

Mangan PS, Sun C, Carpenter M, Goodkin HP, Sieghart W, Kapur J (2005) Cultured Hippocampal Pyramidal Neurons Express Two Kinds of GABAA Receptors. Mol Pharmacol 67:775-788

Marty S, Wehrle R, Sotelo C (2000) Neuronal activity and brain-derived neurotrophic factor regulate the density of inhibitory synapses in organotypic slice cultures of postnatal hippocampus. J Neurosci 20:8087-8095

Melchitzky DS, Lewis DA (2003) Pyramidal neuron local axon terminals in monkey prefrontal cortex: Differential targeting of subclasses of GABA neurons. Cereb Cortex 13:452-460

Melchitzky DS, Lewis DA (2005) Synaptic targets of somatostatin-labeled axon terminals in monkey prefrontal cortex. Soc Neurosci Abstr 675.6

Nusser Z, Sieghart W, Somogyi P (1998) Segregation of different GABAA receptors to synaptic and extrasynaptic membranes of cerebellar granule cells. J Neurosci 18:1693-1703

Pesold C, Impagnatiello F, Pisu MG, Uzunov DP, Costa E, Guidotti A, Caruncho HJ (1998) Reelin is preferentially expressed in neurons synthesizing γ-aminobutyric acid in cortex and hippocampus of adult rats. Proc Natl Acad Sci USA 95:3221-3226

Pesold C, Liu WS, Guidotti A, Costa E, Caruncho HJ (1999) Cortical bitufted, horizontal, and Martinotti cells preferentially express and secrete reelin in perineuronal nets, postsynaptically modulating gene expression. Proc Natl Acad Sci USA 96:3217-3222

Pierri JN, Chaudry AS, Woo T-U, Lewis DA (1999) Alterations in chandelier neuron axon terminals in the prefrontal cortex of schizophrenic subjects. Am J Psychiatry 156:1709-1719

Raghavachari S, Kahana MJ, Rizzuto DS, Caplan JB, Kirschen MP, Bourgeois B, Madsen JR, Lisman JE (2001) Gating of human theta oscillations by a working memory task. J Neurosci 21:3175-3183

Rao SG, Williams GV, Goldman-Rakic PS (2000) Destruction and creation of spatial tuning by disinhibition: GABA$_A$ blockade of prefrontal cortical neurons engaged by working memory. J Neurosci 20:485-494

Reyes A, Lujar R, Rozov BN, Somogyi P, Sakman B (1998) Target-cell-specific facilitation and depression in neocortical circuits. Nat Neurosci 1:279-285

Rujescu D, Bender A, Keck M, Hartmann AM, Ohl F, Raeder H, Giegling I, Genius J, McCarley RW, Moller HJ, Grunze H (2006) A pharmacological model for psychosis based on N-methyl-D-aspartate receptor hypofunction: molecular, cellular, functional and behavioral abnormalities. Biol Psychiatry 59:721-729

Sawaguchi T, Matsumura M, Kubota K (1989) Delayed response deficits produced by local injection of bicuculline into the dorsolateral prefrontal cortex in Japanese macaque monkeys. Exp Brain Res 75:457-469

Schmiedt C, Brand A, Hildebrandt H, Basar-Eroglu C (2005) Event-related theta oscillations during working memory tasks in patients with schizophrenia and healthy controls. Brain Res Cogn Brain Res 25:936-947

Silberberg G, Markram H (2007) Disynaptic inhibition between neocortical pyramidal cells mediated by Martinotti cells. Neuron 53:735-746

Silver H, Feldman P (2005) Evidence for sustained attention and working memory in schizophrenia sharing a common mechanism. J Neuropsychiatry Clin Neurosci 17:391-398

Stein V, Nicol RA (2003) GABA generates Excitement. Neuron 37:375-378

Straub RE, Lipska BK, Egan MF, Goldberg TE, Callicot JH, Mayhew MB, Vakkalanka RK, Kolachana BS, Kleinman JE, Weinberger DR (2007) Allelic variation in GAD1 (GAD67) is associated with schizophrenia and influences cortical function and gene expression. Mol Psychiatry in press:1-16

Szabadics J, Varga C, Molnar G, Olah S, Barzo P, Tamas G (2006) Excitatory effect of GABAergic axo-axonic cells in cortical microcircuits. Science 311:233-235

Torrey EF, Barci BM, Webster MJ, Bartko JJ, Meador-Woodruff JH, Knable MB (2005) Neurochemical markers for schizophrenia, bipolar disorder, and major depression in postmortem brains. Biol Psychiatry 57:252-260

Volk DW, Austin MC, Pierri JN, Sampson AR, Lewis DA (2000) Decreased glutamic acid decarboxylase67 messenger RNA expression in a subset of prefrontal cortical gamma-aminobutyric acid neurons in subjects with schizophrenia. Arch Gen Psychiatry 57:237-245

Volk DW, Austin MC, Pierri JN, Sampson AR, Lewis DA (2001) GABA transporter-1 mRNA in the prefrontal cortex in schizophrenia: Decreased expression in a subset of neurons. Am J Psychiatry 158:256-265

Volk DW, Pierri JN, Fritschy J-M, Auh S, Sampson AR, Lewis DA (2002) Reciprocal alterations in pre- and postsynaptic inhibitory markers at chandelier cell inputs to pyramidal neurons in schizophrenia. Cereb Cortex 12:1063-1070

Wang XJ, Tegner J, Constantinidis C, Goldman-Rakic PS (2004) Division of labor among distinct subtypes of inhibitory neurons in a cortical microcircuit of working memory. Proc Natl Acad Sci U S A 101:1368-1373

Wei W, Zhang N, Peng Z, Houser CR, Mody I (2003) Perisynaptic localization of delta subunit-containing GABA(A) receptors and their activation by GABA spillover in the mouse dentate gyrus. J Neurosci 23:10650-10661

Weickert CS, Hyde TM, Lipska BK, Herman MM, Weinberger DR, Kleinman JE (2003) Reduced brain-derived neurotrophic factor in prefrontal cortex of patients with schizophrenia. Mol Psychiatry 8:592-610

Weickert CS, Ligons DL, Romanczyk T, Ungaro G, Hyde TM, Herman MM, Weinberger DR, Kleinman JE (2005) Reductions in neurotrophin receptor mRNAs in the prefrontal cortex of patients with schizophrenia. Mol Psychiatry 10:637-650

Whittington MA, Traub RD (2003) Interneuron diversity series: inhibitory interneurons and network oscillations in vitro. Trends Neurosci 26:676-682

Woo RS, Li XM, Tao Y, Carpenter-Hyland E, Huang YZ, Weber J, Neiswender H, Dong XP, Wu J, Gassmann M, Lai C, Xiong WC, Gao TM, Mei L (2007) Neuregulin-1 enhances depolarization-induced GABA release. Neuron 54:599-610

Woo T-U, Walsh JP, Benes FM (2004) Density of glutamic acid decarboxylase 67 messenger RNA-containing neurons that express the N-methyl-D-aspartate receptor subunit NR2A in the anterior cingulate cortex in schizophrenia and bipolar disorder. Arch Gen Psychiatry 61:649-657

Woo T-U, Whitehead RE, Melchitzky DS, Lewis DA (1998) A subclass of prefrontal gamma-aminobutyric acid axon terminals are selectively altered in schizophrenia. Proc Natl Acad Sci U S A 95:5341-5346

Yamada MK, Nakanishi K, Ohba S, Nakamura T, Ikegaya Y, Nishiyama N, Matsuki N (2002) Brain-derived neurotrophic factor promotes the maturation of GABAergic mechanisms in cultured hippocampal neurons. J Neurosci 22:7580-7585

Yau HJ, Wang HF, Lai C, Liu FC (2003) Neural development of the neuregulin receptor ErbB4 in the cerebral cortex and the hippocampus: preferential expression by interneurons tangentially migrating from the ganglionic eminences. Cereb Cortex 13:252-264

Chapter 11

Increased Cortical Excitability as a Critical Element in Schizophrenia Pathophysiology

Patricio O'Donnell

Abstract. The identification of schizophrenia predisposing genes critical for brain development and synaptic transmission has highlighted the developmental aspects of this disorder. Yet, symptoms typically emerge after adolescence, and environmental factors (both perinatally and during adolescence) also play a role in schizophrenia. As cortical interneurons have repeatedly been identified as abnormal in schizophrenia and these neurons and their dopamine innervation mature late (i.e., during adolescence), it is possible that early insults may affect these neurons but symptoms appear when interneurons should have matured. Abnormal interneuron activity, in particular in the prefrontal cortex, would increase cortical network activity, impairing information processing and the filtering of weak inputs (increasing "noise"). The specific profile of NMDA glutamate receptors cortical interneuron express may make them vulnerable to the effects of non-competing antagonists; in that way, the glutamate and GABA hypotheses of schizophrenia may converge by blockade of NMDA receptors in interneurons being responsible for symptoms. This specific distribution of NMDA receptor subtypes may allow the search of interneuron-targeting novel pharmacological tools, opening the possibility to restore balance in cortical networks.

11.1 Introduction

Important clues are slowly being revealed about brain changes in schizophrenia, a disorder once described as "the graveyard of pathology" because of the absence of identifying pathological changes. On one hand, the existence of predisposing genes, which had been anticipated for decades, has now been definitely established (Mirnics et al. 2001; Weinberger et al. 2001; Harrison and Weinberger 2004). On the other hand, several groups have unveiled pathological alterations in post-mortem studies, revealing structural and functional changes in diverse cortical and subcortical brain regions (Harrison 2004; Lewis et al. 2004). Despite this wealth of new information, a unifying view of pathophysiological mechanisms leading to this devastating disorder is still missing. Many attempts have been made at establishing a *systems* model of what goes wrong in the brain of those affected with schizophrenia, and each model has indeed advanced our understanding of some aspect of this disease. However, we are still short of an integrative picture identifying critical pathophysiological processes that could eventually lead to new therapeutic approaches. Here, I offer an attempt of a comprehensive hypothesis by linking most brain systems affected in schizophrenia within a single conceptual model, drawing from both preclinical and clinical findings, which emphasizes increased cortical excitability due to alterations in cortical interneurons as a key element that drives pathophysiological changes.

11.2 Susceptibility Genes in Schizophrenia

Important clues are slowly being revealed about brain changes in schizophrenia, a disorder once described as "the graveyard of pathology" because of the absence of identifying pathological changes. On one hand, the existence of predisposing genes, which had been anticipated for decades, has now been definitely established (Mirnics et al. 2001; Weinberger et al. 2001; Harrison and Weinberger 2004). On the other hand, several groups have unveiled pathological alterations in post-mortem studies, revealing structural and functional changes in diverse cortical and subcortical brain regions (Harrison 2004; Lewis et al. 2004). Despite this wealth of new information, a unifying view of pathophysiological mechanisms leading to this devastating disorder is still missing. Many attempts have been made at establishing a *systems* model of what goes wrong in the brain of those affected with schizophrenia. Each model has indeed advanced our understanding of some aspect of this disease; however, we are still short of an integrative picture identifying critical pathophysiological processes that could eventually lead to new therapeutic approaches. Here, I offer an attempt of a comprehensive hypothesis by linking most brain systems affected in schizophrenia within a single conceptual model, drawing from both preclinical and clinical findings, which emphasizes increased cortical excitability due to alterations in cortical interneurons as a key element that drives pathophysiological changes.

Genes related to dopamine (DA) and the primary excitatory and inhibitory neurotransmitters, glutamate and gamma-amino butyric acid (GABA), have also been found with some relation to schizophrenia. For example, polymorphisms in a region close to GAD1, the gene encoding the GABA-synthesizing enzyme glutamic acid decarboxylase-67 (GAD67), are associated with childhood-onset schizophrenia (Addington et al. 2005). In addition, single-nucleotide polymorphisms have been identified for the DA-inactivating enzyme catechol-O-methyl transferase (COMT) (Chen et al. 2004), and whether a val or met aminoacid is present in position 158 confers high or low efficacy to this enzyme (Chen et al. 2004). It is more likely to find the val-val combination in patients suffering schizophrenia and their relatives than in the rest of the population (Weinberger et al. 2001; Harrison and Weinberger 2004). Other susceptibility genes recently uncovered include that for a subunit of the metabotropic glutamate receptor (mGluR3) (Egan et al. 2004) and many others reviewed in the preceding chapters of this book. Thus, in addition to having provided strong support for the theories emphasizing a neurodevelopmental component in schizophrenia, the newly identified predisposing genes highlight the potential for anomalies in interactions among glutamate, GABA and DA in synaptic transmission.

11.3 Temporal Lobe Alterations

Diverse brain structures exhibit abnormal morphological features in schizophrenia, and one of the most frequently mentioned regions is the hippocampus in the temporal lobe. There have been reports indicating cellular disarray (Kovelman and Scheibel 1984; Altschuler et al. 1987; Arnold et al. 1991), although other groups failed to replicate these findings (Christison et al. 1989). Synapse-related markers, including GAP43 (Eastwood and Harrison 1998; Chambers et al. 2005), synaptophysin (Eastwood et al. 1995; Chambers et al. 2005), calcineurin (Eastwood et al. 2005), complexin (Harrison and Eastwood 1998) and spinophilin (Law et al. 2004a), have been found abnormal. In addition, morphometric studies revealed loss of dendritic spine volume in CA3 pyramidal neurons (Kolomeets et al. 2005). There have also been reports of deficits in hippocampal GABA interneurons in schizophrenia (Benes 1999). Some of the genetic markers associated with schizophrenia may be related to this limbic brain region. For example, Neuregulin is typically expressed by hippocampal pyramidal neurons and interneurons (Law et al. 2004b), and DISC-1 is expressed in the hippocampus (Austin et al. 2004). Furthermore, BDNF is important for adult hippocampal function (Monteggia et al. 2004). A common aspect of these findings is that they suggest an early developmental alteration that may have caused either abnormal cell migration or differentiation. Thus, some of the predisposing genetic factors could provide risk for schizophrenia by affecting hippocampal cell differentiation.

11.4 Neocortical alterations

Anomalies have also been reported in neocortical regions. These include loss of a number of GABA markers in the PFC and cingulate cortex (see (Lewis et al. 2005) for review). Synaptic markers abnormally expressed in the PFC of schizophrenia patients include the vesicular glutamate transporter I, complexin (Eastwood and Harrison 2005), the vesicle-associated membrane protein (VAMP) (Halim et al. 2003), and BDNF (Weickert et al. 2003). The findings suggest deficits in both glutamatergic and GABAergic cortical transmission. The latter is further strengthened by observations of an abnormal population of parvalbumin-containing interneurons in the PFC of schizophrenia patients (Beasley and

Reynolds 1997). These cells express lower levels of parvalbumin and GAD67 (Hashimoto et al. 2003), probably causing dysfunctional interneurons and impaired inhibitory activity in these areas. There have also been reports of loss of chandelier cells, which are part of the parvalbumin-positive population (Lewis et al. 1999). PFC GABA transporter is decreased in schizophrenia (Volk et al. 2002), and there is an increase in GABA-A receptor α_2 subunits, which has been interpreted as compensation for decreased GABA innervation (Volk and Lewis 2002). Diminished density of GABA neurons has been reported in the cingulate cortex and hippocampus (Benes 1999). In addition, brains from schizophrenia patients have reduced length of tyrosine hydroxylase (TH) terminals (Akil et al. 1999), indicating a decreased DA innervation, and reduced PFC spine density (Glantz and Lewis 2000), which indicates abnormal innervation in general. These losses are not accompanied by gliosis, suggesting they are not the result of a degenerative process. Instead, they may have been present from earlier stages in life.

How did these early deficits fail to alter cognition or to yield symptoms before adolescence? This is a crucial question to understand the pathophysiology of schizophrenia, a disorder with clear genetic predisposition and perinatal contributing factors but yet without unequivocal symptoms until after adolescence. Cortical GABA interneurons do exhibit a significant reorganization during adolescence. Chandelier cell synapses on pyramidal axons reach stable adult levels after adolescence (Anderson et al. 1995), and electrophysiological evidence indicates that PFC interneurons acquire the ability of being excited by D_2 DA agonists after adolescence in rats (Tseng and O'Donnell 2004). Local PFC circuits, including interneuron connections, mature during adolescence in monkeys (Woo et al. 1997). Amygdala projections to the PFC also mature at that age (Cunningham et al. 2002). Thus, it is possible that prefrontal inputs as well as local interneuronal function acquire adult functional levels during late adolescence. Imaging studies suggest abnormal post-pubertal development as an important element in schizophrenia (Pantelis et al. 2005). It can be speculated that when DA fibers establish an adult level of contact with interneurons, just after adolescence, they may encounter an abnormal population of GABA neurons in a brain predisposed towards schizophrenia (Benes 1997). This may yield insufficient interneuron activation, which may not be able to filter excessive cortical activity. Thus, the periadolescent maturation of these interneurons could be responsible for the delayed emergence of symptoms.

11.5 Hypofrontality

The PFC has been proposed to be hypoactive in schizophrenia. This "hypofrontality" was determined with imaging studies during tasks that normally engage the PFC (e.g., working memory tasks) (Berman et al. 1986; Carter et al. 1998; Ragland et al. 1998). However, recent findings have suggested that the PFC of schizophrenia patients may already be at a high baseline level of activity and working memory deficits could be derived from a reduced capacity of the system

(Manoach 2003). Transcranial-magnetic stimulation studies have also provided evidence of increased cortical excitability (or loss of inhibition) in the PFC of schizophrenia patients (Daskalakis et al. 2002; Eichhammer et al. 2004). Thus, although there is a clear prefrontal impairment in schizophrenia, the widely held concept of hypofrontality may be better defined as inability to engage the PFC at required instances; whether this is the result of a "quiet" or an "overly active" PFC remains controversial.

11.6 Dopamine Systems

A DA alteration has been considered a hallmark of schizophrenia pathophysiology for decades. An abnormal DA system is evident by all clinically effective antipsychotics being DA antagonists and the correlation between clinical efficacy and their affinity for D_2 receptors (Seeman 1987). Furthermore, to achieve clinical effects, antipsychotic drugs need to occupy 60 to 80% of D_2 receptors (Kapur and Seeman, 2001). However, attempts to measure DA in schizophrenia yielded conflicting results, with some studies failing to detect the expected increase in DA levels (van Kammen 1991) and others providing evidence of increased DA activation in striatal and cortical regions (Abi-Dargham et al. 1998; Laruelle 2000). DA synthesis in the midbrain is correlated with the val/met COMT polymorphism (Meyer-Lindenberg et al. 2005), suggesting that the PFC (a region where COMT is critical) controls DA neuron activity. Thus, although it is clear that DA projections are altered (and likely activated in excess when mesocortical projections are required – i.e., by stressful or salient conditions), it is hard to envision DA system alterations as the primary event in pathophysiological changes that may occur in schizophrenia. More likely, changes in DA physiology may be a late link in a chain of events that may have started with the deleterious effects of predisposing genes and their interaction with environmental conditions on cortical glutamate and GABA transmission.

11.7 Schizophrenia Hypotheses

Several hypotheses on the pathophysiology of schizophrenia have been advanced over the years. The DA hypothesis has been the most popular as pharmacological evidence strongly indicates an over-active DA system. However, it is difficult to assign causality to a DA hyperfunction. Furthermore, several authors have tried to account for data inconsistent with this idea, trying to incorporate changes in physiological patterns of activity of DA systems. It has been proposed that DA projections are hypoactive at rest in schizophrenia, resulting in decreased tonic DA release in cortical and subcortical areas. This would cause less autoreceptor activation, thereby enhancing release evoked by a burst of DA cell action potentials

(Grace 1991). The elegance of this hypothesis is without question, and it still may account for important pieces of this puzzle.

Cortical glutamate is also a central player in the current thinking of schizophrenia pathophysiology, with a deficit in this excitatory aminoacid being another popular idea (Deutsch et al. 1989; Olney and Farber 1995; Meador-Woodruff and Healy 2000). This is primarily due to the fact that blockade of NMDA receptors with phencyclidine (PCP) or ketamine may cause psychosis or relapse of symptoms in schizophrenia patients in remission (Luby et al. 1959; Lahti et al. 1995). An actual decrease in glutamate transmission in schizophrenia has also been suggested by imaging studies (Callicott et al. 2000a). However, there is also evidence that PCP can elicit glutamate release in the PFC (Moghaddam et al. 1997), which is probably dependent on the increased pyramidal neuron firing this drug causes (Jackson et al. 2004). Furthermore, in schizophrenia, poor working memory performance is observed along with greater magnitude of PFC fMRI activation, a finding indicating exaggerated and inefficient PFC activity (Callicott et al. 2000b). In an open study with an NMDA agonist, most patients worsened (Campion et al. 1994), suggesting that a hypo-NMDA model may not be adequate. A MRS study revealed increased glutamine levels in the medial PFC of schizophrenia patients (Bartha et al. 1997). This finding was interpreted as indicating decreased glutamate levels, but it is more likely to reflect increased glutamatergic transmission because glutamate is converted into glutamine after uptake into glial cells (Haydon 2001). These apparent inconsistencies could be explained by either a dynamic interplay between basal and evoked glutamate similar to that described for DA, or by a selective participation of interneurons in the effects of non-competitive NMDA antagonists (see below).

11.8 What have we learned from animal models?

The first approaches at animal research related to schizophrenia pathophysiology were, not surprisingly, pharmacological in nature. The ability of amphetamines to induce psychosis (Snyder 1973) was in line with the DA hypothesis of schizophrenia. Subsequent developments highlighted that while amphetamine psychosis showed little resemblance to schizophrenia, other agents such as LSD or PCP caused changes more akin to the disorder (Luby et al. 1959; Javitt and Zukin 1991). Although these initial research lines provided important insight, it became clear that the developmental aspects of schizophrenia were not addressed.

Diverse developmental models emerged over the last couple of decades. While all have strengths and weaknesses and it is clear that they lack face validity, they may provide valuable information related to plausible pathophysiological scenarios. These models range from subtle environment-based manipulations (i.e., raising animals in social isolation (Paulus et al. 1998), prenatal malnutrition or stress (Rehn et al. 2004)) to more aggressive tactics such as impairing cell division during gestation (Flagstad et al. 2004) or lesioning a critical brain area in the early postnatal period (Lipska et al. 1993).

One of the most comprehensively studied developmental models is the neonatal ventral hippocampal lesion. This manipulation does not reproduce hippocampal changes in schizophrenia, as patients do not present a lesion. However, the absence of proper hippocampal innervation at a critical developmental period may have long-term consequences on PFC neural circuits. Rats with a neonatal hippocampal lesion exhibit a variety of behavioral anomalies that emerge after puberty, including enhanced locomotion (Sams-Dodd et al. 1997; Al-Amin et al. 2001), altered reaction to stress and amphetamines (Lipska et al. 1993; Kato et al. 2000), sensorimotor gating deficits (Lipska et al. 1995; Swerdlow et al. 1995), and deficits in social behavior in adult rats (Sams-Dodd et al. 1997). Lesioned animals show working memory deficits (Chambers et al. 1996; Lipska et al. 2002) and enhanced liability for cocaine self-administration (Chambers and Self 2002). Some studies have reproduced similar deficits in monkeys with a neonatal hippocampal lesion, which show abnormal social behaviors at adult ages (Bachevalier et al. 1999). The relevance of these findings to schizophrenia is highlighted by their reversal with antipsychotic drugs (Lipska and Weinberger 1994; Le Pen and Moreau 2002). These animals also exhibit anatomical and neurochemical changes. In the PFC, there have been reports of reduced BDNF levels (Ashe et al. 2002), decreased GAD67 mRNA (Lipska et al. 2003) and changes in spine density and spine length (Flores et al. 2005). Also, the DA response to stress is attenuated in the nucleus accumbens (Brake et al. 1999) and D_2 mRNA is decreased in the striatum (Lipska et al. 2003). Furthermore, some of the behavioral deficits were prevented with a lesion to the PFC (Lipska et al. 1998). These findings indicate that the PFC is critically affected in this model, and suggest that DA deficits may arise as a consequence.

Further characterization of DA alterations in these animals was obtained using electrophysiological methods. Electrical stimulation of the ventral tegmental area (VTA), the source of DA innervation to the PFC, results in a suppression of firing in pyramidal neurons in normal animals (Lewis and O'Donnell 2000). In lesioned rats, VTA stimulation evoked a dramatic increase in firing (O'Donnell et al. 2002), which can be interpreted as underlying a loss of filtering of weak or unwanted information that could disrupt goal-directed behavior and attention. VTA stimulation also produces an abnormally high cell firing in the nucleus accumbens in adult, not prepubertal rats (Goto and O'Donnell 2002), which can also be prevented by haloperidol (Goto and O'Donnell 2002) or a prefrontal cortical lesion (Goto and O'Donnell 2004). As all these abnormal responses are observed only after adolescence, this suggests that periadolescent maturation of prefrontal cortical circuits is affected by the absence of proper hippocampal innervation.

The increased PFC cell firing suggests a state of hyperexcitability in these animals. Indeed, fast-spiking interneurons in the medial PFC mature around adolescence (Tseng and O'Donnell 2007) and fail to acquire the ability to become excited by D_2 dopamine receptors in lesioned animals (Tseng and O'Donnell 2006). Some of the predisposing genes have an impact on cortical interneurons. For example, NRG1 downregulates GABA-A receptors in the hippocampus, suggesting that this factor is important for the early postnatal pruning of GABA synapses (Okada and Corfas 2004). An interesting recent publication reported anomalies in cortical GABA interneurons in transgenic mice expressing soluble neural cell adhesion molecule (NCAM) (Pillai-Nair et al. 2005), which has also been found at increased levels in schizophrenia patients (Vawter et al. 2000). The absence of proper hippocampal inputs to the PFC during early development may result, in addition to abnormal electrical activity, in a reduction of PFC BDNF (Ashe et al. 2002), a trophic factor that modulates GABA receptors (Jovanovic et al. 2004) and has receptors expressed at high levels in parvalbumin-positive interneurons (Gorba and Wahle 1999). Reduced BDNF mRNA has also been observed in the PFC of schizophrenia patients (Hashimoto et al. 2005). In NVHL animals, PFC GABA transmission is impaired, as revealed by reduced GAD67 mRNA (Lipska et al. 2003) and increased levels of GABA-A receptor subunits (Mitchell et al. 2005). Thus, it is conceivable that abnormal interneurons are among the key elements underscoring behavioral, chemical and electrophysiological anomalies emerging during adolescence in animals with a neonatal hippocampal lesion.

11.9 A Systems perspective

All the elements reviewed above could be synthesized into a unique, yet complex, set of pathophysiological conditions (Fig. 11.1). Development and synaptic transmission-related genes may predispose towards schizophrenia, most likely when combined with a perinatal environmental insult (Cannon et al. 2003). This could lead to widespread changes in cortical areas. As cortical circuits and their DA innervation do not mature until late adolescence (Rosenberg and Lewis 1995), there are not obvious early physiological or behavioral consequences. However, the post-adolescent period may be marked by the emergence of labile cortical circuits, with improper modulation of neural activity derived from abnormal interneurons. Cortical (and most importantly, prefrontal) circuits can cope with regular demands, but they may break down with increasing loads. It is possible that an improper peripubertal maturation of PFC circuitry can render the system into a state with reduced flexibility. In fact, GABA interneurons can be driven by DA (Gao and Goldman-Rakic 2003; Tseng and O'Donnell 2004), require network activity to develop, and NMDA blockade could be critically affect them (de Lima et al. 2004). So, conditions that normally place intense demands on re-wiring and plasticity, coupled perhaps with stress, may extenuate interneuronal ability to keep pyramidal activity under control.

Stress does place a high demand on DA systems, and it is conceivable that a stressful event may throw precarious PFC circuits into a largely irreversible pattern of activity. Indeed, stress is almost invariably present in the onset of schizophrenia symptoms (Thompson et al. 2004). Once pyramidal cells become exceedingly active, the system adapts, settling into a new steady state with higher basal activity and less room to further increase (which would be needed whenever working memory or response selection are required). Thus, a gene-environment interaction could derive in abnormal maturation of cortical circuits.

The glutamatergic hypothesis of schizophrenia can be reformulated taking interneurons into consideration. There is evidence suggesting that non-competing NMDA antagonists increase glutamate levels and pyramidal cell firing in the PFC (Moghaddam et al. 1997; Jackson ct al. 2004). Imaging studies have also revealed that ketamine can increase PFC metabolic activity (Vollenweider et al. 1997). It is possible that PCP and other non-competitive NMDA antagonists could be psychotomimetic because of their higher affinity for NMDA receptors located in interneurons. Cortical and hippocampal interneurons express NMDA receptors with NR2C and NR2D subunits, which are not seen in pyramidal neurons (Standaert et al. 1996; Scherzer et al. 1998). Different levels of NR2C and NR2D have been reported in brains from schizophrenia patients (Akbarian et al. 1996) and single-nucleotide polymorphisms in the NR2D gene have been associated with schizophrenia in a Japanese population (Makino et al. 2005). NR2D/NR1 NMDA receptors have a distinct pharmacological profile: they are less sensitive to competitive antagonists and highly sensitive to psychosis-prone non-competing antagonists (Buller et al. 1994). They are also less sensitive to magnesium blockade (Kuner and Schoepfer 1996), which makes them relatively easy to become activated (perhaps not needing AMPA-induced depolarization). NR2D receptors are also glycine-dependent (Williams 1995), and this may contribute to the potential success of strategies enhancing NMDA function by blocking glycine transporter sites (Coyle and Tsai 2004). In addition, NR2D receptors change during postnatal development. They are in high numbers during early development and decrease later. They also exhibit unique topographical distribution, being abundant in the adult thalamus and sparse in the cortex, measured either as mRNA (Wenzel et al. 1996) or protein (Dunah et al. 1996). NR2D receptors peak later than other NMDA receptor subunits during postnatal development (Monyer et al., 1994). The elevated NR2D (more "excitable") mRNA in schizophrenia PFC may be an attempt to compensate for hypoactive interneurons. Thus, the view of schizophrenia as related to epilepsy (Stevens 1992) can be reformulated by emphasizing impaired interneuronal activation in selected cortical areas.

11.10 New potential therapeutic approaches

If an abnormal periadolescent maturation of cortical interneurons is critical for schizophrenia pathophysiology, agents that improve GABA function may prove efficacious. A restoration of interneuron activation that should occur during DA bursts could be obtained by selectively enhancing NMDA receptors located in interneurons. In this regard, the attempts at increasing NMDA function by augmenting glycine (a modulator of the NMDA receptor) (Coyle and Tsai 2004) could be successful. However, if loss of interneuron function is responsible for increased pyramidal neuron excitability, a further excitation of pyramidal neurons would not be desirable. Thus, targeting NMDA receptors located in interneurons could be obtained by designing agonists that have a high affinity for NR2C or NR2D-containing NMDA receptors. Interestingly, haloperidol (the most commonly used antipsychotic) has effects on NMDA receptors fitting that profile (Ilyin et al. 1996). As the primary drawback of this agent is the motor side-effect derived from excessive blockade of D_2 DA receptors, it would be worth searching for an agent with a similar profile on NMDA receptors and attenuated antidopaminergic activity. Alternatively (or in conjunction), D_2 DA innervation of interneurons could be selectively targeted. Perhaps the relative efficacy of one the newest players in schizophrenia treatment, aripiprazole, derives from its partial D_2 agonism (Shapiro et al. 2003). It is conceivable that a combination of partial D_2 agonism with selective NMDA agonism at the NR2C or 2D subunits (or even antagonism at NMDA receptors that do not contain those subunits) could offer advantages in the treatment of this devastating disorder.

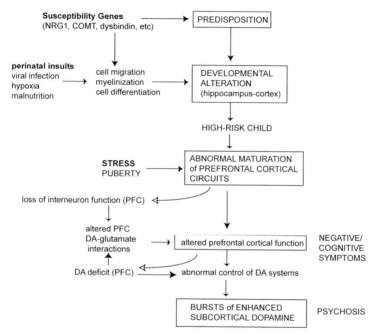

Fig. 11.1. Hypothetical sequence of events leading to schizophrenia symptoms. A combination of predisposing genes with environmental factors may render an individual susceptible for the disease during early developmental stages. However, as PFC function matures during adolescence, florid symptom onset will not take place until this late maturation puts PFC information processing into evidence after adolescence.

As the mechanisms reviewed here imply a predisposed state in a vulnerable population, another important future therapeutic development should be prevention. The critical issue in this regard will be identifying the population at risk, not an easy task when none of the genetic markers confer vulnerability alone.

11.11 Conclusions

A combination of genetic predisposition, environmental factors and abnormal post-pubertal cortical development may be the key to understanding the complexity of schizophrenia pathophysiology. As cortical interneurons mature dramatically during adolescence, they are likely the target of late developmental insults and perhaps involved in the emergence of symptoms. Abnormal cortical interneurons could render cortical areas into a hyperactive state in which there is a reduced functional capacity. Unique features of glutamatergic and dopaminergic modulation of this neuronal population make GABA interneurons a potential target for novel therapeutic approaches.

Acknowledgements

Work supported by MH57683, MH60131, and a NARSAD Independent Investigator Award.

References

Abi-Dargham A, Gil R, Krystal J, Baldwin R, Seibyl J, Bowers M, van Dyck C, Charney D, Innis R, Laurelle M (1998) Increased striatal dopamine transmission in schizophrenia: confirmation of a second cohort. Am J Psychiat 155:761-767.

Addington AM, Gornick M, Duckworth J, Sporn A, Gogtay N, Bobb A, Greenstein D, Lenane M, Gochman P, Baker N, Balkissoon R, Vakkalanka RK, Weinberger DR, Rapoport JL, Straub RE (2005) GAD1 (2q31.1), which encodes glutamic acid decarboxylase (GAD(67)), is associated with childhood-onset schizophrenia and cortical gray matter volume loss. Mol Psychiatry 10:581-588.

Akbarian S, Sucher NJ, Bradley D, Tafazzoli A, Trinh D, Hetrick WP, Potkin SG, Sandman CA, Bunney WE, Jr., Jones EG (1996) Selective alterations in gene expression for NMDA receptor subunits in prefrontal cortex of schizophrenics. J Neurosci 16:19-30.

Akil M, Pierri JN, Whitehead RE, Edgar CL, Mohila C, Sampson AR, Lewis DA (1999) Lamina-specific alterations in the dopamine innervation of the prefrontal cortex in schizophrenic subjects. Am J Psychiatry 156:1580-1589.

Al-Amin HA, Shannon Weickert C, Weinberger DR, Lipska BK (2001) Delayed onset of enhanced MK-801-induced motor hyperactivity after neonatal lesions of the rat ventral hippocampus. Biol Psychiatry 49:528-539.

Altschuler LL, Conrad A, Kovelman JA, Scheibel A (1987) Hippocampal pyramidal cell orientation in schizophrenia. Arch Gen Psychiat 44:1094-1098.

Anderson SA, Classey JD, Conde F, Lund JS, Lewis DA (1995) Synchronous development of pyramidal neuron dendritic spines and parvalbumin-immunoreactive chandelier neuron axon terminals in layer III of monkey prefrontal cortex. Neuroscience 67:7-22.

Arnold SE, Hyman BT, Van Hoesen GW, Damasio AR (1991) Some cytoarchitectural abnormalities of the entorhinal cortex in schizophrenia. Arch Gen Psychiat 48:625-632.

Ashe PC, Chlan-Fourney J, Juorio AV, Li XM (2002) Brain-derived neurotrophic factor (BDNF) mRNA in rats with neonatal ibotenic acid lesions of the ventral hippocampus. Brain Res 956:126-135.

Austin CP, Ky B, Ma L, Morris JA, Shughrue PJ (2004) Expression of Disrupted-In-Schizophrenia-1, a schizophrenia-associated gene, is prominent in the mouse hippocampus throughout brain development. Neuroscience 124:3-10.

Bachevalier J, Alvarado MC, Malkova L (1999) Memory and socioemotional behavior in monkeys after hippocampal damage incurred in infancy or in adulthood. Biol Psychiat 46:329-339.

Bartha R, Williamson PC, Drost DJ, Malla A, Carr TJ, Cortese L, Canaran G, Rylett RJ, Neufeld RW (1997) Measurement of glutamate and glutamine in the medial prefrontal cortex of never-treated schizophrenic patients and healthy controls by proton magnetic resonance spectroscopy. Arch Gen Psychiatry 54:959-965.

Beasley CL, Reynolds GP (1997) Parvalbumin-immunoreactive neurons are reduced in the prefrontal cortex of schizophrenics. Schizophr Res 24:349-355.

Benes FM (1997) The role of stress and dopamine-GABA interactions in the vulnerability for schizophrenia. J Psychiatr Res 31:257-275.

Benes FM (1999) Evidence for altered trisynaptic circuitry in schizophrenic hippocampus. Biol Psychiatry 46:589-599.

Berman KF, Zec RF, Weinberger DR (1986) Physiologic dysfunction of dorsolateral prefrontal cortex in schizophrenia. II. Role of neuroleptic treatment, attention, and mental effort. Arch Gen Psychiat 43:126-135.

Brake WG, Sullivan RM, Flores G, Srivastava L, Gratton A (1999) Neonatal ventral hippocampal lesions attenuate the nucleus accumbens dopamine response to stress: an electrochemical study in the adult rat. Brain Res 831:25-32.

Buller AL, Larson HC, Schneider BE, Beaton JA, Morrisett RA, Monaghan DT (1994) The molecular basis of NMDA receptor subtypes: native receptor diversity is predicted by subunit composition. J Neurosci 14:5471-5484.

Callicott JH, Bertolino A, Egan MF, Mattay VS, Langheim FJ, Weinberger DR (2000a) Selective relationship between prefrontal N-acetylaspartate measures and negative symptoms in schizophrenia. Am J Psychiatry 157:1646-1651.

Callicott JH, Bertolino A, Mattay VS, Langheim FJ, Duyn J, Coppola R, Goldberg TE, Weinberger DR (2000b) Physiological dysfunction of the dorsolateral prefrontal cortex in schizophrenia revisited. Cereb Cortex 10:1078-1092.

Campion D, d'Amato T, Bastard C, Laurent C, Guedj F, Jay M, Dollfus S, Thibaut F, Petit M, Gorwood P, Babron MC, Waksman G, Martinez M, Mallet J (1994) Genetic study of dopamine D_1, D_2, and D_4 receptors in schizophrenia. Psychiat Res 51:215-230.

Cannon TD, van Erp TG, Bearden CE, Loewy R, Thompson P, Toga AW, Huttunen MO, Keshavan MS, Seidman LJ, Tsuang MT (2003) Early and late neurodevelopmental influences in the prodrome to schizophrenia: contributions of genes, environment, and their interactions. Schizophr Bull 29:653-669.

Carter CS, Perlstein W, Ganguli R, Brar J, Mintun M, Cohen JD (1998) Functional hypofrontality and working memory dysfunction in schizophrenia. Am J Psychiatry 155:1285-1287.

Chambers JS, Thomas D, Saland L, Neve RL, Perrone-Bizzozero NI (2005) Growth-associated protein 43 (GAP-43) and synaptophysin alterations in the dentate gyrus of patients with schizophrenia. Prog Neuropsychopharmacol Biol Psychiatry 29:283-290.

Chambers RA, Self DW (2002) Motivational responses to natural and drug rewards in rats with neonatal ventral hippocampal lesions: an animal model of dual diagnosis schizophrenia. Neuropsychopharmacology 27:889-905.

Chambers RA, Moore J, McEvoy JP, Levin ED (1996) Cognitive effects of neonatal hippocampal lesions in a rat model of schizophrenia. Neuropsychopharmacol 15:587-594.

Chen J, Lipska BK, Halim N, Ma QD, Matsumoto M, Melhem S, Kolachana BS, Hyde TM, Herman MM, Apud J, Egan MF, Kleinman JE, Weinberger DR (2004) Functional analysis of genetic variation in catechol-O-methyltransferase (COMT): effects on mRNA, protein, and enzyme activity in postmortem human brain. Am J Hum Genet 75:807-821.

Christison GW, Casanova MF, Weinberger DR, Rawlings R (1989) A quantitative investigation of hippocampal pyramidal cell size, shape and variability of orientation in schizophrenia. Arch Gen Psychiat 46:1027-1032.

Coyle JT, Tsai G (2004) The NMDA receptor glycine modulatory site: a therapeutic target for improving cognition and reducing negative symptoms in schizophrenia. Psychopharmacology (Berl) 174:32-38.

Cunningham MG, Bhattacharyya S, Benes FM (2002) Amygdalo-cortical sprouting continues into early adulthood: implications for the development of normal and abnormal function during adolescence. J Comp Neurol 453:116-130.

Daskalakis ZJ, Christensen BK, Chen R, Fitzgerald PB, Zipursky RB, Kapur S (2002) Evidence for impaired cortical inhibition in schizophrenia using transcranial magnetic stimulation. Arch Gen Psychiatry 59:347-354.

de Lima AD, Opitz T, Voigt T (2004) Irreversible loss of a subpopulation of cortical interneurons in the absence of glutamatergic network activity. Eur J Neurosci 19:2931-2943.

Deutsch SI, Mastropaolo J, Schwartz BL, Rosse RB, Morihisa JM (1989) A "glutamatergic hypothesis" of schizophrenia. Rationale for pharmacotherapy with glicine. Clin Neuropharmacol 12:1-13.

Dunah AW, Yasuda RP, Wang YH, Luo J, Davila-Garcia M, Gbadegesin M, Vicini S, Wolfe BB (1996) Regional and ontogenic expression of the NMDA receptor subunit NR2D protein in rat brain using a subunit-specific antibody. J Neurochem 67:2335-2345.

Eastwood SL, Harrison PJ (1998) Hippocampal and cortical growth-associated protein-43 messenger RNA in schizophrenia. Neuroscience 86:437-448.

Eastwood SL, Harrison PJ (2005) Decreased expression of vesicular glutamate transporter 1 and complexin II mRNAs in schizophrenia: further evidence for a synaptic pathology affecting glutamate neurons. Schizophr Res 73:159-172.

Eastwood SL, Burnet PWJ, Harrison PJ (1995) Altered synaptophisin expression as a marker of synaptic pathology in schizophrenia. Neuroscience 66:309-319.

Eastwood SL, Burnet PW, Harrison PJ (2005) Decreased hippocampal expression of the susceptibility gene PPP3CC and other calcineurin subunits in schizophrenia. Biol Psychiatry 57:702-710.

Egan MF, Straub RE, Goldberg TE, Yakub I, Callicott JH, Hariri AR, Mattay VS, Bertolino A, Hyde TM, Shannon-Weickert C, Akil M, Crook J, Vakkalanka RK, Balkissoon R, Gibbs RA, Kleinman JE, Weinberger DR (2004) Variation in GRM3 affects cognition, prefrontal glutamate, and risk for schizophrenia. Proc Natl Acad Sci USA 101:12604-12609.

Eichhammer P, Wiegand R, Kharraz A, Langguth B, Binder H, Hajak G (2004) Cortical excitability in neuroleptic-naive first-episode schizophrenic patients. Schizophr Res 67:253-259.

Flagstad P, Mork A, Glenthoj BY, van Beek J, Michael-Titus AT, Didriksen M (2004) Disruption of neurogenesis on gestational day 17 in the rat causes behavioral changes relevant to positive and negative schizophrenia symptoms and alters amphetamine-induced dopamine release in nucleus accumbens. Neuropsychopharmacology 29:2052-2064.

Flores G, Alquicer G, Silva-Gomez AB, Zaldivar G, Stewart J, Quirion R, Srivastava LK (2005) Alterations in dendritic morphology of prefrontal cortical and nucleus accumbens neurons in post-pubertal rats after neonatal excitotoxic lesions of the ventral hippocampus. Neuroscience 133:463-470.

Gao WJ, Goldman-Rakic PS (2003) Selective modulation of excitatory and inhibitory microcircuits by dopamine. Proc Natl Acad Sci USA 100:2836-2841.

Glantz LA, Lewis DA (2000) Decreased dendritic spine density on prefrontal cortical pyramidal neurons in schizophrenia. Arch Gen Psychiatry 57:65-73.

Gorba T, Wahle P (1999) Expression of TrkB and TrkC but not BDNF mRNA in neurochemically identified interneurons in rat visual cortex in vivo and in organotypic cultures. Eur J Neurosci 11:1179-1190.

Goto Y, O'Donnell P (2002) Delayed mesolimbic system alteration in a developmental animal model of schizophrenia. J Neurosci 22:9070-9077.

Goto Y, O'Donnell P (2004) Prefrontal lesion reverses abnormal mesoaccumbens response in an animal model of schizophrenia. Biol Psychiatry 55:172-176.

Grace AA (1991) Phasic versus tonic dopamine release and the modulation of dopamine system responsivity: A hypothesis for the etiology of schizophrenia. Neuroscience 41:1-24.

Halim ND, Weickert CS, McClintock BW, Hyde TM, Weinberger DR, Klcinman JE, Lipska BK (2003) Presynaptic proteins in the prefrontal cortex of patients with schizophrenia and rats with abnormal prefrontal development. Mol Psychiatry 8:797-810.

Harrison PJ (2004) The hippocampus in schizophrenia: a review of the neuropathological evidence and its pathophysiological implications. Psychopharmacology (Berl) 174:151-162.

Harrison PJ, Eastwood SL (1998) Preferential involvement of excitatory neurons in medial temporal lobe in schizophrenia. Lancet 352:1669-1673.

Harrison PJ, Weinberger DR (2004) Schizophrenia genes, gene expression, and neuropathology: on the matter of their convergence. Mol Psychiatry.

Hashimoto T, Volk DW, Eggan SM, Mirnics K, Pierri JN, Sun Z, Sampson AR, Lewis DA (2003) Gene expression deficits in a subclass of GABA neurons in the prefrontal cortex of subjects with schizophrenia. J Neurosci 23:6315-6326.

Hashimoto T, Bergen SE, Nguyen QL, Xu B, Monteggia LM, Pierri JN, Sun Z, Sampson AR, Lewis DA (2005) Relationship of brain-derived neurotrophic factor and its receptor TrkB to altered inhibitory prefrontal circuitry in schizophrenia. J Neurosci 25:372-383.

Haydon PG (2001) GLIA: listening and talking to the synapse. Nat Rev Neurosci 2:185-193.

Ilyin VI, Whittemore ER, Guastella J, Weber E, Woodward RM (1996) Subtype-selective inhibition of N-methyl-D-aspartate receptors by haloperidol. Mol Pharmacol 50:1541-1550.

Jackson ME, Homayoun H, Moghaddam B (2004) NMDA receptor hypofunction produces concomitant firing rate potentiation and burst activity reduction in the prefrontal cortex. Proc Natl Acad Sci U S A 101:8467-8472.

Javitt DC, Zukin SR (1991) Recent advances in the phenciclidine model of schizophrenia. Am J Psychiat 148:1301-1308.

Jovanovic JN, Thomas P, Kittler JT, Smart TG, Moss SJ (2004) Brain-derived neurotrophic factor modulates fast synaptic inhibition by regulating GABA(A) receptor phosphorylation, activity, and cell-surface stability. J Neurosci 24:522-530.

Kapur S, Seeman P (2001) Does fast dissociation from the dopamine D_2 receptor explain the action of atypical antipsychotics?: A new hypothesis. Am J Psychiatry 158:360-369.

Kato K, Shishido T, Ono M, Shishido K, Kobayashi M, Suzuki H, Nabeshima T, Furukawa H, Niwa S (2000) Effects of phencyclidine on behavior and extracellular levels of dopamine and its metabolites in neonatal ventral hippocampal damaged rats. Psychopharmacology (Berl) 150:163-169.

Kolomeets NS, Orlovskaya DD, Rachmanova VI, Uranova NA (2005) Ultrastructural alterations in hippocampal mossy fiber synapses in schizophrenia: A postmortem morphometric study. Synapse 57:47-55.

Kovelman JA, Scheibel AB (1984) A neurohistological correlate of schizophrenia. Biol Psychiat 19:1601-1917.

Kuner T, Schoepfer R (1996) Multiple structural elements determine subunit specificity of Mg2+ block in NMDA receptor channels. J Neurosci 16:3549-3558.

Lahti AC, Koffel B, LaPorte D, Tamminga CA (1995) Subanesthetic doses of ketamine stimulate psychosis in schizophrenia. Neuropsychopharmacology 13:9-19.

Laruelle M (2000) The role of endogenous sensitization in the pathophysiology of schizophrenia: implications from recent brain imaging studies. Brain Res Rev 31:371-384.

Law AJ, Weickert CS, Hyde TM, Kleinman JE, Harrison PJ (2004a) Reduced spinophilin but not microtubule-associated protein 2 expression in the hippocampal formation in schizophrenia and mood disorders: molecular evidence for a pathology of dendritic spines. Am J Psychiatry 161:1848-1855.

Law AJ, Shannon Weickert C, Hyde TM, Kleinman JE, Harrison PJ (2004b) Neuregulin-1 (NRG-1) mRNA and protein in the adult human brain. Neuroscience 127:125-136.

Le Pen G, Moreau JL (2002) Disruption of prepulse inhibition of startle reflex in a neurodevelopmental model of schizophrenia: reversal by clozapine, olanzapine and risperidone but not by haloperidol. Neuropsychopharmacology 27:1-11.

Lewis BL, O'Donnell P (2000) Ventral tegmental area afferents to the prefrontal cortex maintain membrane potential 'up' states in pyramidal neurons via D_1 dopamine receptors. Cerebral Cortex 10:1168-1175.

Lewis DA, Volk DW, Hashimoto T (2004) Selective alterations in prefrontal cortical GABA neurotransmission in schizophrenia: a novel target for the treatment of working memory dysfunction. Psychopharmacology (Berl) 174:143-150.

Lewis DA, Hashimoto T, Volk DW (2005) Cortical inhibitory neurons and schizophrenia. Nat Rev Neurosci 6:312-324.

Lewis DA, Pierri JN, Volk DW, Melchitzky DS, Woo T-UW (1999) Altered GABA neurotransmission and prefrontal cortical dysfunction in schizophrenia. Biol Psychiat 46:616-626.

Lipska B, al-Amin H, Weinberger D (1998) Excitotoxic lesions of the rat medial prefrontal cortex. Effects on abnormal behaviors associated with neonatal hippocampal damage. Neuropsychopharmacology 19:451-464.

Lipska BK, Weinberger DR (1994) Subchronic treatment with haloperidol and clozapine in rats with neonatal excitotoxic hippocampal damage. Neuropsychopharmacology 10:199-205.

Lipska BK, Jaskiw GE, Weinberger DR (1993) Postpuberal emergence of hyperresponsiveness to stress and to amphetamine after neonatal excitotoxic hippocanpal damage: a potential animal model of schizophrenia. Neuropsychopharmacology 90:67-75.

Lipska BK, Aultman JM, Verma A, Weinberger DR, Moghaddam B (2002) Neonatal damage of the ventral hippocampus impairs working memory in the rat. Neuropsychopharmacology 27:47-54.

Lipska BK, Lerman DN, Khaing ZZ, Weickert CS, Weinberger DR (2003) Gene expression in dopamine and GABA systems in an animal model of schizophrenia: effects of antipsychotic drugs. Eur J Neurosci 18:391-402.

Lipska BK, Swerdlow NR, Geyer MA, Jaskiw GE, Braff DL, Weinberger DR (1995) Neonatal excitotoxic hippocampal damage in rats cause post-pubertal changes in prepulse inhibition of startle and its disruption by apomorphine. Psychopharmacol 132:303-310.

Luby ED, Cohen BD, Rosenbaum G, Gottlieb JS, Kelly R (1959) Study of a new schizophrenomimetic drug - Sernyl. Arch Neurol Psychiat 81:363-369.

Makino C, Shibata H, Ninomiya H, Tashiro N, Fukumaki Y (2005) Identification of single-nucleotide polymorphisms in the human N-methyl-D-aspartate receptor subunit NR2D gene, GRIN2D, and association study with schizophrenia. Psychiatr Genet 15:215-221.

Manoach DS (2003) Prefrontal cortex dysfunction during working memory performance in schizophrenia: reconciling discrepant findings. Schizophr Res 60:285-298.

Meador-Woodruff JH, Healy DJ (2000) Glutamate receptor expression in schizophrenic brain. Brain Res Rev 31:288-294.

Meyer-Lindenberg A, Kohn PD, Kolachana B, Kippenhan S, McInerney-Leo A, Nussbaum R, Weinberger DR, Berman KF (2005) Midbrain dopamine and prefrontal function in humans: interaction and modulation by COMT genotype. Nat Neurosci 8:594-596.

Mirnics K, Middleton FA, Lewis DA, Levitt P (2001) Analysis of complex brain disorders with gene expression microarrays: schizophrenia as a disease of the synapse. Trends Neurosci 24:479-486.

Mitchell CP, Grayson DR, Goldman MB (2005) Neonatal lesions of the ventral hippocampal formation alter GABA-A receptor subunit mRNA expression in adult rat frontal pole. Biol Psychiatry 57:49-55.

Moghaddam B, Adams B, Verma A, Daly D (1997) Activation of glutamatergic neurotransmission by ketamine: a novel step in the pathway from NMDA receptor blockade to dopaminergic and cognitive disruptions associated with the prefrontal cortex. J Neurosci 17:2921-2927.

Monteggia LM, Barrot M, Powell CM, Berton O, Galanis V, Gemelli T, Meuth S, Nagy A, Greene RW, Nestler EJ (2004) Essential role of brain-derived neurotrophic factor in adult hippocampal function. Proc Natl Acad Sci U S A 101:10827-10832.

Monyer H, Burnashev N, Laurie DJ, Sakmann B, Seeburg PH (1994) Developmental and regional expression in the rat brain and functional properties of four NMDA receptors. Neuron 12:529-540.

O'Donnell P, Lewis BL, Weinberger DR, Lipska BK (2002) Neonatal hippocampal damage alters electrophysiological properties of prefrontal cortical neurons in adult rats. Cereb Cortex 12:975-982.

Okada M, Corfas G (2004) Neuregulin1 downregulates postsynaptic GABAA receptors at the hippocampal inhibitory synapse. Hippocampus 14:337-344.

Olney JW, Farber NB (1995) Glutamate receptor dysfunction and schizophrenia. Arch Gen Psychiat 52:998-1007.

Pantelis C, Yucel M, Wood SJ, Velakoulis D, Sun D, Berger G, Stuart GW, Yung A, Phillips L, McGorry PD (2005) Structural brain imaging evidence for multiple pathological processes at different stages of brain development in schizophrenia. Schizophr Bull 31:672-696.

Paulus MP, Bakshi V, Geyer MA (1998) Isolation rearing affects sequential organization of motor behavior in post-pubertal but not pre-pubertal Lister and Sprague-Dawley rats. Behav Brain Res 94:271-280.

Pillai-Nair N, Panicker AK, Rodriguiz RM, Gilmore KL, Demyanenko GP, Huang JZ, Wetsel WC, Maness PF (2005) Neural cell adhesion molecule-secreting transgenic mice display abnormalities in GABAergic interneurons and alterations in behavior. J Neurosci 25:4659-4671.

Ragland JD, Gur RC, Glahn DC, Censits DM, Smith RJ, Lazarev MG, Alavi A, Gur RE (1998) Frontotemporal cerebral blood flow change during executive and declarative memory tasks in schizophrenia: a positron emission tomography study. Neuropsychology 12:399-413.

Rehn AE, Van Den Buuse M, Copolov D, Briscoe T, Lambert G, Rees S (2004) An animal model of chronic placental insufficiency: relevance to neurodevelopmental disorders including schizophrenia. Neuroscience 129:381-391.

Rosenberg DR, Lewis DA (1995) Postnatal maturation of the dopaminergic innervation of monkey prefrontal and motor cortices: a tyrosine hydroxylase immunohistochemical analysis. J Comp Neurol 358:383-400.

Sams-Dodd F, Lipska BK, Weinberger DR (1997) Neonatal lesions of the rat ventral hippocampus result in hyperlocomotion and deficits in social behavior in adulthood. Psychopharmacol 132:303-310.

Scherzer CR, Landwehrmeyer GB, Kerner JA, Counihan TJ, Kosinski CM, Standaert DG, Daggett LP, Velicelebi G, Penney JB, Young AB (1998) Expression of N-methyl-D-aspartate receptor subunit mRNAs in the human brain: hippocampus and cortex. J Comp Neurol 390:75-90.

Seeman P (1987) Dopamine receptors and the dopamine hypothesis of schizophrenia. Synapse 1:133-152.

Shapiro DA, Renock S, Arrington E, Chiodo LA, Liu LX, Sibley DR, Roth BL, Mailman R (2003) Aripiprazole, a novel atypical antipsychotic drug with a unique and robust pharmacology. Neuropsychopharmacology 28:1400-1411.

Snyder SH (1973) Amphetamine psychosis: a model of schizophrenia mediated by catecholamines. Am J Psychiat 130:61-67.

Standaert DG, Landwehrmeyer GB, Kerner JA, Penney JB, Jr., Young AB (1996) Expression of NMDAR2D glutamate receptor subunit mRNA in neurochemically identified interneurons in the rat neostriatum, neocortex and hippocampus. Mol Brain Res 42:89-102.

Stevens JR (1992) Abnormal reinnervation as a basis for schizophrenia: a hypothesis. Arch Gen Psychiat 49:238-243.

Swerdlow NR, Lipska BK, Weinberger DR, Braff DL, Jaskiw GE, Geyer MA (1995) Increased sensitivity to the sensorimotor gating-disruptive effecs of apomorphine after lesions of medial prefrontal cortex of ventral hippocampus in adult rats. Psychopharmacology 122:27-34.

Thompson JL, Pogue-Geile MF, Grace AA (2004) Developmental pathology, dopamine, and stress: a model for the age of onset of schizophrenia symptoms. Schizophr Bull 30:875-900.

Tseng KY, O'Donnell P (2004) Dopamine-glutamate interactions controlling prefrontal cortical pyramidal cell excitability involve multiple signaling mechanisms. J Neurosci 24:5131-5139.

Tseng KY, O'Donnell P (2006) Peri-adolescent maturation of dopamine modulation of interneuron activity in normal animals and in a developmental animal model of schizophrenia. Biol Psychiat 59:196S.

Tseng JY, O'Donnell P (2007) Dopamine modulation of prefrontal cortical interneurons changes during adolescence. Cereb Cortex 17:1235-1240.

van Kammen DP (1991) The biochemical basis of relapse and drug response in schizophrenia: review and hypothesis. Psychological Medicine 21:881-895.

Vawter MP, Frye MA, Hemperly JJ, VanderPutten DM, Usen N, Doherty P, Saffell JL, Issa F, Post RM, Wyatt RJ, Freed WJ (2000) Elevated concentration of N-CAM VASE isoforms in schizophrenia. J Psychiatr Res 34:25-34.

Volk DW, Lewis DA (2002) Impaired prefrontal inhibition in schizophrenia: relevance for cognitive dysfunction. Physiol Behav 77:501-505.

Volk DW, Pierri JN, Fritschy JM, Auh S, Sampson AR, Lewis DA (2002) Reciprocal alterations in pre- and postsynaptic inhibitory markers at chandelier cell inputs to pyramidal neurons in schizophrenia. Cereb Cortex 12:1063-1070.

Vollenweider FX, Leenders KL, Scharfetter C, Antonini A, Maguire P, Missimer J, Angst J (1997) Metbolic hyperfrontality and psychopathology in the ketamine model of psychosis using emission tomography (PET) and [18F]fluorodeoxyglucose (FDG). Eur Neuropsychopharmacol 7:9-24.

Weickert CS, Hyde TM, Lipska BK, Herman MM, Weinberger DR, Kleinman JE (2003) Reduced brain-derived neurotrophic factor in prefrontal cortex of patients with schizophrenia. Mol Psychiatry 8:592-610.

Weinberger DR, Egan MF, Bertolino A, Callicott JH, Mattay VS, Lipska BK, Berman KF, Goldberg TE (2001) Prefrontal neurons and the genetics of schizophrenia. Biol Psychiatry 50:825-844.

Wenzel A, Villa M, Mohler H, Benke D (1996) Developmental and regional expression of NMDA receptor subtypes containing the NR2D subunit in rat brain. J Neurochem 66:1240-1248.

Williams K (1995) Pharmacological properties of recombinant N-methyl-D-aspartate (NMDA) receptors containing the epsilon 4 (NR2D) subunit. Neurosci Lett 184:181-184.

Woo TU, Pucak ML, Kye CH, Matus CV, Lewis DA (1997) Peripubertal refinement of the intrinsic and associational circuitry in monkey prefrontal cortex. Neuroscience 80:1149-1158.

Index

Printed in the United States of America